ENVIRONMENTAL RESEARCH TRENDS

ENVIRONMENTAL RESEARCH TRENDS

MARVIN A. CATO
EDITOR

Nova Science Publishers, Inc.
New York

For permission to use material from this book please contact us:
Telephone 631-231-7269; Fax 631-231-8175
Web Site: http://www.novapublishers.com

LIBRARY OF CONGRESS CATALOGING-IN-PUBLICATION DATA
Environmental research trends / Marin A. Cato (editor).
p. cm.
Includes index.
ISBN-13: 978-1-60021-556-8 (hardcover)
ISBN-10: 1-60021-556-4 (hardcover)
1. Pollution--Environmental aspects. 2. Environmental monitoring. I. Cato, Marin A.
QH545.A1E587 2006
363.73--dc22 2006101093

Published by Nova Science Publishers, Inc. ✦ New York

CONTENTS

PREFACE

The environment consists of the surroundings in which an organism operates, including air, water, land, natural resources, flora, fauna, humans and their interrelation. It is this environment which is both so valuable, on the one hand, and so endangered on the other. And it is people which are by and large ruining the environment both for themselves and for all other organisms. This book reviews the latest research in this field which is vital for everyone.

Expert Commentary - It is aimed to introduce biological monitoring which has been carried out national-widely in Japan for 15 years and its future direction. Biological monitoring is an assessment of overall exposure to chemicals that are present in our environment including workplaces through measurement of the appropriate determinants in biological specimens collected from humans. The measurement of lead concentration in blood (Pb-B) and *delta*-aminolevulinic (ALA) acid in urine of workers who handle lead, and urinary metabolites of workers who handle organic solvents (xylene, *N,N*-dimehyleformamide, styrene, tetrachloroethylene, 1,1,1-trichloroethane, trichloroethylene, toluene, and *n*-hexane) have been one of dispensable items of occupational health examinations since 1989. During the past 15 years, the total numbers of Pb-B and ALA analyses decreased about 10 – 20 %. On the other hands, annual analyses of urinary metabolites have been constantly 500,000 to 600,000. Among urinary metabolites of organic solvents, the analyses of total trichlorocompounds and trichloroacetic acid have decreased, in contrast, N-methylformamide and 2,5-hexanedione analyses have increased. The measurement results are classified into one of the three categories, that is, distribution 1, 2 and 3. The percentages of distribution 2 (more than 1/3 - 1/2 of the biological exposure indices (BEI) and less than BEI) and distribution 3 (over BEI) have deceased for the past 15 years. The increase of out side order of measurement resulted in oligopoly by a few laboratories specializing in clinical chemistry. TLV-TWA (ACGIH) and OEL (JSOH) has been altered during the past 15 years. Development of new analytical methods affected the measurement result. Especially, the change of ALA analysis from a colorimetric method to a HPLC method lowered measurement levels. Determinants of occupational exposure should be specific to the occupational exposure itself. Therefore, chemical substances itself and its metabolites are useful. It has been attempted to expand the number of subjected chemical substances, and more than 15 determinants are recently nominated. On the other hand, nonspecific early biological health effects which reflect combined exposure or long time exposure are suitable for determinants of environmental exposure. Hemoglobin adducts, Comet assay and urinary

8-hydoroxyguanine have been reported to show some significances in the exposure of 7, 20 and 9 chemical substances, respectively, in humans.

Chapter 1 - This chapter will provide a comprehensive review of research to date on the use of composting for bioremediation of contaminated soils and waste during the last twenty years.

Environmental pollution is of current environmental concern. In an effort to find solutions, bioremediation techniques have shown promising results in the treatment of contaminated soils and waste. Composting approaches as a bioremediation technology to treat contaminated waste were first reported in the 1980s. Currently, new sustainable technologies are demanded by the composting industry due to recent regulatory changes moving waste management away from landfill towards more integrated, practical, sustainable and economic schemes.

Thus, in-vessel composting is presented as a sustainable bioremediation technology to treat contaminated soils amended with biodegradable municipal solid waste. Special attention will be given to a case study of the disappearance of polycyclic aromatic hydrocarbons (PAHs) present in an aged coal-tar contaminated soil from a former manufactured gas plant site during in-vessel composting using green waste as an amendment. This chapter will discuss in detail a set of operational parameters to maximize contaminant disappearance and microbial activity towards the production of a mature and stable final compost using in-vessel composting as a sustainable bioremediation technology. The effect of temperature, soil to amendment ratio, moisture content and type of amendment on the fate of contaminants, organic matter and microbial communities during in-vessel composting using a multi-parameter factorial design and advanced analytical and microbiological techniques will be presented.

In summary, this chapter will determine optimal operation conditions during in-vessel composting of contaminated soils and waste, used as a sustainable bioremediation technology, will identify environmental indicators, and will include discussion of technological, regulatory and sustainable aspects.

Chapter 2 - The paper elucidates a comprehensive account of heavy metals (Ba, Cr, Cu, Fe, Mn, Ni, Pb, Zn, Cd, As and Hg) and organochlorine pesticide residues in sediments from Sunderban mangrove wetland, West Bengal, India, one of the luxuriant mangrove ecosystems and biosphere reserves in the world. Sedimentary organic carbon, ph and textural properties were worked out in order to assess their role on metal distribution and pesticide residues. The sampling sites have been selected covering various geomorphological areas of intertidal zones that are under variable stresses both from natural and anthropogenic sources. The coastal processes of wave-tide-current climate together with riverine hydrodynamics act as the primary control of sedimentation in these areas. Sandflats (2-4 φ), mixed flats (sand+mud) and mudflats (silt+clay; $\varphi < 4$) characterize the intertidal zone depending on the energy regimes and availability of the material. The heavy metallic pollutants as well as the organochlorine pesticides, as measured with the help of ICPAES and GC equipped with electron capture detector respectively, have accumulated in the sediment matrices from multifarious sources. Distribution of metals in the sediment cores is characteristically different from each other with elevated values either in surface or deeper layers. Regarding organochlorine pesticides, an elevated level of ΣHCH, ΣDDT and metabolites of HCH and DDT, β-HCH, pp'-DDT and pp'-DDE were found to be dominant. The sources of contaminants are closely related to human activities, such as domestic and industrial discharges, agricultural chemical applications and soil erosion due to deforestation. The resulting compositional dataset was subjected to principal component analysis, cluster analysis, analysis of variance, correlation matrix and regression analysis as and when required. Pollution Load Index (PLI) values range from 0.35 to 0.76 and index of geoaccumulation (I_{geo}) exhibited higher values in case of Zn and Cd revealing moderately contaminated status. The study is compared to other estuarine environment in India and abroad. The present data will serve as a baseline against which future anthropogenic effects may be assessed.

Chapter 3 - The development of innovative monitoring strategies, able to provide detailed information on temporal and spatial evolution of contaminants in superficial soil, combining fast, not invasive and low cost acquisition procedures for in situ monitoring pollution, is a crucial task for environmental research. In this context, electromagnetic parameter measurements represent suitable tools for identifying innovative experimental procedures. Particularly, magnetic susceptibility measurements of superficial soil may be used as a simple, rapid, cheap and non-destructive proxy variable to evaluate heavy metal patterns in industrial, urban and agricultural areas. We developed an experimental procedure to carry out field surveys for characterizing heavy metal patterns in superficial soil. This procedure is based on in situ soil magnetic susceptibility measurements supported by analytical determinations of metal concentrations. The well defined experimental protocol allows to reduce intrinsic environmental biases and to standardize the measurement procedure making simpler data interpretation.

In this chapter we present a review of case studies concerning the characterization of contamination patterns in different industrial areas of Basilicata region in Southern Italy. Basilicata is an inner region, rather unpolluted, in which the main land use is agriculture The investigated sites show different geomorphologic and anthropogenic features, so they represent optimal sites for studying the potentiality of this kind of approach in different environmental conditions.

Chapter 4 - Water contamination by toxic metals has become a serious problem threatening the entire ecosystem including man. Metals released into the environment by technological activities tend to persist indefinitely, circulating and eventually accumulating throughout the food chain. Heavy metal ions in trace quantities are difficult to remove from aqueous streams using conventional methods such as chemical precipitation, membrane separation, ion exchange etc. The process of biosorption appears to be one of the few alternatives available for such situations. It combines the advantages of biotechnology (such as low operating cost, minimization of the volume of chemical and/or biological sludge to be disposed of, high efficiency in detoxifying very dilute effluents, and no nutrient requirements) with the advantages of conventional chemical methods e.g. resistance to the physicochemical properties of heavy metal solutions (very high or low pH, high metal concentrations) because of the use of dead biomass.

The mechanism of biosorption onto the cell surface does not depend on the cell metabolism, but is based on physical and chemical interactions between metal and function groups of the cell wall. While this is a metabolism independent process, metal binding is very quick and mostly reversible. Used biomass has all the characteristics of other sorbents (e.g. active carbon), which presents a great advantage for industrial application of this process.

Experimental results presented in this work deal with kinetics study and partially with the study of mechanisms of biosorption by materials of different origin.

The basic parameters chosen to examine and to compare biosorptive abilities of selected biomaterials were capacity and rate of the process. The process was very fast – metals were removed from the solution within the first 10 minutes by most of the biosorbents. Only some of the biosorbents significantly increased pH value of the solution resulting in metal precipitation. But precipitates can cause difficulties in metal – sorbent separation.

Probable mechanisms responsible for the biosorption is ion-exchange which offers the possibility to remove metal bound on the biosorbent particles by desorption.

The advantage of this process is its selectivity. Two times the capacity was required to remove silver by alga *Chlorella kessleri* than copper and cadmium. The advantage of this process was the difference between pH value necessary for Ag biosorption (2.9) and for copper and cadmium, 5 and 5.6, respectively, helping in the process of selectivity because copper and cadmium are not removed from the solution at low pH value.

Chapter 5 - Efforts devoted to conservation of endangered species may often be hindered by threats occurring at spatially distant areas used by individuals along their seasonal migration routes. Therefore, specific conservation programs may temporarily remain out of the control of regional conservation efforts when individuals move from areas subjected to conservation measures to those areas that are unprotected. The evaluation of the repercussions of different threats at different areas and their role in the overall conservation status and trends represents a current challenge in conservation management of many species. Conservation of migratory birds represent a clear example of these trans-frontier challenges, as conservation effort in a given area may be unsuccessful without the knowledge of threats occurring in other key areas throughout the annual cycle. We evaluated the responsibility of Spain in the contamination of migratory red (*Milvus milvus*) and black kites (*Milvus migrans*) with heavy metals and bacteria resistant to antibiotics acquired in urban and refuse dumps in which these species feed upon livestock remains. Urban and refuse dumps are sources of abundant and predictable food for kites in Spain such that these places may be crucial for the conservation of wintering red kites and floating and migrating black kites from northern and central Europe. These foraging locales may also represent contamination traps when associated with sources of industrial and urban residues and/or when the food on which kites feed upon contain high levels of detrimental contaminants, pathogens, veterinary drugs, and bacteria resistant to these drugs. We assessed the origin of non-breeding kites using refuse dumps in Spain, and evaluated the role of dumps in attracting large numbers of wintering red kites as well as floating and migrating black kites in different Spanish regions. Results indicate the presence of high levels of resistance to antibiotics in faecal bacteria isolated from wintering red kites associated with carcass dumps of livestock treated intensively with antibiotics, especially fattening pigs. On the contrary, low levels of antibiotic resistance were found in areas where wintering red kites mostly feed upon natural prey and free-range livestock carcasses. Nestlings of both species showed high levels of bacterial resistance associated with a diet primarily composed by carrion from stabled livestock treated intensively with antibiotics. Increasing numbers of floating and migrating black kites from northern Europe foraging at a refuse dump associated with a solid-waste incinerator may be increasingly at risk from exposure to airborne contaminants. Detrimental effects of using these contamination traps in Spain may be contributing to the declining trends of local and wintering, floating and migrating kites breeding in distant areas. Global conservation plans for threatened species should consider simultaneously the impact of variable threats at different areas, especially in highly mobile species.

Chapter 6 - Some aspects of influence of industrial pollutant on small mammals were studied in an ecologo-toxicological experiment. Polymetal dust, containing Pb and Cd, mixed at 1 % concentration with conventional animal food, was administrated to laboratory mice of the inbreed line BALB/c. The experiment spanned 90 days. Structural and numerical chromosome changes were studied as characteristics of genotoxicity. Resting metabolism and thermoregulation, examined over the range of ambient temperatures 0–30°C, were considered as parameters of the general physiological status. Also body weight of the mice was established in the context of metal bioaccumulation. On days 15, 40, 60, and 90 bone marrow samples were taken and oxygen consumption, body temperature and body weight were measured. Males had significantly higher chromosome aberration frequency (CAF) than females. The highest CAF observed in males was 25.7% (on day 60), while in females it was 16.3% (on day 40). Chromatide breaks were about 1.8 times more frequent than centromeric fusions. High aneuploidy was found in males (13.5%), 3 times more than in females (4%). Fragments, isochromatide breaks, c/t aberrations and polyploidy were found infrequently. No dose-effect dependence between the frequency of chromosome aberrations and duration of exposure was recorded. This suggests that up to day 90 of the exposure there was no decline in the efficacy of the repair mechanisms. The oxygen consumption, body temperature, and body weight sharply decreased on day 90; at 25 °C they were 59%, 7%, and 28%, respectively, lower compared to the control group. On day 90 of exposure the experimental mice could not thermoregulate and their rate of resting metabolism reduction was 1.35 times higher compared to that in the controls. The results confirm that chromosome alterations are a very sensitive parameter indicating heavy metal burden. Resting metabolism, body temperature, and body weight are not among the most sensitive biomarkers but they could be used for estimation of tolerable metal levels in physiological aspect.

Chapter 7 - Organisms are continuously exposed to a variety of anthropogenic toxicants daily released to the environment and, at the present, the realistic effects on cells are still a challenge. In this chapter we discuss the use of new methods to evaluate the effects of pollutants in cells using *in vitro* and *in vivo* studies. The uses of primary cultured cells allow to evaluate the toxic mechanisms and cell responses in order to investigate the exposure to rational concentrations. The method consists in extracting from hepatic and/or muscle lipids the mixture of pollutants (organochlorine compounds (OCs) or polychlorinated biphenyls (PCBs)), which were chronically bioaccumulated in biota from field, and test their effects on cells *in vitro*. Cells such as hepatocytes are isolated through non-enzymatic perfusion protocol and different concentrations of OCs, PCBs or organo metals have been tested. Cell death, oxidative stress and ultrastructural parameters are evaluated. The following immunological parameters such as, production of nitric oxide, macrophage activity and cell attachment were considered to evaluate the *in vitro* exposure to leukocytes extracted from fish blood and macrophages from peritoneal cavity and head kidney of fish. The study of leukocyte migration into peritoneal cavity of fish is a new approach adapted from mammals to tropical fish specie. This method utilizes the peritoneal exudates after lipopolysaccharide (LPS, E.coli, 0111:B4)-induced leukocyte migration to evaluate the systemic response. These new methods developed have been used in recent investigations to study the effects of pollutants in tropical fish with a very satisfactory result, increasing the knowledge about the toxic mechanisms involved with the exposure to chemicals in aquatic organisms.

Chapter 8 - The paper presents the results of the long term study on the concentrations of natural (^{7}Be, ^{210}Pb) and fission (^{137}Cs) radionuclides in ground level air in an urban area (city of Belgrade) from 1985 to 2005. The periods before the nuclear plant accident at Chernobyl in 1986, during the accident and afterwards are investigated. The seasonal trends (average monthly concentrations, seasonal indices and quarterly seasonal indices, min/max ratios, radionuclides loading indices) are considered due to weather conditions and microclimate on different sampling sites within the city. Activity of the radionuclides was determined on a high purity Germanium (HPGe) detector by standard gamma spectrometry.

The results of the study on concentrations, distribution and transport of suspended PM_{10} and $PM_{2.5}$ particles in air in Belgrade city center are presented, too. Trace metal concentrations in suspended particles were determined by Graphite Furnace Atomic Absorption Spectrometry (GF-AAS).

In: Environmental Research Trends
Editor: Marvin A. Cato, pp. 1-2
ISBN 978-1-60021-556-8

Expert Commentary A

Emerging Nonbacterial Pathogens Call for New Approaches and Techniques for Monitoring Drinking Waters

Malamatenia Arvanitidou*
Laboratory of Hygiene, Aristotle University of Thessaloniki, Greece

Waterborne outbreaks of cryptosporidiosis, giardiasis and viral diseases have been well documented, and occurrence of the respective causative agents, viruses, *Cryptosporidium* and *Giardia*, in water supplies is a critical issue for the water industry. Furthermore, the imperfect relationship between coliform bacteria and pathogens, such as viruses and protozoa, through wastewater treatment has been known for some time. Conventional methods for the detection of nonbacterial pathogens in water supplies are labor- intensive and time- consuming. Rapid and sensitive methods are essential for the water quality industry. The advantages of molecular methods such as FISH and PCR include greater sensitivity, rapid analysis of many samples, relatively low cost, simultaneous detection of multiple pathogens, and the ability to discriminate between species and strains if suitable probes and primers are selected.

The most important criterion in applying this new technology to environmental microbiology is the removal of inhibitors and contaminants from the samples. Although a number of inhibitors and contaminants from various environmental samples, have been identified with possible removal procedures, it is believed that numerous inhibitors may exist which are yet to be identified. The application of PCR- gene probe technology to various environmental samples for detection of pathogens and other microorganisms may be affected severely if a relatively universal method for removal of inhibitors from the environmental samples is not developed.

Lastly, automation of the molecular methods is necessary for the development of on-line devices to track the microbial community composition for the detection of pathogens. Employing well trained technical personell and establishing guidelines for interpretation of signals generated from such systems are still in place. Although a variety of methodologies

* Εργαστήριο Υγιεινής Laboratory of Hygiene, 54124 Θεσσαλονίκη 54124 Thessaloniki, Greece. Τηλ. 2310999143, 999144 Tel. #2310999143, 999144. Fax 2310999701 Fax #2310999701.

are in use, we emphasize the need for continued development and fuller evaluation of the techniques currently being used.

In: Environmental Research Trends
Editor: Marvin A. Cato, pp. 3-17
ISBN 978-1-60021-556-8

Expert Commentary B

BIOLOGICAL MONITORING – NATIONAL-WIDE PRACTICE AND FUTURE DIRECTION

Toshihiro Kawamoto* and Osamu Wada**
*Department of Environmental Health,
University of Occupational and Environmental Health
**The University of Tokyo, Japan

ABSTRACT

It is aimed to introduce biological monitoring which has been carried out national-widely in Japan for 15 years and its future direction. Biological monitoring is an assessment of overall exposure to chemicals that are present in our environment including workplaces through measurement of the appropriate determinants in biological specimens collected from humans. The measurement of lead concentration in blood (Pb-B) and *delta*-aminolevulinic (ALA) acid in urine of workers who handle lead, and urinary metabolites of workers who handle organic solvents (xylene, *N,N*-dimehyleformamide, styrene, tetrachloroethylene, 1,1,1-trichloroethane, trichloroethylene, toluene, and *n*-hexane) have been one of dispensable items of occupational health examinations since 1989. During the past 15 years, the total numbers of Pb-B and ALA analyses decreased about 10 – 20 %. On the other hands, annual analyses of urinary metabolites have been constantly 500,000 to 600,000. Among urinary metabolites of organic solvents, the analyses of total trichlorocompounds and trichloroacetic acid have decreased, in contrast, N-methylformamide and 2,5-hexanedione analyses have increased. The measurement results are classified into one of the three categories, that is, distribution 1, 2 and 3. The percentages of distribution 2 (more than 1/3 - 1/2 of the biological exposure indices (BEI) and less than BEI) and distribution 3 (over BEI) have deceased for the past 15 years. The increase of out side order of measurement resulted in oligopoly by a few laboratories specializing in clinical chemistry. TLV-TWA (ACGIH) and OEL (JSOH) has been altered during the past 15 years. Development of new analytical methods affected the

* Correspondence to : Toshihiro Kawamoto, M.D., Ph, D., D.A.B.T. Department of Environmental Health, University of Occupational and Environmental Health Iseigaoka, Yahatanishi-ku, Kitakyushu 807-8555, Japan. Tel : +81-93-691-7243. Fax : +81-93-691-9341. E-mail : kawamott@med.uoeh-u.ac.jp

measurement result. Especially, the change of ALA analysis from a colorimetric method to a HPLC method lowered measurement levels. Determinants of occupational exposure should be specific to the occupational exposure itself. Therefore, chemical substances itself and its metabolites are useful. It has been attempted to expand the number of subjected chemical substances, and more than 15 determinants are recently nominated. On the other hand, nonspecific early biological health effects which reflect combined exposure or long time exposure are suitable for determinants of environmental exposure. Hemoglobin adducts, Comet assay and urinary 8-hydoroxyguanine have been reported to show some significances in the exposure of 7, 20 and 9 chemical substances, respectively, in humans.

Abbreviations

ACGIH	American Conference of Governmental Industrial Hygienists
ALA	*delta*-aminolevulinic acid
BEI	Biological Exposure Index
HA	Hippuric acid
HD	2,5-Hexanedione
MA	Mandelic Acid
MHA	Methylhippuric Acid
NMF	N-Methylformamide
OEL	Occupational Exposure Level
Pb-B	Lead in blood
JSOH	Japan Society for Occupational Health
TLV-TWA	Threshold Limit Value-Time Weighed Average
TCA	Trichloroacetic Acid
TTC	Total Trichlorocompounds

Introduction

Biological monitoring is an assessment of overall exposure to chemicals that are present in our environment including workplaces through measurement of the appropriate determinants in biological specimens collected from workers or residents.

Partial amendments to the Japanese regulation of the prevention of lead poisoning and Japanese regulation of the prevention of organic solvent poisoninig were made in 1989. As results. Measurement of lead in blood (Pb-B) and *delta*-aminolevulinic acid (ALA) became indispensable items of the occupational health examination for workers who handle lead. Also, the measurement of urinary metaboliteds of workers who handle eight kinds of organic solvents (xylene, *N,N*-dimehyleformamide, styrene, tetrachloroethylene, 1,1,1-trichloroethane, trichloroethylene, toluene, and *n*-hexane) became mandatory. The relationship between chemicals and their determinants are shown in Table 1. The biological monitoring in Japan has some character, national-wide, indispensable, twice a year and duty to submit the results to Labor Standards Inspection Office.

In this article, we introduce the experience of biological monitoring which has been carried out national-widely for 15 years from 1990 to 2004 in Japan and its future direction.

Table 1. Indispensable biological monitoring in Japan

Chemicals	Sample	Detarminants
Lead	blood	Lead (Pb-B)
Lead	urine	Delta aminolevulinic acid (ALA)
Xylene	urine	Methylhippuric acid (MHA)
N,N-Dimethylformamide	urine	N-methylformamide (NMF)
Styrene	urine	Mandelic acid (MA)
Tetrachloroethylene	urine	Trichloroacetic acid (TCA) or Total tichlorocompounds (TTC)
1,1,1-trichlororthane	urine	Trichloroacetic acid (TAC) or Total tichlorocompounds (TTC)
Trichloroethylene	urine	Trichloroacetic acid (TCA) or Total tichlorocompounds (TTC)
Toluene	urine	Hippuric acid (HA)
Normal hexane	urine	2,5-Hexanedione (HD)

Biological Monitoring in Japan

1. Number of Measurement Samples

The measurement numbers of lead in blood (Pb-B) and ALA in urine were around 140,000 in 1991. However, the annual numbers have decreased about 10 to 20% during the last 15 years (Figure 1).

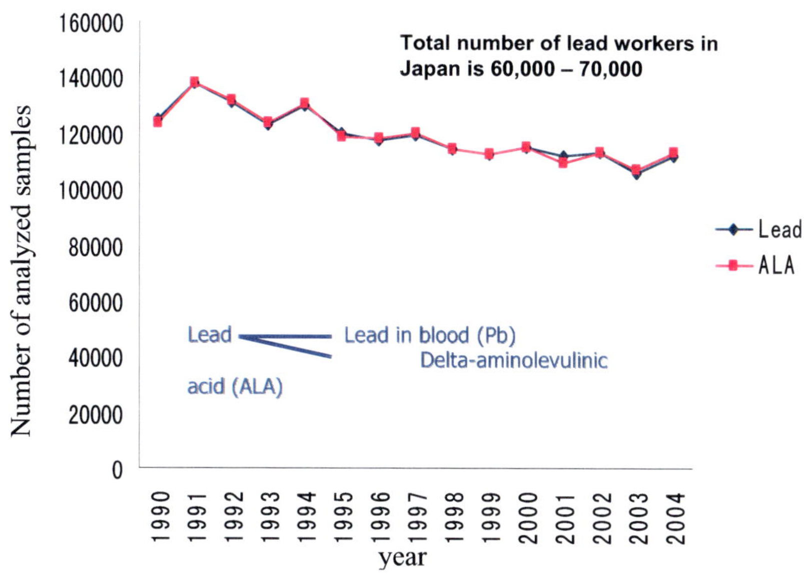

Figure 1. Secular changes of the number of analyzed samples (lead).

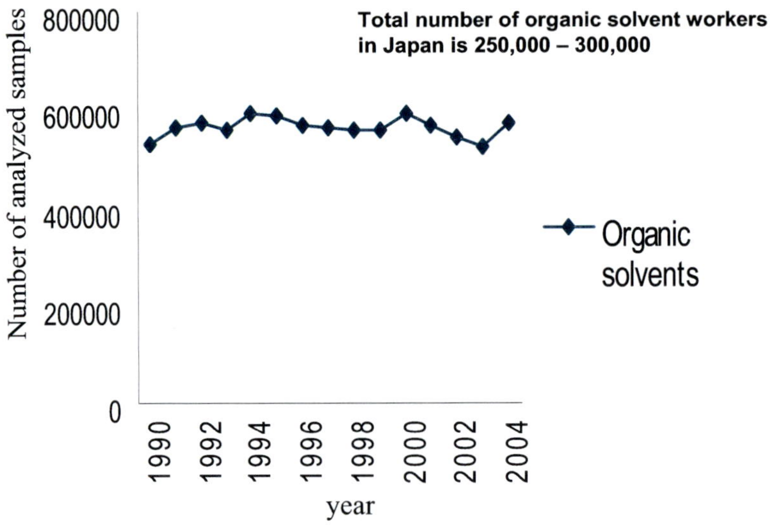

Figure 2. Secular changes of the number of analyzed samples (organic solvents).

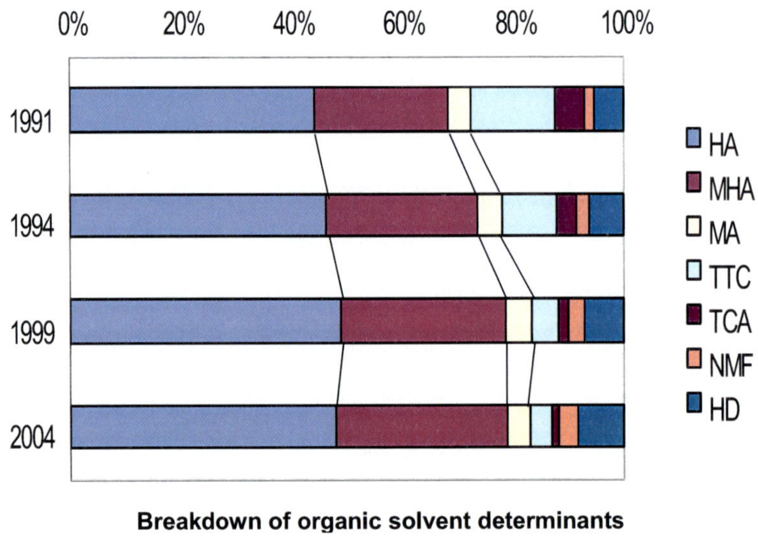

Figure 3. Share of analyzed samples of each metabolites in 1991, 1994, 1999 and 2004.

The annual change of both analyzing numbers have been corresponding because lead in blood and urinary ALA is measured at the same time as indispensable items in the occupational health examination in workplaces. As workers take biological monitoring twice a year, the number of workers who handle lead is calculated to be 55,000 to 60,000 in 2004.

On the other hands, the total measurement number of the metabolites of organic solvents (HA, MHA, MA, TTC, TCA, NMF and HD) have been constant around 500,000 to 600,000 for the last 15 years (Figure 2). Among the urinary metabolites, the analyses of TTC and TCA have decreased, but MHA, NMF and HD have increased (Figure3). One of the biggest reasons why TTC and TCA measurements decreased is the reduction of usage by Montreal Protocol [1].

2. Measurement Results

The determinants levels are classified into one of the three categories, that is, distribution 1, 2 and 3 according to values set by Ministry of Health, Labor and Welfare. Figure 4 shows the relationship between exposure concentrations to a chemical and levels of its biological monitoring determinant. BEI represents the level of a determinant that is most likely to be observed in specimens collected from healthy workers who have been exposed to the chemical to the same extent as workers with inhalation exposure at the Threshold Limit Value (TLV). The criteria classifying the distribution 3 from the distribution 2 are set corresponding values to the BEIs recommended by the American Conferences of Governmental Industrial Hygienist (ACGIH) in 1989 except that lead in blood is set 40 ug/*d*l of blood. The value classifying the distribution 2 form the distribution 1 are set from 1/3 to 1/2 of the BEIs or the value for classification of the distribution 2 and 3 [2]. Determinants and their criteria for alssification in the biological monitoring of lead and organic solvents are shown in Table 2 and 3. Trends of the measurement results of biological monitoring determinants with time are shown in Figure 5.

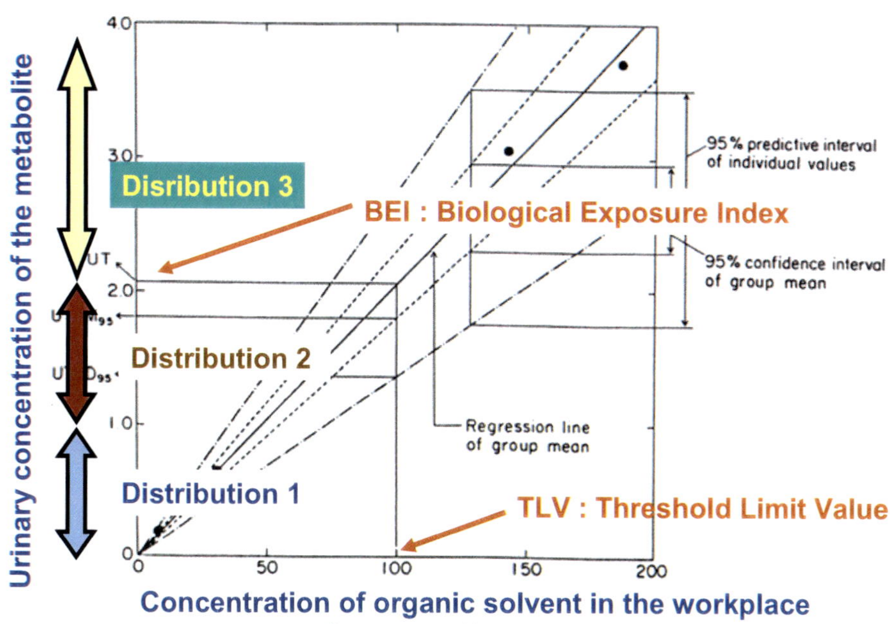

Lines fitted to the means, confidence limits of the means, and predictive limits of individual samples from the groups (UT, 2.05; UTLM$_{95}$, 1.80; UTLD$_{95}$, 1.38, g MHA/g creat)

Figure 4. Relationship between exposure concentrations to a chemical and levels of its biological monitoring determinant.

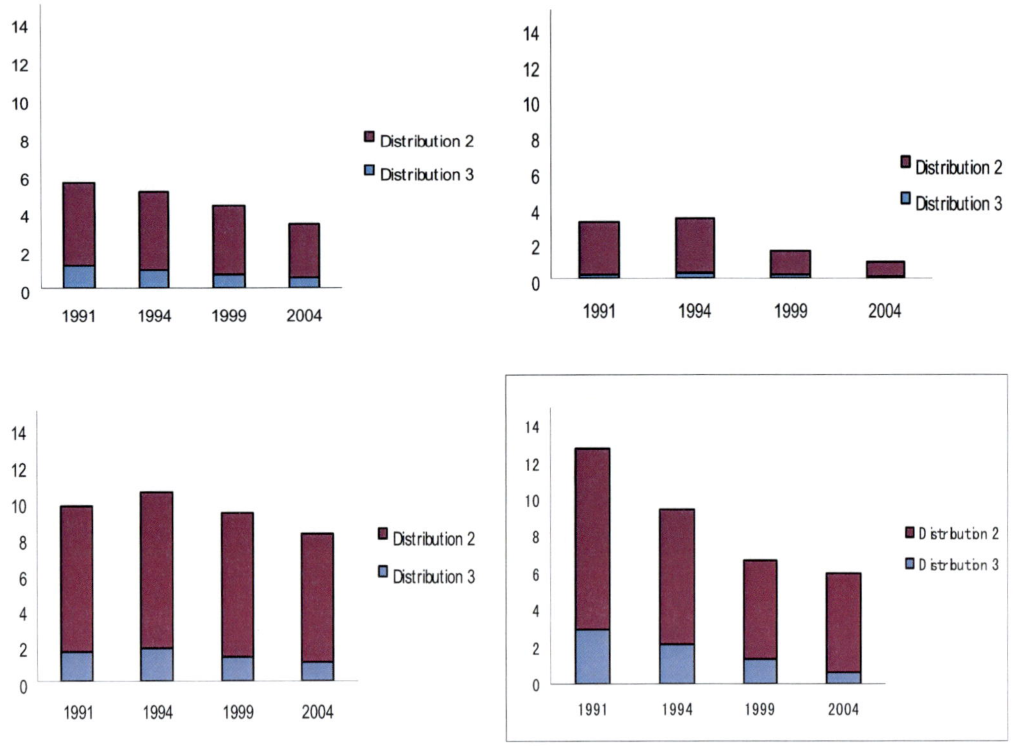

Figure 5. Frequency of distribution 2 and 3.

Table 2. Determinants and their criteria for classification in the biological monitoring of lead

Determinant	Unit	Distribution 1	Distribution 2	Distribution 3
Blood lead	μg/dl	≦ 20	20 - 40	> 40
Urinary ALA*	mg/ℓ	≦ 5	5 - 10	> 10
Protoporphyrin in erythrocyte	μg/dl	≦100	100 - 250	>250

*ALA: δ - aminolevulinic acid.

All determinants except HA reduced percentage of the distribution 2 and 3 from 1991 to 2004. The decrease of the distribution 2 and 3 usually indicates that the number of workers with high exposure has decreased. The percentages of distribution 2 (more than 1/3 - 1/2 of the biological exposure indices (BEI) and less than BEI) and distribution 3 (over BEI) have deceased for the past 15 years. It can be concluded that occupational exposure to lead or the organic solvents has been decreased after the commencement of national-wide biological monitoring.

Table 3. Determinants and their criteria for classification in the biological monitoring of organic solvents

Solvent	Determinant	Unit	Distribution		
			1	2	3
Xylene	Urinary methylhippuric acid	g/ℓ	≦0.5	0.5 - 1.5	> 1.5
N,N-Dimethylformamide	Urinary *N*-methylformamide	mg/ℓ	≦ 10	10 - 40	> 40
Styrene	Urinary mandelic acid	g/ℓ	≦0.3	0.3 - 1	> 1
Tetrachloroethylene	Urinary trichloroacetic acid	mg/ℓ	≦ 3	3 - 10	> 10
	Urinary total trichloro-compounds	mg/ℓ	≦ 3	3 - 10	> 10
1,1,1-Trichloroethane	Urinary trichloroacetic acid	mg/ℓ	≦ 3	3 - 10	> 10
	Urinary total trichloro-compounds	mg/ℓ	≦ 10	10 - 40	> 40
Trichloroethylene	Urinary trichloroacetic acid	mg/ℓ	≦ 30	30 - 100	>100
	Urinary total trichloro-compounds	mg/ℓ	≦100	100 - 300	>300
Toluene	Urinary hippuric acid	g/ℓ	≦ 1	1 - 2.5	> 2.5
normal Hexane	Urinary 2,5-hexanedione	mg/ℓ	≦ 2	2 - 5	> 5

3. Increase of Outside Order of Measurement

Figure 6 shows two typical scheme of biological monitoring in Japan. Institute for occupational health examination visit workplaces and collect samples. Some institutes analyze samples in their own laboratories. The other institutes have samples analyzed in laboratories specializing in clinical chemistry (outside order). In 1999, more than 20% of the institutes for occupational health examination analyze sample for biological monitoring in their own laboratories. However, the rates reduced to only 10% after 10 years later, and around 90% of the institutes for occupational health examination externalize the analyses (Figure 7).

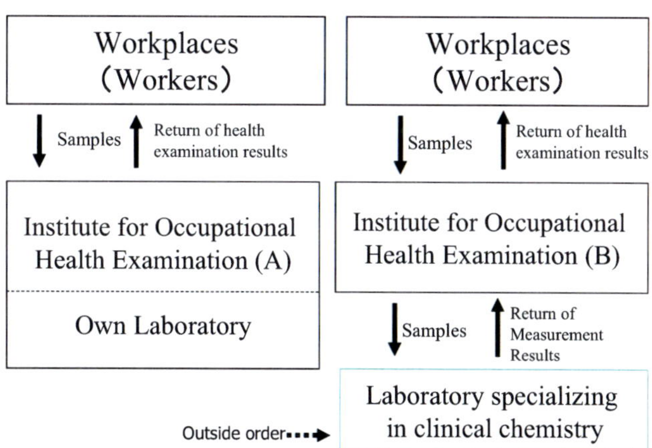

Figure 6. Two typical scheme of biological monitoring in Japan.

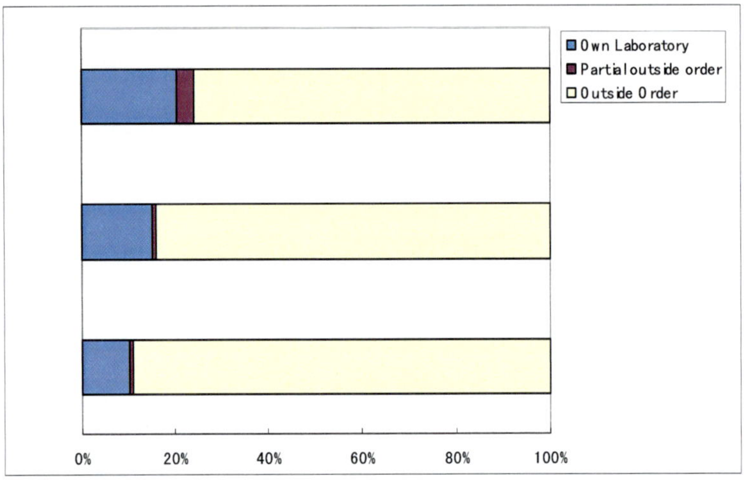

Figure 7. Increase of outside order (Pb-B measurement).

However, The increase of outside order of measurement resulted in oligopoly by a few laboratories specializing in clinical chemistry. Figure 8 shows the rates of outside orders by each determinant. The outside order rates of HD, NMF, TTC and TCA are higher than those of HA, MHA, MA, ALA and Pb-B.

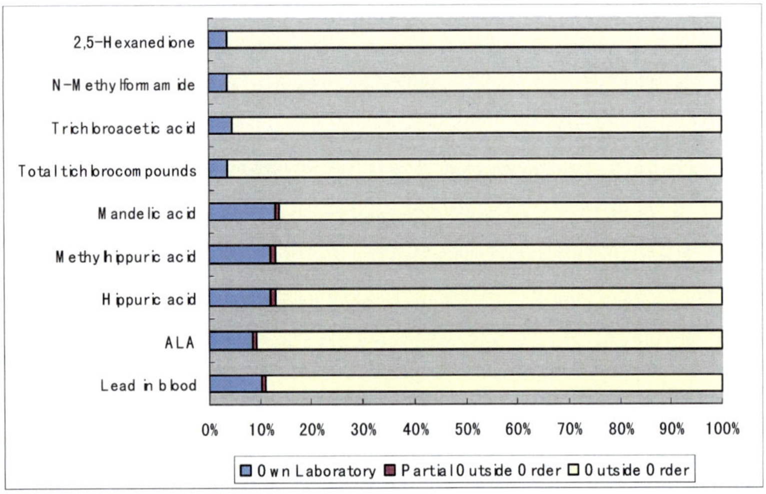

Figure 8. Rate of outside order by each determinant in 2004.

4. Development of New Analytical Methods

During 15 years, new analytical methods has been developed. The biggest change is the alteration of ALA measurement from colorimetric method to HPLC methods. Only one laboratory specializing in clinical chemistry out of seven big ones [3] analyzed ALA by HPLC methods in 1990, however, all of the seven laboratories used HPCL method in 1999

(Figure 9). This alteration of analytical method caused a big problem in measurement results. Colorimetric methods measures aminoacetone etc. as ALA.

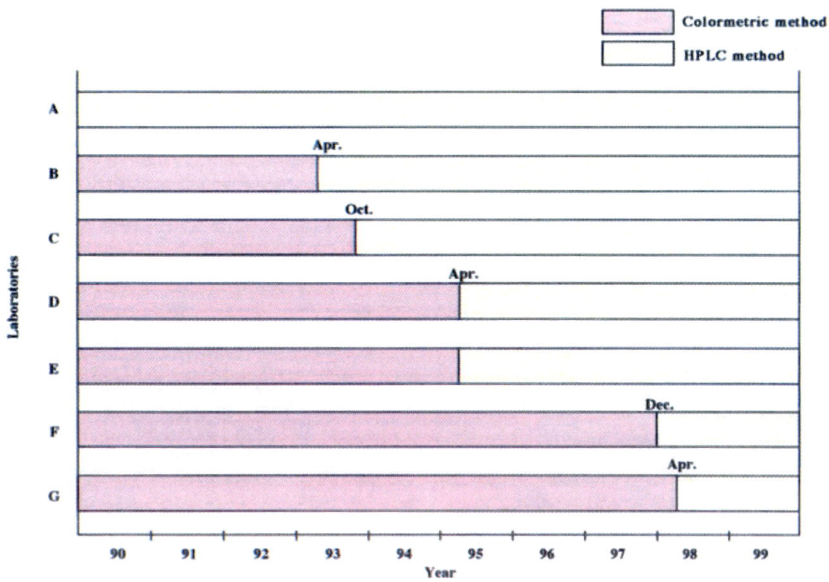

Figure 9. Transition of the analytical methods for urinary ALA form cororimetric methods to HPLC methods.

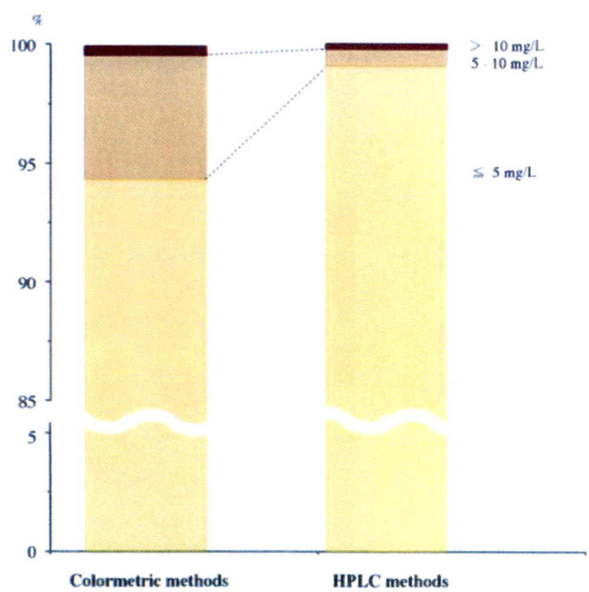

Figure 10. Comparison of ALA measurement results between colorimetric methods and HPLC methods.

However, HPLC methods measure only ALA. As a result, the measurement results by the colorimetric method always higher than those by the HPLC method. Figure 10 shows the differences of measurement results by both methods. BEI for ALA have been set by some field data measured by the colorimetric method. Therefore, a new BEI for the HPLC method should be set in order not to apply the old BEI to the evaluation of lead workers. However, it has not been set for long time. The number of urine with high ALA concentration decreased drastically during the 18 years. The reason seems to be not only the improvement of working environment but also the alteration of analytical methods form colorimetric one to HPLC.

5. Alteration of TLV-TWA (ACGIH) and OEL (JSOH)

After starting national-wide biological monitoring, some chemicals were altered in their TLVs-TWA by ACGIH or Occupational Exposure Levels (OEL) by Japan Society for Occupational Health (Table 4). Although the criteria classifying the distribution 1, 2 and 3 are set corresponding values to the BEIs, which usually decreased in conjunction with decrease TLV-TWA, it is decided not to change the criteria classifying distribution 1, 2 and 3. The one reason is that the criterion classifying distribution 1 an2 is useful even though BEI is decreased. The other reason is that the criteria should be constant in order to compare the measurement results in the same scale for long years. Therefore, the distribution 1, 2 and 3 became now not correspond to TLVs-TWA or OELs, but a classifying standard for statistics.

Table 4. Decrease of TLV

Toluene		
TLV-TWA (ACGIH)	50 ppm	(No change)
OEL (JSOH)	100 ppm	→50 ppm (135 mg/m3) ('97)
Xylene		
TLV-TWA (ACGIH)	100 ppm	(No change)
OEL(JSOH)	100 ppm	→50 ppm (217 mg/m3) ('01)
Trichloroethylene		
TLV-TWA (ACGIH)	50 ppm	(No change)
OEL(JSOH)	50 ppm	→25 ppm (135 mg/m3) ('97)
Styrene		
TLV-TWA (ACGIH)	50 ppm	→20 ppm
OEL(JSOH)	50 ppm	→20 ppm (85 mg/m3) ('99)

However, The increase of outside order of measurement resulted in oligopoly by a few laboratories specializing in clinical chemistry. Figure 8 shows the rates of outside orders by each determinant. The outside order rates of HD, NMF, TTC and TCA are higher than those of HA, MHA, MA, ALA and Pb-B.

FUTURE DIRECTION OF BIOLOGICAL MONITORING

There are three directions in biological monitoring in the future. One is expansion of the subjecting chemicals of biological monitoring. The next is countermeasure of the problems of running biological monitoring. The last one is to expand biological monitoring to environmental exposure.

1. Expansion of the Subjecting Chemicals of Biological Monitoring

Japan Industrial Safety and Health Association (JISHA : Subordinate Institute of Ministry of Health, labor and Welfare) reevaluated health examination for organic solvents from 2002 to 2004 and submitted a reported in 2005 [4]. In this reports, biological determinants of 15 organic solvents were listed. This review has been carried out based on evidences of articles which indicate the validity of the determinants. Table 5 show the international comparison of biological monitoring. The operation of biological monitoring varies according to countries. In order to set forward better biological monitoring, international collaboration is very important.

Table 5. International Comparison of Biological Monitoring

PRTR No.	Chemical substances	CAS-No	Korea	Laws (Japan)	JSOH	ACGIH	DFG
15	Aniline	62-53-3				O	O
40	Ethylbenzene	100-41-4				O	
44	Ethyleneglycol monoethylether	110-80-5	choice				O
60	Cadmium and its compounds	-	●	▲		O	
63	Xylene	1330-20-7	●	●		O	O
69	Chromium(VI) compounds	-	choice			O	
93	Chlorobenzene	108-90-7	choice			O	O
99	Vanadium pentaoxide	1314-62-1	choice	▲		O	
100	Cobalt and compounds	-				O	
112	Carbon tetrachloride	56-23-5					O
120	3,3'-Dichloro-4,4'-diaminodiphenylmethane	101-14-4			O		
139	*o*-dichlorobenzene	95-50-1					O
140	*p*-dichlorobenzene	106-46-7					O
145	Dichloromenthane	75-09-2	choice			O	
172	N,N-Dimethylformamide	68-12-2	●	●		O	O
175	Mercury and its compounds	-	●	▲	O	O	O
177	Styrene	100-42-5	●	●		O	O
200	Tetrachloroethylene	127-18-4	●	●		O	

●: Indispensable item, ▲: Further examination item, O: BEI is recommended

Table 5. Continued

PRTR No.	Chemical substances	CAS-No	Korea	Laws (Japan)	JSOH	ACGIH	DFG
209	1,1,1-Trichloroethane	71-55-6	●	●			○
211	Trichloroethylene	79-01-6	●	●	○	○	
227	Toluene	108-88-3	●	●	○	○	○
230	Lead and its compounds	-	●	●	○	○	○
231	Nickel	7440-02-0	choice	▲			
232	Nickel compounds	-		▲			
235	Nitroglycol	628-96-6		▲			
236	Nitroglycerin	55-63-0					○
237	*p*-Nitorochlorobenzene	100-00-5	choice	▲			
240	Nitrobenzene	98-95-3		▲		○	○
241	Carbon disulfide	75-15-0				○	○
252	Arsenic and its compounds	-	choice	▲			
266	Phenol	108-95-2	recommend	▲		○	○
283	Hydrogen fluoride	-	choice	▲		○	
294	Beryllium and its compounds	-		▲			
299	Benzene	71-43-2	choice			○	
303	Pentachlorophenol	87-86-5	choice	▲		○	
311	Manganese and its compounds	-	choice	▲			○

●: Indispensable item, ▲: Further examination item, ○: BEI is recommended

Table 6. Biological half-time of each determinant

Determinants	Specimen	Half-time
Lead	blood	900 hrs
Cadmium	Urine	20 yrs
Hippuric acid	Urine	1.5 hrs
Methylhippuric acid	Urine	3 hrs
Mandelic acid	Urine	4 hrs
Trichloroacetic acid	Urine	75 hrs

2. Countermeasure of the Problems of Running Biological Monitoring

National-wide biological monitoring in Japan has many problems in the operation. The biggest problem is the sampling time of specimens from workers. Table 6 shows biological half-limes of some biological monitoring determinants. Heavy metals have long half-times and no problems in sampling time, that is, we can collect specimens at any time. However, urinary determinants of organic solvents have very short half-time. Short half-time implies that sampling time is limited. For example, it is meaningless to analyze morning urine more than 12 hours after finishing exposure. Actually some immoderate institutes of occupational health examination collect urine samples in the morning (before work-shift) and disrupt biological monitoring operation in Japan. In order to resolve such disadvantage of urinary metabolites, it is important to develop novel determinants which are independent from sampling time and reflect relatively long time exposure periods.

3. Expansion of Biological Monitoring to Environmental Exposure

Environmental exposure is different form occupational exposure as shown in Table 7 One of important purpose of biological monitoring to occupational exposure is to clarify whether an occupational factor is involved in a certain health effect of workers.

Table 7. Difference of occupational exposure and environmental exposure

	Occupational Exposure	Environmental Exposure
Number of exposed substances	Limited (1 – some)	Numerous
Targeting substances	Known	Unknown Secondary pollutants (ex. Oxidants)
Exposure time	Known	Unknown
Dose	High	Low

So biological determinants for occupational exposure are better to be specific to chemicals, i.e., parent chemicals, their metabolites and specific health effects. On the other hand, determinants for environmental exposure should reflect on nonspecific, combined, and long time exposure. A working group got started in 2002 under Ministry of Environment in order to develop new biological monitoring systems for environmental exposure. The group focused three determinants, that is, hemoglobin adducts [5], Comet assay and 8-hydorxyguanine. The results of literature reviews are shown in Table 8. The validity of the determinants was reported in some chemical exposure. However, all significances are recognized only in workers.

Table 8.

PRTR No.	Chemical substances	CAS No.	Hb adducts	Comet assay	8-OHdG
2	Acryamide	79-06-1	○		
7	Acrylonitrile	107-13-1	○	○	
25	Antimony and its compounds	-		○	
26	Asbestos	1332-21-4			○
42	Ethylene oxide	75-21-8	○		
56	1,2-Epoxypropane (Propione oxide)	75-56-9	○		
60	Cadmium and its compounds	-		○	
68	Chromium and chromium(III) compounds	-		○	
69	Chromium (VI) compounds	-			○
77	Chloroethylene (Vinyl chloride)	75-01-4		○	
82	2-Chloro-2',6'-diethy-N-(methoxymethyl)acetanilide (alachlor)	15972-60-8		○	
100	Cobalt and its compounds	-		○	
131	2,4-Dichlorophenoxyacetic acid (2,4-D; 2,4-PA)	94-75-7		○	

○: Significance has been reported in human studies.

PRTR No.	Chemical substances	CAS No.	Hb adducts	Comet assay	8-OHdG
175	Mercury and its compounds	-			○
177	Styrene	100-42-5	○	○	○
179	Dioxin and derivatives	-		○	
199	Tetrachloroisophthalonitrile (chlorothalonil; TPN)	1897-45-6		○	
227	Toluene	108-88-3		○	
230	Lead and its compounds	-		○	○
231	Nickel	7440-02-0		○	○
241	Carbon disulfide	75-15-0		○	
252	Arsenic and its compounds	-		○	○
268	1,3-Butadiene	106-99-0	○	○	
299	Benzene	71-43-2		○	○
306	Polychlorinated biphenyl (PCB)	1336-36-3		○	
311	Manganese and its compounds	-			○
329	1-Naphthyl N-methylcarbamate (carbaryl; NAC)	63-25-2		○	
340	4,4-Methylene dianiline	101-77-9	○		

○: Significance has been reported in human studies.

REFERENCES

[1] Albert, LA, El-Sebae, AH, Fairhurst, S; Gilbert, B; Jensen, AA; Kawamoto, T; Nielsen IR; Wahlstrom, B; Walentowics, R; Wood, G. Environmental Health Criteria 136. 1,1,1-*Trichloroethylene*. Geneva: World Health Organization; 1992.

[2] Ogata, M; Numano, T; Hosokawa, M; Michitsuji, H. Large scale biological monitoring in Japan. *Kawasaki Journal of Medical Welfare*. 1996:2(1), 1-15.

[3] Kawamoto, T; Kodama, Y; Kohno, K. Interlaboratory quality control and status of n-hexane biological monitoring in Japan. *Arch. Environ. Contam. Toxicol.* 1995:28, 529-536.

[4] Sakurai, H; *Retrace of specific health examination for organic solvents.* The 33rd Annual Meeting of Oargnic Solvebnt Poisoning, Nagoya, 2006.

[5] Ogawa, M; Oyama, T; Isse, T; Yamaguchi, T; Murakami, T; Endo, Y; Kawamoto, T. Hemoglobin adducts as a marker of exposure to chemical substances, especially PRTR class I designated chemical substances. *J. Occup. Health.* in press.

In: Environmental Research Trends
Editor: Marvin A. Cato, pp. 19-61
ISBN: 978-1-60021-556-8

Chapter 1

IN-VESSEL COMPOSTING AS A SUSTAINABLE BIOREMEDIATION TECHNOLOGY OF CONTAMINATED SOILS AND WASTE

B. Antizar-Ladislao* and N. J. Russell
Imperial College London, Wye campus, Wye,
Ashford, Kent, TN25 5AH, UK

ABSTRACT

This chapter will provide a comprehensive review of research to date on the use of composting for bioremediation of contaminated soils and waste during the last twenty years.

Environmental pollution is of current environmental concern. In an effort to find solutions, bioremediation techniques have shown promising results in the treatment of contaminated soils and waste. Composting approaches as a bioremediation technology to treat contaminated waste were first reported in the 1980s. Currently, new sustainable technologies are demanded by the composting industry due to recent regulatory changes moving waste management away from landfill towards more integrated, practical, sustainable and economic schemes.

Thus, in-vessel composting is presented as a sustainable bioremediation technology to treat contaminated soils amended with biodegradable municipal solid waste. Special attention will be given to a case study of the disappearance of polycyclic aromatic hydrocarbons (PAHs) present in an aged coal-tar contaminated soil from a former manufactured gas plant site during in-vessel composting using green waste as an amendment. This chapter will discuss in detail a set of operational parameters to maximize contaminant disappearance and microbial activity towards the production of a mature and stable final compost using in-vessel composting as a sustainable bioremediation technology. The effect of temperature, soil to amendment ratio, moisture content and type of amendment on the fate of contaminants, organic matter and microbial communities during in-vessel composting using a multi-parameter factorial design and advanced analytical and microbiological techniques will be presented.

* Current address: University of Cantabria, Department of Water and Environment Science and Technology, Bulevar Ronda Rufino Peón, 254, Torrelavega, Cantabria, 39316 Spain. E-mail: blanca.antizar@unican.es

In summary, this chapter will determine optimal operation conditions during in-vessel composting of contaminated soils and waste, used as a sustainable bioremediation technology, will identify environmental indicators, and will include discussion of technological, regulatory and sustainable aspects.

I. Composting

Composting is an aerobic process where organic materials are biologically decomposed. Heat produced during this process leads to elevations in temperature characteristic of composting. Conventional composting processes typically comprize four major microbiological stages in relation to temperature: mesophilic, thermophilic, cooling and maturation, during which the structure of the microbial community also changes [1-3].

Composting systems are generally divided into three categories: windrow, static pile, and in-vessel. In the windrow approach, the solid waste mixtures are composted in long rows and aerated by convective air movement and diffusion. The mixtures are mechanically turned periodically to expose the organic matter to ambient oxygen. This approach to composting leads to a characteristic fluctuating temperature regime as core temperatures will fall on turning due to introduction of cooler masses of air and release of entrapped heated air, followed by a subsequent rise in temperature and a new temperature peak as maximum microbial activity resumes. Over time the peak temperatures fall as the level of substrates declines. Recording of temperature fluctuations results in the characteristic jagged tooth thermal pattern of windrowing. Windrowing is a simple mechanical system with no effective control of either oxygen or temperature levels within the organic mass undergoing decomposition. The technique, though popular commercially due to its ease of implementation and relative low cost of installation, has been criticized microbiologically since considerable periods of time elapse when either oxygen or temperature or both will become limiting for microbial diversity and decomposition.

In the static pile approach, piles of solid waste mixture, often with bulking agents (wood chips, straw) as a matrix improver, are aerated by using a forced-aeration system, which is installed under the piles to maintain a minimum oxygen level throughout the compost mass. The forced aeration system may be in either the positive or negative mode which leads to maximum core temperatures slightly above or below the central point respectively. Aeration may be on a timed basis [4] leading to high core temperatures and a severely restricted microbial diversity or by temperature feedback control [5]. This latter approach is based on a greater understanding of the microbial ecology involved and in which the control of air flow is dictated by reference to an upper temperature limit above which the fans blow continuously until heat loss through latent heat of vaporization of water (seen as steaming) reduces the core temperature to an acceptable operational level. In both the Epstein [4] and the Finstein *et al.* [5] static systems there are substantial variations in temperature throughout the composting mass and the cooler edges are routinely covered with a thick layer of finished compost acting as a thermal blanket to ensure that these edges reach regulatory temperature levels. Although the Finstein *et al.* [5] system allows for the control of maximum core temperature, and in theory supplies excess of oxygen for decomposition, both static approaches suffer from this gradation of temperature (and oxygen) throughout the organic waste mass. In static systems active composting will occur over a three to four week period depending on the nature of the substrate being processed. This active phase is often followed by a two to three month period of maturation.

In-vessel composting takes place in a partially or completely enclosed container in which environmental conditions can be controlled. Enclosed vessels more closely approximate a laboratory incubator where the organic mass and its associated microflora should be exposed to a more even temperature profile. Control of temperature in in-vessel systems is usually achieved through recycling of exhaust gases with intermittent mixing of fresh air to maintain an agreed temperature [6-8]. In theory such systems should allow for excellent control of temperature within the vessel and considerably less variation of temperature through the composting mass. However in-vessel systems have serious limitations in general composting due to limited throughputs and the high installation costs [9]. Because of this, they are often used as a form of pre-treatment "bioreactor" for up to 5 days prior to conventional composting (usually windrowing), where further decomposition, stabilization, and degassing takes place [4,10,11].

The capability of microorganisms to biodegrade specific contaminants may not differ significantly from the ambient soil environment to that of composting but the transformation potential differs for several reasons [12]. Firstly, elevated temperatures of composting (> 50^0C) will increase the rate of enzyme action involved in biodegradation processes. Secondly, co-oxidation may be enhanced due to the range of alternative substrates present. Thirdly, modifications in the physical and chemical microenvironments within the composting mass can serve to increase the diversity of the microflora to which the contaminant is exposed. Lastly, high temperatures will typically increase the solubility and mass transfer rates of the contaminant, thereby making them more available to metabolism. However, some of these positive attributes listed above may be in conflict with the overall impact of temperature and microbial activity.

Over the past two decades there has been much discussion concerning the appropriate levels of temperature for maximizing decomposition rates in composting [5,13]. Reviews of the literature during this period established unequivocally that lower temperatures favour more efficient composting [14]. In practice however composting processes are often subject to regulatory levels of temperature (e.g. 70^0C during 1 hr; 55^0C during 72 hrs) in order to meet United Kingdom national levels of pathogen reduction [15,16]. Emphasis is therefore placed on pathogen reduction above that of composting process efficiency. The impact of this approach to composting is important when considering organic wastes that are likely to be non-pathogen or low pathogen containing materials and whose contaminant substrates are often complex requiring multiple enzyme systems for degradation. Research into bioremediation of PAH contaminated wastes has clearly shown that the most effective degraders of such substrates belong to the fungal class [10,17-22]. More specifically the majority of these belong to the *Basidiomycota* which contain the fungal genera responsible for wood decomposition and hence possess the necessary complex array of ligninolytic enzymes and non-enzymatic mechanisms for the degradation of lignin whose basic chemical structures are similar to the PAHs. Lignin degraders have been shown to be amongst the most potent degraders of PAHs, as epitomized by *Phanaerochaete chrysosporium* [22].

The relationship of these organisms to temperature in the context of composting as a bioremediation technology therefore becomes critical. Microorganisms exhibit a wide range of temperature adaptation and evolution. Microorganisms are found in the coldest and hottest regions of Earth [23-25]. At the extreme ends of this range are a very few highly specialized genera whose wall and cell membrane modifications allow them to survive and multiply in such conditions. As has been pointed out above, current composting approaches and technologies tend to emphasize the higher end of the range (> 70^{0}C) in order to meet regulatory requirements for pathogen control. Such temperatures, which will be easily reached in uncontrolled operations, i.e. windrowing [4], severely inhibit the microbial diversity and hence enzymatic potential of the system. Those organisms of particular interest in PAH degradation will be eliminated above approximately 45^{0}C.

Some speculation has been given to the temperature range at which microbial diversity is maximized. What little research has been done suggests that the range between 40-45^{0}C allows maximum diversity to be expressed for both, bacteria, actinomycetes and fungi [13,26]. It is of interest to note that this range represents an overlap point at which both mesophilic and moderately thermophilic microorganisms will be viable and active. A strategy of temperature control at this range would seem therefore to be appropriate for the bioremediation of PAH contaminated waste.

II. Polycyclic Aromatic Hydrocarbons (PAH)-Contaminated Soils and Waste

PAHs are a class of organic chemicals consisting of two or more benzene rings fused in a linear, angular or cluster arrangement. Their occurrence in the environment is partly the result of natural processes including forest fires and volcanic eruptions, and partly due to anthropogenic activities including the incomplete combustion of fossil fuels, accidental discharge during transport, use and disposal of petroleum products, and incineration of refuse and wastes [27-29]. Thus, in areas of high population density and industrial activity PAH releases to soils and the wider environment have led to higher concentrations of these contaminants than would be expected from natural processes alone. The problem is exacerbated because PAHs have accumulated in soils, sediments and animals after their release to the environment because of their hydrophobicity (Table 1). This gives cause for concern because they have been implicated in many adverse effects on wildlife and human health – most notably carcinogenesis [30-32]. As a result there is a need to identify and clean up sites that have become heavily contaminated so that they do not pose unnecessary risks to health. A further incentive for clean up is the presence of many highly contaminated sites on the premises of former manufactured gas plants which were typically found on land close to the center of towns and cities which is often now a valuable capital asset if it can be cleaned up to a saleable condition [33]. Thus, the potential of using physical, chemical or biological technologies (or hybrid combinations of these) to remediate PAH-contaminated sites has received much attention in recent years [34-36]. Since the 1970s, research on the biological degradation of PAHs has demonstrated that bacteria, fungi and algae possess catabolic abilities that may be used for the bioremediation of PAH-contaminated waste and water [18,37,38]. Bioremediation technologies such as phytoremediation [39,40], land farming [41] and composting [10,42,43] have been used for biodegradation of PAH-contaminated wastes.

Table 1. Structure and physico-chemical properties of 16 USEPA regulated PAHs

PAH	Molecular weight	Formula	log K_{ow}	Solubility ($mmol \cdot l^{-1}$)	Henry constant ($atm.m^3 \cdot mol^{-1}$)
Naphthalene	128	$C_{10}H_8$	3.00-4.00	$2.4x10^{-1}$	$4.5x10^{-3}$
Acenaphthylene	152	$C_{12}H_8$	3.70	-	-
Acenaphthene	154	$C_{12}H_{10}$	3.92-5.07	$2.9x10^{-2}$	$2.4x10^{-4}$
Fluorene	166	$C_{13}H_{10}$	4.18	$1.2x10^{-2}$	$7.4x10^{-5}$
Anthracene	178	$C_{14}H_{10}$	4.46-4.76	0.07	-
Phenanthrene	178	$C_{14}H_{10}$	4.45	$7.2x10^{-3}$	$2.7x10^{-4}$
Fluoranthene	202	$C_{16}H_{10}$	4.90	$1.3x10^{-3}$	$1.9x10^{-3}$
Pyrene	202	$C_{16}H_{10}$	4.90	$7.2x10^{-4}$	$1.3x10^{-5}$
Benzo[a]anthracene	228	$C_{18}H_{20}$	5.61-5.70	-	$1.2x10^{-6}$
Chrysene	228	$C_{18}H_{20}$	5.61	$1.3x10^{-5}$	$6.7x10^{-7}$
Benzo[b]fluoranthene	252	$C_{20}H_{12}$	6.57	-	-
Benzo[k]fluoranthene	252	$C_{20}H_{12}$	6.84	-	-
Benzo[a]pyrene	252	$C_{20}H_{12}$	6.04	$1.5x10^{-5}$	$2.7x10^{-7}$
Dibenzo[a,h]Anthracene	278	$C_{22}H_{14}$	5.80-6.50	$1.8x10^{-6}$	$2x10^{-9}$
Indeno[1,2,3-cd]pyrene	276	$C_{22}H_{12}$	7.66	-	-
Benzo[g,h,i]perylene	276	$C_{22}H_{12}$	7.23	$2x10^{-5}$	$2x10^{-7}$

Source: International Society for Polycyclic Aromatic compounds [44].

III. Chronological Evolution of Composting of PAH-Contaminated Soils and Waste as a Bioremediation Technology

Composting bioremediation approaches consist of the addition of compost ingredients to a contaminated soil or waste, where the compost matures in the presence of the contaminated soil or waste. Composting is a relatively new sustainable bioremediation technology, so few investigations have been conducted and most of them have been performed at the laboratory scale [6]. In this section, the application of composting as a bioremediation technology to treat PAHs contaminated soils and waste is presented, with emphasis on the progressive development of this technology on the treatment of PAHs either as single compound or a mixture of pollutants.

A. Initial Studies

Crawford *et al.* [45] were one of the earliest research groups to report on a composting bioremediation approach to remove PAHs from a contaminated soil. They reported a controlled composting study in the USA at the beginning of the 1980s, although they do not give details on the technology, scale or conditions used. In their study, naphthalene, pyrene and benzo[a]anthracene, at initial concentration of 500 $mg \cdot kg^{-1}$, showed varying degrees of degradation. Naphthalene was completely removed during the first 7 days of composting; benzo[a]anthracene showed 25% removal during the first 7 days and 42% total removal over

30 days of composting; pyrene showed no decrease during the first 7 days of composting and a moderate removal (not percentage removal available) during the last 23 days of experiment.

Racke and Frink [46] studied the fate of phenanthrene during sewage sludge composting at laboratory scale, using radiolabel phenanthrene (1.3-1.6 $mg \cdot kg^{-1}$ dry weight), which allowed them to study the fate of phenanthrene. After 18-20 days of composting, between 10-11% of the phenanthrene was degraded and between 15-17% of unextractable phenanthrene metabolites remained in the compost, either bound to organic matter or incorporated into microorganisms.

Adenuga *et al.* [47] investigated the biodegradation of pyrene (13 $mg \cdot kg^{-1}$) using in-vessel composting technology at the laboratory scale. They amended a spiked soil with composted sewage sludge (40 g mixture), in quantities ranging from 0-50% dry weight basis and the moisture was adjusted to 40%. The mixture was placed in amber-flasks connected to an airflow system, in a water bath. Then the temperature was raised from 20^0C to 60^0C over a period of 21 days, probably trying to stimulate the mesophilic and thermophilic microbiological stages characteristic of composting. The results of their preliminary studies showed that pyrene could be degraded in the composting of soil/sludge mixture although the rate and extent were not fully described.

Crawford *et al.* [45] reported a pilot study to assess the feasibility of treating 2-, 3- and 4-ring PAH-contaminated soils by composting leaves. The study used small windrows (about 19 m^3 each) of varied ratios of soils from a former industrial site contaminated with low levels of 2-, 3- and 4-ring PAHs (about 100 $mg \cdot kg^{-1}$) and other semi-volatile compounds (less than 10 $mg \cdot kg^{-1}$). During their study, it was observed that temperature, moisture contents and ratio of carbon to nitrogen were deficient for optimal composting operation, which indicated the need for larger windrows and increased process control. Complete removal of these PAHs due to mesophilic degradation, abiotic breakdown, volatilization or a combination of them, occurred within 150 days with most losses during the first 63 days. They found that the amendment ratio did not affect the extent of degradation of the PAHs, and appeared to only slightly decrease the rate of degradation of semi-volatile compounds with increasing soil content.

After the completion of these preliminary studies, it was observed that disappearance of PAHs from contaminated waste was feasible using composting approaches, although its optimization required adequate oxygen supply, sufficient nutrients, suitable pH, temperature and moisture for the microbial activity. The ratio of PAH-contaminated waste in the composting mixture needed to be optimized as well to avoid toxicity effects. Attempts to study the fate of the target PAHs were already observed, with special emphasis on the entrapment of PAHs in the solid matrix or incorporation into the microbial biomass.

B. Towards Optimization of Physico-Chemical Parameters in Composting

Civilini [48] described a laboratory scale in-vessel-composting process (2 kg mixture) operated at a constant temperature of 45^0C during 15 days following the Finstein approach [5]. Civilini [48] used municipal solid wastes and fertilizer, to clean up PAHs present in creosote-contaminated soil, using an optimal ratio of starting material to creosote [80% compost, 5% fresh organic matter, 5% fresh organic mater mixed with fertilizer (N:P:K 20:50:5), 8% creosote-contaminated soil, 2% fertilizer (N:P:K 20:50:5), 0.2% creosote] to avoid toxicity effects. Water moisture and airflow were continuously controlled, and samples were taken at 0, 5, 10 and 15 days. He investigated the fate of 2-, 3- and 4-ring PAHs (naphthalene, acenaphthene, fluorene, phenanthrene, anthracene, fluoranthene, pyrene, benzo[a]anthracene and chrysene). He reported a PAH removal between 81.63% (benzo[a]anthracene) to 98.63% (fluorene) after 15 days treatment, although the initial PAH concentration was not given. The author accounted for volatilization in this study, which was found to be less than 10% of total losses for all the PAHs with the exception of acenaphthene, which was found to be approximately 54%. The lack of thermophilic phase must be considered in this study, since it allows the development of important degraders of PAHs. Civilini [48] observed a selection of microorganisms during days 5 to 10, in which only sporogenic aerobic and/or facultative Gram-positive bacteria increased and all other groups decreased, maybe caused by the reduction of water-soluble PAHs. During days 10 to 15, a pathogenic problem was identified (*Escherichia coli*) which confirmed the unsatisfactory sanitization activity of the process at 45^{0}C [11].

So far, there was no report of the bioremediation of PAH-contaminated wastes using composting approaches where the mixture was inoculated. By the mid-1990s, research on biodegradation had shown that special groups of microorganisms (i.e. white-rot fungi) have a remarkable potential to degrade PAHs. These fungi naturally degrade lignin to obtain the cellulose that is inside wood fibre, but the non-specific mechanisms that enable them to degrade lignin also allow them to degrade a wide range of pollutants.

C. Towards Optimization of Microbiological Parameters in Composting

McFarland and co-workers [49-51] investigated the removal of PAH in a contaminated soil using fungi. They followed the fate of benzo[a]pyrene in a silt loam soil at laboratory scale under an in-vessel-composting regime (reactor volume, 125 ml) in the presence and absence of *P. chrysosporium* [51]. The soil spiked with benzo[a]pyrene (150 mg·kg^{-1}) was amended with corn cobs (primary growth substrate) using a soil to amendment ratio of 2:1 (dry weight) and the reactor head space was periodically purged with humidified oxygen to keep aerobic conditions and water moisture. PAHs were also monitored in $HgCl_2$ (4%) treated systems to compare the impact of biotic and abiotic processes. Information on the temperature profile during composting was not given. Samples were taken after 1, 7, 14, 21, 28, 35, 84, 91 and 95 days. This study showed that although the benzo[a]pyrene appeared to be removed, there was no appreciable difference between the uninoculated and inoculated systems with 65.6 and 62.8% removal, respectively, after 95 days, although initial rates of removal were faster in the inoculated incubations. During poison test conditions, removal of benzo[a]pyrene was observed, which suggested the possibility of irreversible adsorption of benzo[a]pyrene to compost materials or a lack of complete microbial inhibition by the $HgCl_2$. Substantial concentration of *P. chrysosporium* (> 1×104 CFU·g^{-1}) was found in both the inoculated and uninoculated compost systems at the end of the treatment period, which explained the similarity in removal in both systems. This suggests that amending soils with suitable fungal substrates may be sufficient to encourage the growth of *P. chrysosporium* populations already present in soil. Other studies have found that autochthonous microflora, with no introduction of foreign microorganisms, offers the greatest potential for PAH degradation in contaminated soils when an organic substrate is added [19-21].

McFarland and Qiu [51] reported the loss of benzo[a]pyrene by first order kinetic during composting with rate constants of 0.003 day^{-1}, 0.08 day^{-1} and 0.06 day^{-1} for poisoned control, fungal unamended and fungal amended treatments respectively. Analysis of gaseous traps indicated that there was no loss through volatilization or mineralization and that nearly 100% of the benzo[a]pyrene removed was attributable to bound residues as the parent compound (approx. 60%) or as chemical intermediates. Furthermore, this study highlighted that the presence of the fungi increased the rate of bound residue formation in the first 30 days of the composting study, where the rate increased from 0.73 $mg \cdot kg^{-1} \cdot day^{-1}$ in the absence of the fungi to 1.58 $mg \cdot kg^{-1} \cdot day^{-1}$ in the presence of the fungi. The authors concluded that the bioaugmentation of a soil-composting system with *P. chrysosporium* was ineffective in degrading benzo[a]pyrene during 95 days of incubation. However, in terms of 'locking up' the PAH within the compost matrix, this technique proved very successful, although the long-term implications for the fate of benzo[a]pyrene are unknown. In addition, when bioaugmentation is practised the rates of nutrient uptake may increase due to a higher metabolism of the target contaminant, which might result in nutrient-limiting conditions leading to a reduction in microbial activity during long-term soil treatment.

Similarly, Joyce *et al.* [52] investigated the fate of a mixture of 3- and 4-ring PAHs (fluorene, anthracene, phenanthrene, pyrene, benzo[a]anthracene) under laboratory scale in-vessel composting conditions of spiked simulated solid municipal waste, monitored over a 60 day period (30 days of active composting followed by 30 days of compost maturation). PAHs were also monitored in $HgCl_2$ (2%) treated systems to compare the impact of biotic and abiotic processes. The moisture of the compost was kept to 50-60% during the first 30 days and to 30% during the last 30 days. The reactors were aerated and/or stirred periodically to maintain aerobic conditions; temperature was kept at 50^0C during the first 30 days and nutrients (0.8% ammonia-nitrogen, 2.3% nitrate, 32.9% urea) were added only during the first 30 days. The results of this study showed that the loss processes occurred during the active phase of composting (first 30 days), when all investigated PAHs showed some abiotic losses in addition to their biological removal from the compost. Anthracene, phenanthrene and pyrene were removed during the composting process by a combination of biotic and abiotic mechanisms, but biotic processes were predominant. However, fluorene was readily lost with abiotic processes accounting for approximately 75% of the removal of this PAH. Benzo[a]anthracene was resistant to biodegradation throughout composting but 40-50% was lost abiotically. These results suggest that the PAHs present in the contaminated soil should be carefully screened, considering the potential for volatilization losses to prevent the process from becoming an air-stripping process.

D. Using Different Amendments

Kirchmann and Ewnetu [53] investigated the biodegradation of petroleum-based oil wastes in 280 litre-aerated-composting bins using horse manure as amendment. PAH concentrations in oil sludge and petroleum residues was 16,800 and 78,500 $mg \cdot kg^{-1}$ dry matter respectively. They studied 4 treatments, using 1.8% oil sludge (5,400 mg $PAH \cdot kg^{-1}$ dry matter), 2.1% petroleum residues (6,100 mg $PAH \cdot kg^{-1}$ dry matter), 7.1% petroleum residues (1,000 mg $PAH \cdot kg^{-1}$ dry matter) and 7.0% paraffin oil (5,300 mg $PAH \cdot kg^{-1}$ dry matter). During all the treatments, the temperature increased from 15^0C to 30^0C during the first 12-17 days and then decreased to 22-25^{0}C. In the treatment using 7.1% petroleum residues, additions of horse manure were applied repeatedly, which resulted in repeated increases in the temperature following each addition. At the start of the composting period, the dominant PAH was naphthalene, which accounted for more than 50% of the measured PAHs. After 135 days of treatment, Kirchmann and Ewnetu [53] reported that the majority of the PAHs were removed with measurable quantities close to or below the detection limit of 0.1 mg·kg^{-1} dry matter, but pyrene, chrysene and dibenzo[a,h]anthracene were only partially removed (10%) and thus still present at concentrations of 0.2-0.8 mg·kg^{-1} dry matter. Thus, Kirchmann and Ewnetu [53] proved that composting PAH-oil wastes with horse manure was a suitable environmental approach. Horse manure is high in lignocellulosic residues and at the temperatures cited considerable impact would have been expected from the natural mycoflora including members of the *Basidiomycotina*.

Loser *et al.* [54] investigated the removal of PAHs during composting of wood containing PAHs with liquid pig manure. They compared the removal of PAHs in artificially contaminated pine-wood (1,000 mg phenanthrene·kg^{-1} and 1,000 mg pyrene·kg^{-1} aged by autoclaving) and in real PAH-polluted waste wood (5,485 mg PAH·kg^{-1}) in a pilot-scale percolator system. They inoculated both kinds of wood with 50 g decomposed wood·kg^{-1}, and mixed with 26 litres of liquid pig manure each. After 31 days of composting treatment, the PAH concentration of the real polluted waste wood was higher than the concentration of the artificially contaminated pine-wood (1,470 mg·kg^{-1} to 170 mg·kg^{-1}) which was probably due to the lower bioavailability of PAHs in the 'naturally' polluted waste and thus slower biodegradation as compared to the artificially contaminated wood. In addition, in pine-wood 93% of phenanthrene and 90% of pyrene were removed, but in 'naturally' polluted waste wood 86% and 32% of phenanthrene and pyrene respectively were degraded. Another possible reason for the higher residual hydrocarbon ratio in the 'naturally' polluted soil could be the presence of higher molecular weight PAHs. In the 'naturally' polluted soil 2- and 3-ring PAHs were decreased by 75 to 98% and 4-ring PAHs by 40 to 45%, but 5- and 6-ring PAHs were reduced by 15% only. Loser *et al.* [54] concluded that remediation of PAH-polluted waste wood by means of microorganisms is possible. Despite using a different PAH-contaminated waste to enhance composting, Loser *et al.* [54] corroborated the suitability of composting approaches to bioremediate highly PAH-contaminated wastes.

Potter *et al.* [42] investigated the degradation of 19 PAHs from a Reilly soil (creosote manufacturing and wood preserving) during in-vessel composting at the laboratory scale. Each of the test conditions in their experiments utilized a 70% soil and 30% corn cob mixture on a dry weight basis, amended with cow manure, a modified OECD fertilizer or activated sludge to adjust the nutrient content (C:N:P 100:5:1). Moisture content in each 208 litre compost-reactor was adjusted to 30-35%, and aerobic conditions facilitated by a continuous vertical air flow. Samples for analyses were taken after 7, 14, 28, 56 and 84 days of treatment. Temperatures in all reactors increased to the upper mesophilic and lower thermophilic ranges (41-53^{0}C) during the first 15 days of treatment and subsequently decreased to ambient temperature, which confirmed aerobic conditions. Compost reactors amended with sludge sustained higher biomass concentrations than those amended with cow manure during the first 28 days. In addition, the greatest amount of biomass appeared between the 15 days of composting corresponding to the highest temperatures, and thus greatest aerobic activity. Following 56 days of composting, all compost-reactors contained similar amounts of biomass. Starting concentrations of total PAHs ranged from 1,606 to 4,445 mg·kg^{-1}, and final concentrations ranged from 888 to 1,556 $mg \cdot kg^{-1}$ in the reactors. They reported that 2- and 3-ring PAHs were removed by an average of 87% in all composters after 84 days of treatment, 4-ring PAHs were reduced by an average of 61%, however, none of the amendment conditions appeared to be effective in degrading 5- to 6-ring PAHs. Most of the concentration reduction occurred within the first 28 days of treatment with a plateau forming by 56 days, suggesting first order kinetics. Potter *et al.* [42] reported removal rates of small PAHs ranging from 0.012 day^{-1} (5% activated sludge) to 0.081 day^{-1} (OECD corn cobs + 1% cow manure), and removal rates of medium PAHs ranging from 0.004 day^{-1} (5% autoclaved sludge) to 0.033 day^{-1} (1% cow manure) during the first 28 days. The results of Potter *et al.* [42] showed a general similarity of final PAH concentrations across all treatments, which might reflect the recalcitrance of PAHs during the composting-bioremediation process whereby different types of amendments did not significantly alter the final results.

Ahtiainen *et al.* [55] constructed 2 pilot compost piles (5 m^3) with soil from a sawmill area heavily contaminated with creosote oil (23,600 mg $PAHs \cdot kg^{-1}$ fresh weight soil) and metals (As, Cr, Cu) and they added spruce bark chips (mixture rate not available) as a bulking agent. One pile was inoculated with *Mycobacterium*, and the other pile was left uninoculated. At the start of composting most of the PAHs (about 78%) constituted small PAHs and hence rather quickly biodegradable. However, the total amount of PAHs was high enough to cause potential inhibition of biological activity. Their data shows that soil native microbes were able to degrade PAHs under composting conditions. Furthermore, inoculation of the compost with known PAH degraders did not speed up the process, in contrast to the findings reported by McFarland and Qiu [51], who used *P. chrysosporium* as the inoculum. The increase in certain PAHs was probably due to the heterogeneous nature of the compost. These results support the recommendations given by Guerin [41] regarding the importance of soil homogenization previous to composting treatment.

Ahtiainen *et al.* [55] also investigated the removal of PAHs during composting at a larger scale, constructing one compost pile (100 m^3) with soil from an old wood-preserving facility heavily contaminated with creosote oil (10,960 mg $PAHs \cdot kg^{-1}$ fresh weight soil) and metals (As, Cr, Cu). The soil was pre-treated with 50% hydrogen peroxide to speed up breakdown of the recalcitrant medium and large PAHs. After the hydrogen peroxide was completely degraded, a microbial inoculum of PAH degraders, nutrients and spruce bark chips were added. Ahtiainen *et al.* [55] found that by pre-treating the soil with hydrogen peroxide, they could achieve similar removal rates of about 96% of small-PAHs, and 57% medium and large PAHs, in shorter periods of time.

Amir *et al.* [56] investigated the removal of PAHs present in lagoon-derived sewage sludge (165 kg) during composting with straw (20 kg) in a heap turned every 15 days to provide aeration, during 180 days of treatment. The initial total extractable PAH concentration after mixing the sludge with straw was 0.21 $mg \cdot kg^{-1}$. They reported an 88% decrease in the concentration of total extractable PAH, and a decrease in the content of organic matter from 30.6% to 24.6% following the composting treatment. During the same period of time, the amount of humic substances increased from 17.5 to 20.8 $mg \cdot g^{-1}$. They suggested two mechanisms of removal, namely biodegradation and adsorption of PAH. These mechanisms depended on the hydrophobicity, number of aromatic rings and molecular structure of the various PAHs present, and possible PAH bioavailability, although the latter finding was not conclusive.

Oleszczuk [57] investigated the effect of different amendments (fly ash and sawdust) on the disappearance of PAHs during sewage sludge composting in 100 dm^3 plastic containers keeping humidity at a constant level (55 – 60%). The initial concentration of PAHs was 10,385 $\mu g\ kg^{-1}$ and 5.4 $\mu g\ kg^{-1}$ in sewage sludge and fly ash respectively. Sawdust did not contain PAHs. Three-ring PAHs were predominant (48.7%) in the sewage sludge. Following composting treatment for 353 days, Oleszczuk reported a decrease in the total PAH load of 87.5, 83.4, 82.9 and 88.1% after composting with sludge alone, and with additions of fly ash (20% and 30% w/w dry weight) and sawdust (30% w/w dry weight), respectively. These results indicated that the composting of sewage sludge alone and the treatment with addition of sawdust were not statistically different.

E. Treatment of PAH Contaminated Soils and Waste with Compost

Mature compost can also be added to contaminated waste for bioremediation purposes. The composts sustain populations of microorganisms with the potential to degrade a variety of organic contaminants [58] and they can improve the contaminated soil environment for indigenous or introduced microorganisms by changing the soil pH, nutrient status, aeration and moisture retention characteristics. However, one major concern of using compost as a bioremediation approach is the problem of mixing non-contaminated material with contaminated soil resulting in a greater quantity of contaminated material if the treatment does not succeed. In fact, dilution has frequently been practised as a simple means of getting contaminated sites within regulatory limits. Research in using compost as a bioremediation approach is limited and most of it has been done in the laboratory scale.

Mahro and Kästner [59] investigated the fate of pyrene in soil and soil-compost mixtures over a period of 100 days, finding that the degradation of pyrene was enhanced significantly with the addition of mature compost with > 80% pyrene removal after 20 days in the presence and < 5% pyrene removed in the absence of the compost. Further, Kästner *et al.* [60] investigated the impact of mature compost addition on the fate of ^{14}C-labeled anthracene in soil at laboratory scale. They used 3-litre volume closed compost reactors incubated at 21 ± 2^0C, continuously aerated with humidified air to keep 60% moisture. In soil-compost incubations, 23% of the ^{14}C-labelled anthracene was mineralized to $^{14}CO_2$ and 42% was irreversibly sequestered/bound to the soil-compost matrix after 103 days (the authors suggested biogenic binding). However, in soil-only incubations, approximately 88% of the PAH was recoverable by solvent extraction with the formation of bound residues being less significant.

Following this study, an interest in the understanding of the interaction between the PAHs and the soil-compost matrix on biodegradation and binding mechanisms increased. Thus, Kästner and Mahro [61] continued this work by investigating the degradation of naphthalene (500 $mg{\cdot}kg^{-1}$), phenanthrene (100 $mg{\cdot}kg^{-1}$), anthracene (100 $mg{\cdot}kg^{-1}$), fluoranthene (100 $mg{\cdot}kg^{-1}$) and pyrene (100 $mg{\cdot}kg^{-1}$) in soil (Ah horizon of a para brown soil at a non-contaminated, rural area) and soil-compost (3:1, w/w) incubations (25^0C, 60% of the water-holding capacity) at laboratory scale during 100 days. The authors found that the presence of the compost enhanced the removal of the PAHs and that the presence of the organic matrix of the compost was essential for enhanced degradation. In contrast to the pure soil, naphthalene, phenanthrene and anthracene were degraded after 20 days in the soil-compost mixture. Fluoranthene and pyrene showed lag phases of 10 days, and complete degradation occurred after 35 days. The authors suggested that the stimulating effect on the PAH degradation was a function of the organic matrix of the compost (humic substances) and the ecological niches of the compost, which have to be colonized by the respective microorganisms. Kästner and Mahro [61] suggested that the addition of compost to contaminated soils may enhance the biodegradation and bioavailability of PAHs, and retain certain volatile PAHs as well as reduce the sorptive effects in soils that prevent the compounds from analytical detection.

Kästner *et al.* [62] investigated the fate of [^{14}C]anthracene (100 $mg{\cdot}kg^{-1}$) in soil (particle size < 2 mm) and in soil-compost mixtures (particle size < 4 mm) in a continuously aerated bioreactor at laboratory scale during 176 days. Soil and compost were mixed at a ratio of 4:1 (dry wt), with a moisture content adjusted to 60%. They reported complete transformation of the parent compound (anthracene). Although the amount of organic carbon, which might act as an additional binding substrate, was larger in the soil-compost mixture (12.7%) than in the native soil (1%), Kästner *et al.* [62] reported less formation of bound residues from [^{14}C]anthracene and a higher mineralization in the soil-compost mixture (67.2% mineralized, 20.7% transformed into bound residues) than in the native soil (43.8% mineralized, 45.4% transformed into bound residues).

Haderlein *et al.* [63] also investigated the impact of humic matter present in the compost on the mineralization of PAH-contaminated wastes. They prepared mixtures of 640 g of a silty soil contaminated with aliphatic hydrocarbons (TPH = 40,000 $mg{\cdot}kg^{-1}$) and PAH (630 $mg{\cdot}kg^{-1}$), 250 g maple leaves, 750 g alfalfa and 80 g (w/w) $CaCO_3$ with a moisture content about 50% (w/w), incubated at 55^{0}C, 50% moisture and aerated either continuously or intermittently during 35 days and left to mature for 90 days at ambient temperature. This mature composted PAH-contaminated soil had a pyrene concentration of about 16 $mg{\cdot}kg^{-1}$. Then Haderlein *et al.* [64] mixed this resulting mature compost with PAH-contaminated soil (80% soil, 20% compost) and left the mixtures for further composting during 100 days. Abiotic controls contained 0.4% NaN_3. Pyrene was rapidly mineralized (> 50% mineralization after 15 days) whereas mineralization in unamended soil was limited to < 3% during the same period. Abiotic controls had a maximum total mineralization of 0.7% of the initially added pyrene. Handerlein *et al.* [64] focused their efforts to elucidate the link between PAH mineralization and humic matter based on their preliminary studies where they found that pyrene mineralization potential during the composting of contaminated soil increases with time, as does humification. The addition of humic acid (previously extracted from the mature compost) to the soil-compost mixture enhanced pyrene mineralization, reaching an increase of 18±14% mineralization values at the end of the experiment. However the addition of fulvic acid (previously extracted from the mature compost) inhibited pyrene mineralization, probably due to the high content on mineral salts remaining in the fulvic acid after extraction and high pH (8-10). This study corroborated the conclusion that humic acid is a major but not sole reason why the addition of compost stimulates PAH mineralization. This is probably due to the increased bioavailability of contaminants sorbed to mineral-humic acid complexes as suggested by Laor *et al.* [65] and the fact that sorption of the microorganisms and PAHs to the colloidal surfaces of humic matter stimulates PAH biodegradation.

Wischmann and Steinhart [66] investigated the removal of PAH/N-PAH in soil-compost mixtures (415 g mixture) as compared to unamended soil (Ah/Al horizon soil material; 400 g) in a 1 litre bioreactor at laboratory scale for 180 days. Soil and soil-compost mixtures were spiked with a PAH/N-PAH standard solution in dichloromethane, resulting in concentrations from 28 to 181 mg·kg^{-1} dry weight. The moisture of the mixture was kept to 50% during the length of the treatment. The compost used in this investigation had a degree of maturity V. Abiotic removal of PAH / N-PAH was investigated in poisoned controls (soil autoclaved 35 min at 130^{0}C, 1.7 bar, 1 g·kg^{-1} $HgCl_2$). Samples were taken after 1, 3, 8, 14, 21, 28, 42, 54, 77, 105 and 180 days. In unamended soils, only PAHs with up to 3-rings were degraded over 105 days. In the soil-compost mixture there was a lag phase of 8 days, and the 2-ring PAHs depleted within the following 49 days. The 3-ring PAHs were eliminated to < 3% during 105 days. Of the PAHs with more than 3-ring PAHs only fluoranthene and pyrene were almost completely transformed within 105 days. The residual concentration of benzo[a]anthracene, chrysene and benzo[a]pyrene decreased to 2, 3 and 27% compared to the initial estimated amounts within 180 days of treatment. PAH removal in the poisoned soil showed similar elimination rates to those in the unamended soil, which suggested abiotic losses predominated in the unamended soil. Microbiological analyses indicated non-sterile conditions in the poisoned soil-compost mixture that suggested potential biotic losses, although they occurred to a lesser extent as compared to the soil-compost mixtures. Longer treatment times were required in the investigation by Wischmann and Steinhart [66] than in the investigation by Kästner and Mahro [61] for complete removal of 2- and 3-ring PAHs, although both investigations used mature compost and not aged-soils. These different results might be explained by the use of different operational parameters and different origins of soil and compost.

Carlstrom and Tuovinen [67] investigated the mineralization of phenanthrene in domestic-waste compost in biometers to assess the impact of the origin of the compost on the mineralization of PAHs. The compost was collected in four different sampling events from the interior thermophilic zone and from the exterior mesophilic zone of compost piles between 3 to 6 months old. Replicates of 5g compost were spiked with phenanthrene (100mg·kg-1) and incubated in the dark at 22±2 or 60±2^{0}C during 90 days. Carlstrom and Tuovinen [67] reported a dominant effect due to the heterogeneity of the yard-waste samples, which obscured the possible effect of surfactant addition, particle size, and moisture content. Nevertheless they reported an effect of the origin of the compost on the mineralization of phenanthrene. Yard- waste compost samples collected from the 50-60^{0}C thermophilic interior zone and incubated at 60±2^{0}C yielded 1-2% mineralization, whereas their incubation at 22 ± 2^{0}C resulted in 17% average mineralization of phenanthrene. These results suggested that phenanthrene-mineralizing microorganisms present in the biometer were not thermophiles, but were able to survive in the thermophilic zone of the compost, probably spore-formers. Carlstrom and Tuovinen [67] were one of the first laboratories to report thermophilic biomineralization of PAHs. Samples collected from an outer 30-40^{0}C mesophilic zone and incubated at 60±2^{0}C showed negligible mineralization, while parallel sub-samples incubated at 22±2^{0}C yielded about 8% mineralization. The removal of phenanthrene was attributed to microbial activity because sterile samples showed negligible $^{14}CO_2$ evolution (< 1%). The rate constants were in the range of 0.083-0.033 day^{-1} for the mineralizable fraction of phenanthrene.

Šašek *et al.* [10] investigated the bioremediation of an MGP site soil contaminated with PAHs by amending with mushroom compost in mid-phase I (consisting of wheat straw, chicken manure), and gypsum during 54 days in a thermally insulated composting chamber (approx. 1,000 kg mixture) followed by a further 100 days of maturation in windrows. The total concentration of twelve US EPA PAHs in the soil was 610 mg PAH·kg^{-1} dry mass of soil. The mixture comprised 64% soil and 36% compost on a dry weight basis, and the moisture content was maintained at 64%. Changes in the temperature of compost were monitored, showing an initial increase of temperature up to 70^{0}C, and after 12 days of composting, the temperature progressively decreased, indicating the different stages of composting. The degradation of individual PAHs was in the range of 20-60% at the end of 54 days of composting followed by further PAH removal (37-80% maximum) after another 100 days of maturation. During composting, the outgoing air was passed through a filter and the filter was analyzed for possible volatilization losses of PAHs. The amount of PAHs retained in the filter was below detection limits, indicating that the removal of PAHs during composting was either due to compost microflora metabolism or irreversible sorption to the compost matrix.

Šašek *et al.* [68] investigated the same composting/compost approach in the treatment of a soil collected from an area of a former-tar-producing plant with a total concentration of 16 PAH listed by US EPA at a higher concentration of 2,832 mg·kg^{-1}. The mixture comprized approximately 47% soil and 53% compost on a dry weight basis, and the mixture moisture was maintained at 64%. Their results showed 42-68% removal of 3- and 4-ring PAHs, and 35-57% removal of 5- and 6-ring PAHs after 42 days of composting. However, an additional 100 days of compost maturation in open-air did not result in a further decrease of PAH concentrations.

Lau *et al.* [69] investigated the effect of temperature on the biodegradation of naphthalene, phenanthrene, benzo[a]pyrene and benzo[g,h,i]perylene using a compost approach in the laboratory. Sterilized garden soil (1g) was spiked with 1 ml of acetone containing the PAHs and mixed with a straw spent-mushroom compost (*Pleurotus pulmonarius*), which is a combination of wheat straw, dried blood, horse manure and ground chalk composted together. The moisture content of the mixture was adjusted to 60%, and the sample was incubated at 4 to 80^{0}C at 200 rpm. Removal of the PAH under investigation occurred due to biodegradation and sorption mechanisms. Under the experimental conditions of 1% spent mushroom compost treating 100 mg PAH·l^{-1} at room temperature, the removal of PAHs varied between 82% naphthalene and 59% phenanthrene. The highest sorption removal (46%) was with phenanthrene. Lau *et al.* [69] reported an increase in PAH removal as temperature was increased. At 50^{0}C, 3 PAHs except phenanthrene were completely removed. At 80^{0}C, 5% of the spent mushroom compost completely degraded the 4 PAHs at 200 mg·kg^{-1}soil. This mechanistic study indicated that increasing the temperature during bioremediation of PAH-contaminated waste using compost enhanced the removal of PAHs, with an optimal removal at 80^{0}C.

Atagana [70] investigated the removal of PAHs during composting of soil contaminated with > 30,000 mg·kg^{-1} creosote with poultry manure for 19 months. He treated a 350 kg mixture (PAH contaminated soil first homogenised and mixed with wood chips in a ratio 1:1 (v/v) and then mixed with solid poultry manure in a ratio 4:1 v/v) set up as static pile compost heap on a wood pallet overlaid with nylon fibre sheet in the open yard, and covered with hay for insulation, and compared PAH removal with the control (mixture of contaminated soil and wood chips without poultry manure). The pH of the compost for most of the experiment remained between 6.2 and 7.6. Temperature in the compost rose to 60^{0}C in the second month of treatment before decreasing to 45^{0}C in the fourth month, and thereafter remained between 35 and 45^{0}C. In general, 2- and 3-ring PAHs were removed below 1 mg·kg^{-1} by the third and fourth months, whereas 4- and 5-ring PAHs were removed below the same limit by the tenth and eleventh months. Chrysene was removed after 16 months of treatment. Atagana reported that microbial activity correlated with temperature fluctuations and hydrocarbon degradation. At the beginning of composting, a mixed population of bacteria (*Pseudomonas* sp., *Bacillus* sp., *Rhodococcus* sp., *Mycobacterium* sp.) and fungi (*Rhizopus*, *Fusarium*, *Aspergillus*, *Penicillium*, *Pleurotus*) dominated the compost. After the fourth month the most dominant species were mainly fungi (*Pleurotus*). Additionally, the removal of PAHs from the control experiment was similar to that of the composting system in the first month, but then, it continued very slowly for the length of the treatment to a final value of 56% removal. Atagana [70] concluded that high nitrogen content (C:N 306:1) and high temperatures (60^0C) had limited effects on the microbial degradation capacity of the compost.

Moretto *et al.* [71] investigated the removal of PAHs during in-vessel composting of a mixture of 146 kg of soot and 1,029 kg of active mass (mixture of sewage biomass and yard waste, 1: 2.5, v/v) characterized by high initial pH (pH of soot = 12.53; pH of active biomass = 8.06). Composting occurred in a 1.86 m^3 closed tank under forced aeration regime during the first 60 days and then natural aeration during 70 days. The initial concentration of US PAHs and not-US PAHs in the soot component of the total mass was 157.62 and 60.52 $mg\cdot kg^{-1}$ (dry basis) respectively. Additionally, the mass was turned by shovel every 30 days. The initial moisture content in the mixture was 55-60%, and the initial C/N ratio was 15-30, within the range values for optimal composting. Temperature in the compost rose to 40-55^0C and remained in the thermophilic range during the 55 first days, and then dropped and remained within the range 35-44^0C until the end of the treatment. Moretto *et al.* [71] reported a 50% removal of 16 US PAHs after 60 days of composting treatment, increasing up to 68% after 130 days. They suggested a more effective degradation of lower molecular compounds, and volatilization of some 2-4 ring PAHs due to an increase in temperature during composting. Thus, this study indicated that composting technology can efficiently be applied to treat alkaline soils heavily contaminated by PAHs.

F. Comparing Composting with other Bioremediation Technologies

Guerin [41] investigated the removal of PAHs during bioremediation, comparing the use of mesophilic composting of soil with conventional land treatment or landfarming, both of them at field scale. This study was of special importance because it was amongst the first to quantitatively compare the treatment of a highly PAH-polluted soil using a composting approach with a different bioremediation technology. The treated contaminated soil (4.3 - 6,915 $mg\cdot kg^{-1}$ total PAH) had a silty clay texture, was visually contaminated with tar residues, and was initially blended with commercially based slow-release nutrients. A ratio of green tree waste to manure to soil of 15:5:80 was used, where green tree waste was fresh ($<$ 5 days old) *Eucalyptus* spp. leaf and stem waste. The initial concentration of naphthalene was 180-300 $mg\cdot kg^{-1}$, phenanthrene was present at 70-230 $mg\cdot kg^{-1}$ and benzo[a]pyrene was present at 58-71 $mg\cdot kg^{-1}$. The soil compost mixture (130 m^3) was placed in a windrow 4 m (wide) $\times$ 24 m (long) $\times$ 1.5 m (high) and regularly mixed during 224 days. The moisture was maintained at 60-80% of the water holding capacity of the treatment soils during the course of the field program. Soil composting temperature reached maximum temperature (42^0C) after 35 days, while there was no self-heating of the soil observed in the land treatment. The soil composting process conditions reduced the total PAH concentrations to below the target level given by the regulatory body (500 mg·kg^{-1}) after 224 days and resulted in a final concentration of 120 mg·kg^{-1}, which was lower than that obtained by land treatment. Losses of the low molecular weight PAH from volatilization throughout the treatment period, as determined by a portable Flame Ionization Detector (FID), were not detected. After 224 days of composting treatment, there was complete removal of the lower molecular weight PAH, 90% degradation of medium PAHs and 70% degradation of large PAHs. Indeno[1,2,3-cd]pyrene and benzo[g,h,i]perylene were most resistant to degradation but approximately 50% of each was lost. Guerin [41] also investigated changes in microbial populations during 224 days of composting. Total heterotrophic populations remained in the range 107-109 per gram soil, showing slightly higher values in the composting treatment than in the land treatment. Microorganisms capable of utilizing naphthalene (*Pseudomonas* spp.) remained in the range 107-108 per gram soil during the first 35 days. However after 224 days they considerably decreased as expected due to the reduction in PAH concentration, and in particular naphthalene concentration. In this study, composting proved to be a suitable field-scale environmental technology for PAH-contaminated soil treatment. Operational parameters such as appropriate amendment and ratio of amendment to soil, oxygen supply and moisture were critical factors in achieving effective soil composting. Contaminated soils can vary greatly in distribution of contaminants, thus special attention should be given to the appropriate soil preparation previous to composting treatment.

Atagana [70] investigated the removal of PAHs using co-composting of PAH-contaminated soil with poultry manure, as described above mentioned, and compared it to landfarming. He reported that more that 20% of the 4- and 5-ring PAHs persisted after 11 months landfarming treatment as compared with all PAHs being removed (except chrysene) to below 1 mg·kg^{-1} after 11 months co-composting with poultry manure.

G. Current State of the Art

Most investigations of the bioremediation of PAH-contaminated soils and waste using composting approaches conducted to date have been at laboratory or pilot-plant scale. Furthermore, efforts have focused on operational considerations rather the fundamental physical, chemical and biological mechanisms that underpin bioremediation-composting technologies. Results from these investigations are difficult to compare due to the use of different experimental conditions: (i) temperature, (ii) moisture, (iii) soil/waste to amendment/compost ratio, (iv) aeration, (v) inoculation vs. non-inoculation; different composting technologies: (i) windrows vs. in-vessel approach; and different scales of operation: (i) laboratory vs. pilot vs. field. Further complications are treatment of wastes/soils of different origin with different total PAHs concentration, distribution of PAHs, bioavailability of PAHs, homogenization, and variations in the sources and types of organic wastes.

Most of the reviewed investigations, regardless the composting/compost approach, technology or scale used, agreed on the importance of optimizing the operations conditions during the application of the bioremediation technology. Emphasis was given to temperature, moisture and maintenance of aerobic conditions but particularly to the ratio of waste to amendment used. The majority of the laboratory scale investigations on the use of composting/compost of PAH-contaminated wastes were kept in the mesophilic or lower thermophilic ranges [42,52,53] with a few investigations in the thermophilic phase [67,69]. Moisture differed from one investigation to another, ranging from 40% to 80%, although the majority of the studies emphasized the importance of keeping the moisture of the mixture constant throughout the treatment. Different technologies require different approaches to aeration, but was commonly achieved by aerating the mixture with humidified air. By doing so, both moisture and aerobic conditions were maintained. Although some of the earliest work on composting/compost bioremediation approaches suggested the importance of finding a suitable mixture ratio between the contaminated waste and the amendment/compost [45], there are still a few reports where the optimal mixture ratio has been investigated [48] and in general it does not exceed 80% soil content on a dry weight basis [62,63].

Composting of PAH-contaminated wastes has received more attention than the treatment of PAH-contaminated wastes with compost. One major concern of using compost as a bioremediation technology is the problem of mixing non-contaminated compost with contaminated waste, resulting in a greater quantity of contaminated material if the treatment does not succeed. Nevertheless, if using compost as a bioremediation technology proves to achieve similar removal levels of PAHs as using composting as a bioremediation technology, the application of compost-bioremediation technology may offer important operational advantages (i.e., homogenization) during the application of the technology at a field scale. Consequently, more mechanistic studies comparing the bioremediation of PAH-contaminated waste using composting and compost approaches applying the same technology and scale at optimal operational conditions are necessary. If fresh composts are mixed with the soil, a composting process may occur [10,68], and then a more appropriate definition would be composting/compost approach rather than composting or compost approach.

Composting and the use of compost have both been successfully applied to the bioremediation of PAH-contaminated wastes. The main mechanisms of PAH removal under both bioremediation regimes were mineralization, binding and volatilization. Mineralization was the more significant mechanism of removal reported in most studies to date. Mineralization of PAHs may be enhanced by increasing their availability to microbial attack during bioremediation processes. In order for the pollutants to become available to microbial attack, desorption from the waste-compost matrix may have to take place. In terms of composting, increases in temperature may enhance the rate of PAH desorption from the matrix and thus transfer to the aqueous phase (Pignatello and Xing, 1996). In terms of compost, enhanced PAH mineralization of PAHs present in soil-compost mixtures may occur due to the presence of humic matter in the compost matrix. This is supported by studies where either humic acid and/or fulvic acid have been added to a soil with very low endogenous humic acids/fulvic acids [64,65,72,73] and an increase in the bioavailable fraction has been observed. Another favourable factor inherent to composting approaches is that normally large amounts of organic matter, considered a major factor in the 'locking-up' of organic pollutants are added to the system[50,74]. Thus, by utilizing composting approaches, the bioavailable fraction will ideally be mineralized and the unavailable fraction will be 'locked up' in the soil-compost matrix, reducing the overall risk. However, the long-term fate or the 'locked up' organic contaminants remains uncertain.

IV. In-Vessel Composting of an Aged Coal-Tar Contaminated Soil: A Case Study

This section focuses on the optimization of operational parameters at laboratory scale using an aged naturally PAH-contaminated coal-tar soil, which will facilitate the highest possible removal of the target PAHs. Complementary studies on the fate of organic matter and microbial community dynamics during in-vessel composting are also presented.

A. Materials and Methods

Contaminated Soil

The coal tar contaminated soil was obtained from a manufactured gas plant site commissioned in 1838 at Clitheroe, Lancashire, United Kingdom. An extensive description of the site and the procedures for soil sampling and preparation were reported by Birnstingl [75]. Prior to use in the present study the coal tar contaminated soil was air-dried and homogenized by passing through a 5 mm followed by a 2 mm sieve and stored in the laboratory at room temperature. Before experimentation the coal tar contaminated soil was diluted by homogenizing with silver sand (1:1) to provide a more homogeneous distribution of the coal tar residue. The results presented here refer to post dilution coal tar contaminated soil, and all results are expressed as post dilution. Thus, the soil contained 100.3 mg $PAHs{\cdot}kg^{-1}$ soil, organic matter content was 4.79±0.16% (wt/dry wt) and soil pH_w was 7.3±0.1.

Green Waste

The soil was conditioned with an artificial green waste for the composting studies, which represents green waste to be used at field-scale based studies. Artificial green waste was prepared by mixing foodstuff (mixture of carrots, cucumber, lettuce, onions, potatoes and tomatoes in equal amounts) (3% dw), sawdust (38% dw), leaves (17% dw), grass (27% dw) and wheat straw (14% dw). Foodstuff, sawdust, wheat straw and leaves where separately blended and grass was cut. The composition of the green waste satisfied the nutrient requirement (C:N 40-50) according to the calculations using Cornell's system [76].

Reactor Design

A set of 360 glass composting reactors (200 ml) was constructed, which allowed conditions to be easily monitored and controlled (Figure 1). These fully enclosed bench-scale reactors each held about 65 g total composting mixture. Composting reactors were placed in triplicates in temperature controlled incubators at 38^0C, 55^0C and 70^0C to simulate representative mesophilic and thermophilic microbiological stages during composting processes [13,26]. For each glass composting reactor, the ingredients of the artificial green waste and soil in the particular S:GW ratio under investigation were thoroughly mixed in a glass beaker (500 ml), and then the mixture introduced into the reactor. Initial moisture content of the composting mixture was measured and double distilled water (DDW) was added when needed to reach the desirable moisture content for the different experimental conditions under investigation. Compost moisture content was measured at intervals to ensure that it was maintained at the required level, and amended with DDW when needed.

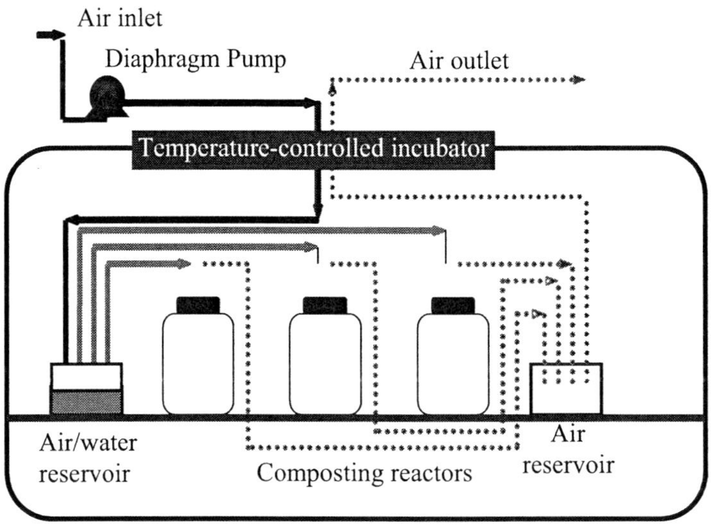

Figure 1. Laboratory set-up of composting reactors.

The reactor units stood vertically with air flowing continuously to avoid oxygen content limitation and vented outdoors to avoid volatiles accumulating in the composting reactors. Airflow up through the composting mixture via a stainless steel air-delivery tube inserted into the bottom of the composting reactors was provided by 100% oil free diaphragm pumps (Model PXW-600-DIOV, VP1, Fisher Scientific). The air inlet was bubbled through a DDW reservoir to avoid excessive water evaporation during aeration.

Sample Analysis

Destructive sampling, in triplicate, for each experimental treatment was executed periodically. The contents of each composting reactor were thoroughly mixed with a stainless steel spatula in a glass beaker (500 ml), and sub-samples were collected for analysis.

Total Organic Matter (TOM) in Composting Mixtures

Ash content was determined using a loss-on-ignition procedure [77]. Triplicate 5 g samples were dried at 110^0C for 24 h (moisture content) and then transferred to a muffle furnace at 550^0C for 12 h. Ash content was calculated from the ratio of pre- and post-ignition sample weights. The residual moisture of the samples was determined to produce the results on a dry matter basis (110^0C).

PAHs in Composting Mixtures

The extraction of 16-E.P.A. Priority PAHs from composting mixtures and soil was done by Accelerated Solvent Extraction (ASE^{TM} 200) and defined in this investigation as total PAHs for simplicity. Briefly, glass fibre disks were placed at the outlet end of the extraction cells and a 7g sample of compost was mixed with 3 g of sodium sulphate and 7 g of Hydromatrix^{TM} and introduced into each extraction cell. Surrogate standards (1-fluoronaphthalene, 2-fluorobiphenyl, purity > 97%, Greyhound Chromatography and Allied Chemicals, UK) were added to the cells prior to extraction to monitor PAH losses. Extraction cells were placed into the auto-sampler tray with copper turnings to remove sulphur. ASE^{TM}

200 conditions for PAH extraction were: 14 MPa (2000 psi), 100^0C, oven heat-up time = 5 min, static time = 5 min, solvent dichloromethane/acetone (1:1), (v/v), flush volume = 60% of extraction cell volume, nitrogen purge = 1 MPA (150 psi) for 60 s. The extracts were cleaned up on a column containing 1 g of activated-florisil (SiO_2, 84.0%; MgO, 15.5%; Na_2SO_4, 0.5%; 60/100 mesh; 130^0C; 12 h) and 2 g of anhydrous sodium sulphate. To remove hydrophobic impurities the column was initially washed with 10 ml dichloromethane before eluting 5 ml of extract and then leaving to dry for 1 min. The PAHs were then eluted with 10 ml dichloromethane.

Internal standards (napahthalene-d8, acenaphthene-d10 in a mixture with chrysene-d12, 1,4-dichlorobenzene-d4, perylene-d12, phenanthrene-d10, purity > 97%, Greyhound Chromatography and Allied Chemicals, UK) were added to the clean extracts prior to analysis. A Hewlett Packard 6890 series gas chromatograph with a 7673 series auto-sampler and a 5973 series mass selective detector was used for the analysis. The GC inlet was operated in pulsed (0.90 min, 30.0 psi) splitless mode at 270^0C with helium as carrier gas. The injection volume was 1 µl and the inlet purged at 50 $ml \cdot min^{-1}$ one minute after injection; the column flow was maintained at 1 $ml \cdot min^{-1}$. Separation was achieved using an HP-5MS column (19091S-433 30 m × 0.25 mm × 0.25 µm). The temperature program comprized 70^0C for 2 min, 10 centigrade degrees per minute to 300^0C where it was held until the end of the analysis (10 min). The MS transfer line was 280^0C providing conductive heating of the MS source to about 230^0C. The instrument was tuned using perfluorotributylamine. The MS was operated in selective ion monitoring (SIM) mode. The 16 US EPA PAHs (quantification ion/confirmation ions), internal standards (quantification ion/confirmation ion) and surrogates (quantification ion/confirmation ions) for SIM GC-MS mode were: naphthalene (128/127, 129, 102), naphthalene-d8 (136/137, 134, 108), 1-fluoronaphthalene (146/120, 125), 2-fluorobiphenyl (172/171, 170), acenaphthylene (152/151, 153, 76), acenaphthene (154/153, 152), acenaphthene-d10 (164/162, 160, 163), fluorene (166/139, 165), phenanthrene (178/165, 163, 82, 176), anthracene (178/179, 176, 89), fluoranthene (202/200, 101, 203), pyrene (202/200, 201, 101, 203), benzo[a]anthracene (228/226, 229), chrysene (228/226, 230, 113), benzo[b]fluoranthene (252/250, 253, 126), benzo[k]fluoranthene (252/253, 250, 126), benzo[a]pyrene (252/207, 253, 250, 126), indeno[1,2,3-c,d]pyrene (276/276, 279, 138), dibenzo[a,h]anthracene (278/279, 139, 276), benzo[g,h,i]perylene (276/138, 137, 277). The GC-MS system was calibrated prior to the analysis of samples using seven calibration standards.

Organic Matter of Composting Mixtures

Contour maps of excitation-emission matrix (EEM) spectra were obtained on water extracts of whole compost (1 g composting mixture; 50 ml DDW) after shaking in an orbital shaker SO1 (Stuart Scientific, UK) for 5 min at 75 $rev \cdot min^{-1}$, and filtration through Whatman No.2 filter paper. The pH of the extract was not adjusted but monitored. A Perkin Elmer LS45 fluorescence spectrophotometer was used, equipped with FL WINLAB Software for data processing. The excitation and emission slits were fixed to 10 nm. The emission (Em) wavelength range was fixed from 280 to 600 nm, whereas the excitation (Ex) wavelength was increased from 200 to 500 nm keeping a constant $\Delta\lambda$ of 20 nm steps at 500 $nm \cdot min^{-1}$.

Biomass

Analysis of bacteria, fungi and actinomycetes were by the dilution and spread-plate method following the "Standard Methods for the Examination of Water and Wastewater" [78] with minor modifications. Briefly, 10 g of the soil green-waste mixture sample were mixed with 90 ml of Ringers' solution and shaken for 10 min. Consecutive 1:10 dilutions were prepared, starting with 1 ml of sample to produce 8 dilutions of each sample. Then 0.1 ml of each dilution was spread onto 5 plates of nutrient agar (with cycloheximide) for bacteria, 5 plates of starch casein (with cycloheximide) for actinomycetes and 5 plates of potato dextrose agar (with rose bengal) for fungi. Cycloheximide was used to inhibit the growth of fungi from the soil, and rose bengal was used to suppress the growth of bacteria. Samples from the soil green-waste mixtures treated at 38^0C, 55^0C and 70^0C were incubated at 38^0C, 55^0C and 70^0C respectively for 72 h. Following incubation, plate colonies were counted.

Additionally, sub-samples of compost (2 g) were spiked with 500 $\mu g \cdot l^{-1}$ ^{13}C-palmitic acid (hexadecanoic acid, internal standard), the total lipid was extracted using the Bligh and Dyer procedure, and the lipid acyl chains and internal standard fatty acid converted to fatty acid methyl esters (FAME) using $MeOH/H_2SO_4$ [79]. The FAME were analysed using the same Hewlett Packard GC/MS system described above. The GC inlet was operated in pulsed (1.40 min, 40.0 psi) splitless mode at 260^0C with helium as carrier gas. The injection volume was 1 µl and the inlet was purged at 50 $ml \cdot min^{-1}$ 20 min after injection; inlet pressure was controlled by electronic pneumatics to maintain a constant column flow of 1 $ml \cdot min^{-1}$. Separation was achieved using an HP-5MS column (19091S-433 30 m × 0.25 mm × 0.25 µm). The temperature program comprised 40^0C for 3 min, 10 centigrade degrees per minute to 150^0C, 3 centigrade degrees per minute to 230^0C, and 30 centigrade degrees per minute to 300^0C, which was maintained for 5 min to allow late eluting peaks to exit the column. The MS transfer line was held at 310^0C, thus providing conductive heating of the MS source to about 230^0C. The MS was operated in selective ion monitoring (SIM) mode, using $m/z = 74$ as the common fragment ion of FAME. To identify the fatty acids, the retention times were compared with those obtained for standard bacterial acid methyl esters (Supelco, UK). The amount of microbial signature acids was calculated using the ^{13}C-16:0 internal standard, which gives a characteristic fragment ion $m/z = 75$ that can be quantified separately from the bulk ^{12}C-FAME in the sample.

The sum of the following fatty acids was used to represent total bacteria: i15:0, a15:0, i16:0, i17:0, cy17:0, 18:1ω7c and cy19:0 [1,80,81]. Gram-positive bacteria were represented by i15:0, a15:0 and i17:0 [82] and Gram-negative bacteria by cy17:0, 18:1ω7c and cy19:0 [81,83]. Thermophilic bacteria (mainly bacilli) were represented by i15:0 and i17:0 [84]. Fungi were represented by 18:2ω6,9 [80].

Data Analysis

Comparison of ΣPAHs disappearance under different composting conditions was investigated using paired *t*-tests. Factor analysis was used to simplify the complex and diverse relationships that exist among the operational parameters and to investigate the structure in the relationship between the operational parameters during in-vessel composting of PAH-contaminated solid waste. Factor analysis was executed, with factors extracted by the principal component method from the correlation matrix and varimax rotated factor loadings. The effect of different operational parameters on suggested indicators was investigated using a two-way multivariable ANOVA analysis and post hoc Tukey test. All statistical tests were performed using the latest version of *StatistiXL* Version 1.5.

B. PAHs Disappearance in an Aged Coal-Tar Contaminated Soil During in-Vessel Composting

Eighteen experimental conditions were tested in triplicate using 360 laboratory-scale composting reactors. The standard composting reactors comprised three temperature levels (T, 38^0C, 55^0C and 70^0C) four soil to green waste ratios (S:GW, 0.6:1, 0.7:1, 0.8:1 and 0.9:1 on a dry weight:weight basis) and three moisture contents (MC, 40%, 60% and 80%). Control reactors consisted of 1:0 S:GW ratio. The logistic approach to pinpoint the optimal operational conditions for maximum degradation of PAHs was, first to investigate the influence of S:GW 0.6-0.9:1 at three temperature levels and MC 60%, then to investigate the influence of MC 40-80% at three temperature levels and optimal S:GW ratio. Using this logistic approach allowed us to investigate eighteen experimental conditions at laboratory-scale rather than thirty six possible combinations.

The initial ΣPAH concentration in the investigated soil was 100.3 mg PAHs·kg^{-1} soil, which is lower than those concentrations of about 450 mg PAH·kg^{-1} soil/sediment reported in a typical manufacturing gas plant site by Erickson *et al.* [85]. However, they were above the Dutch List action level of 40 mg PAH·kg^{-1} air dried soil. The 16 US EPA-listed priority pollutant PAHs were grouped as 2- and 3-ring PAHs (naphthalene, acenaphthylene, acenaphthene, fluorene, anthracene, phenanthrene), 4-ring PAHs (fluoranthene, pyrene, benzo[a]anthracene, chrysene) and 5- and 6-ring PAHs (benzo[b]fluoranthene, benzo[k]fluoranthene, benzo[a]pyrene, dibenzo[a,h]anthracene, indeno[1,2,3-c,d]pyrene, benzo[g,h,i]perylene) and thus defined as small, medium and large PAHs respectively for ease of discussion. The concentrations of the 16 US EPA-listed PAHs investigated in the present study before treatment and after 98 days at three different temperatures (T, 38^0C, 55^0C and 70^0C), four soil to green-waste ratio (S:GW, 0.6:1, 0.7:1, 0.8:1 and 0.9:1) and three different moisture contents (MC,40%, 60% and 80%) as well as the control reactors PAH content are presented in Table 2. The PAH concentration in all reactors remained consistently distributed in the composting mixtures due to an initial extensive blending of soil and green waste. A mechanistic and fundamental investigation was undertaken to optimize the operational parameters during in-vessel composting of PAH-contaminated solid wastes.

Effect of Temperature

The effect of temperature on PAH-contaminated waste degradation was initially investigated at 38, 55 and 70^0C for as S:GW 0.6:1 and MC 60%. Temperature was an important factor affecting the in-vessel composting of this PAH contaminated soil (Table 2). PAH losses were greater at 38^0C (76.4% removed) than at 55^0C or 70^0C, with differences also observed between the two higher temperature treatments (52.2% and 22.3% removed respectively at 55^0C and 77^0C). Small PAHs were reduced by an average of 98.2% at 38^0C as compared to 53.0% and 16.9% at 55 and 70^0C respectively at S:GW 0.6:1 and MC 60%. Medium PAHs were reduced by 76.2%, 53.7% and 22.7% at 38^0C, 55^0C and 70^0C respectively and large PAHs were reduced by an average of 40% at 38^0C, 55^0C and 70^0C, S:GW 0.6:1 and MC 60%.

Table 2. PAH bioavailability and lost ($mg \cdot kg^{-1}$ dry soil) and PAH disappearance (% removal) after 98 days of continuous in-vessel composting treatment

Treatment MC/S:GW	Time (days)	38 ^{0}C PAH bioavailable	38 ^{0}C PAH lost	38 ^{0}C PAH removal	55 ^{0}C PAH bioavailable	55 ^{0}C PAH lost	55 ^{0}C PAH removal	70 ^{0}C PAH bioavailable	70 ^{0}C PAH lost	70 ^{0}C PAH removal
60%/ 0.6:1	21	50 ± 0	48 ± 2		54 ± 0	14 ± 4		46 ± 3	20 ± ±5	
	56	23 ± 2	13 ± 2		36 ± 5	39 ± 3		38 ± 1	2 ± 3	
	98	22 ± 2	15 ± 2	76.4	22 ± 9	0 ± 2	52.2	21 ± 7	0 ± 2	22.3
60%/ 0.7:1	21	52 ± 1	41 ± 11		62 ± 11	45 ± 8		52 ± 8	12 ± 8	
	56	26 ± 13	26 ± 16		66 ± 20	6 ± 11		34 ± 6	11 ± 5	
	98	18 ± 8	16 ± 4	82.0	41 ± 7	17 ± 2	67.7	24 ± 8	23 ± 3	46.0
60%/ 0.8:1	21	32 ± 2	36 ± 16		70 ± 7	37 ± 11		33 ± 2	4 ± 1	
	56	35 ± 8	40 ± 3		74 ± 11	27 ± 5		23 ± 7	19 ± 0	
	98	9 ± 0	2 ± 3	76.7	68 ± 9	0 ± 10	60.6	17 ± 1	33 ± 2	55.7
60%/ 0.9:1	21	32 ± 9	50 ± 15		29 ± 6	37 ± 6		42 ± 10	19 ± 2	
	56	17 ± 0	23 ± 2		25 ± 1	18 ± 2		23 ± 2	0 ± 2	
	98	8 ± 1	0 ± 2	69.1	30 ± 1	0 ± 2	62.4	22 ± 3	0 ± 1	54.0
40%/ 0.8:1	21	47 ± 1	18 ± 14		53 ± 6	10 ± 15		60 ± 1	11 ± 11	
	56	63 ± 2	22 ± 17		72 ± 3	9 ± 2		62 ± 6	2 ± 11	
	98	66 ± 1	0 ± 13	39.2	56 ± 13	0 ± 6	21.9	30 ± 18	0 ± 8	40.4
80%/ 0.8:1	21	28 ± 2	72 ± 7		33 ± 3	45 ± 13		50 ± 0	0 ± 9	
	56	34 ± 1	0 ± 4		48 ±	0 ± 0		43 ± 0	0 ± 12	
	98	24 ± 2	0 ± 8	68.4	65 ± 21	0 ± 7	51.2	46 ± 5	0 ± 10	34.4

PAH concentration at start: 100.3±3.2 $mg \cdot kg^{-1}$ dry soil.

Regarding concentrations of individual PAHs, chrysene, benzo[k]fluoranthene and benzo[a]pyrene were not removed at a S:GW ratio 0.6:1 and MC 60%, thus proving their recalcitrant nature. The final concentration of benzo[g,h,i]perylene was ~55% of the initial concentration following the three temperature levels treatments, which may indicate that the mechanisms of removal of this particular PAH may be due to mechanisms other than biological degradation. Benzo[g,h,i]perylene is expected to be strongly bound to the organic composting matrix due to its high hydrophobicity (log K_{OW} = 7.23) [6] and volatilisation is not expected to be an important fate process, given its Henry constant value ($2 \cdot 10^{-7}$ $atm \cdot m^3 \cdot mol^{-1}$) [6]. In addition, adsorption to the organic composting matrix is expected to attenuate its potential volatilization. However, under the investigated composting conditions, benzo[g,h,i]perylene may partially exist in particulate phase (vapour pressure 6×10^{-8} Pa at 25^0C) [6] or adsorbed onto fine particles, and it may have been transferred to the atmosphere.

Large PAHs are often difficult to biodegrade and different strategies have been investigated in an attempt to find the best conditions to remove these recalcitrant contaminants. Most of these strategies have been developed to overcome the low bioavailability of large PAHs, such as an increase in temperature [86] or use of surfactants [34,87] combined with the use of specialized microorganisms which degrade aromatic hydrocarbons[18,21,51]. Current composting approaches and technologies tend to emphasize the use of high temperatures (> 70^{0}C) in order to meet regulatory requirements for pathogen control [15]. Such temperatures severely inhibit the microbial diversity and hence enzymatic potential of the system [26]. What little research has been done suggests that the range between 40 and 45^{0}C allows for such maximum diversity to be expressed for both, bacteria, actinomycetes, and fungi [13,26]. It is of interest to note that this range represents an overlap point at which both mesophilic and thermophilic microorganisms will be viable and active.

Effect of Soil to Green Waste Ratio

The effect of soil to green waste ratio on PAH-contaminated waste degradation was investigated at 38, 55 and 70^{0}C (Table 2). A temperature of 38^{0}C proved to be the optimal for ΣPAHs removal. Of the 12 experimental conditions (S:GW 0.6-0.9:1, MC 60, T 38-70^{0}C), ΣPAHs removal at 38^{0}C was significantly higher than ΣPAHs removal at 55-70^{0}C ($p < 0.01$) over time; and ΣPAHs removal at 55^{0}C was as well higher than ΣPAHs removal at 70^{0}C ($p < 0.1$). Higher removal rates were apparent at S:GW 0.7-0.8:1, although no significant ($P>0.1$)differences in ΣPAH removal were found for different S:GW at MC 60%. Small PAHs and medium PAHs were removed by an average of 91.3% and 79.4% respectively at 38^{0}C, S:GW 0.7:1, 0.8:1, 0.9:1 and MC 60%. Average removal of large PAHs was 58.0% and 52.3% at S:GW 0.7:1 and 0.8:1 respectively as compared to 25.0% at S:GW 0.9:1, both at 38^{0}C and MC 60%. Increasing the S:GW ratio to 0.9:1 resulted in a lower removal of benzo[b]fluoranthene, benzo[k]fluoranthene and indeno[1,2,3-c,d]pyrene. Crawford *et al.* [45] reported a pilot study to assess the feasibility of treating 2-, 3- and 4-ring PAH-contaminated soils (about 100 mg·kg^{-1}) by composting leaves in small windrows (about 19 m^3 each). They found that the amendment ratio did not affect the extent of degradation of the PAHs. Our study corroborates the results of Crawford *et al.* [45] within the S:GW ratios investigated, but the S:GW ratio may influence the rate of removal of PAHs.

Effect of Moisture Content

The effect of moisture content on PAH-contaminated waste degradation was investigated at 40, 60 and 80% for a S:GW 0.8:1 and T 38, 55 and 70^{0}C (Table 2). The removal of ΣPAHs was significantly lower ($p < 0.05$) at a MC 40% as compared to 60% within the investigated temperature range, although no significant difference was found between MC 60 and 80%. As aforementioned, while removal of PAHs was also observed at 70^{0}C and MC 40-80%, this occurred to a lesser extend than at 38^{0}C or 55^{0}C. Small PAHs were removed by an average of 88.1% at 38^{0}C within the MC range under investigation. Medium PAHs were removed by an average of 75.8% at MC 60% and 80% as compared to 49.5% at MC 40% and 38^{0}C, and large PAHs were removed by 52.3% and 24.9% at MC 60% and 80% as compared to no removal at MC 40% and 38^{0}C. Potter *et al.* [42] investigated the degradation of 19 PAHs from a soil during in-vessel composting at the laboratory scale, using 70% soil and 30% corn-cob mixture on a dry weight basis, amended with cow manure, a modified OECD fertilizer, or activated sludge to adjust the nutrient content (CNP 100:5:1). They maintained a MC of 30-35%, aerobic conditions, and temperature range 41-53^{0}C. Potter *et al.* [42] reported that 2-

and 3-ring PAHs were removed by an average of 87% in all composters after 84 days of treatment, and 4-ring were reduced by an average of 61%; however, none of the amendments conditions appeared effective in degrading 5- and 6-ring PAHs. This investigation is in agreement with Potter *et al.* [42] results, and emphasizes the importance of MC optimisation during in-vessel composting-remediation of contaminated wastes.

Optimal Conditions for PAH Disappearance during in-Vessel Composting

The factor analysis result using 18 operational conditions indicated that 3 factors could explain 99.8% of the total variance [43]. Thus, for a simpler and easier interpretation, factor rotation was performed using varimax rotation after factor extraction, and rotation was normalized. On the basis of these analyses, optimal operational conditions for degradation of PAHs would occur at MC 60%, S:GW 0.8:1 and T 38^{0}C. In this study it appears that 38^{0}C enhances the biological removal of PAHs, which might occur due to promotion of the native microbial population and activity [2,26]. Temperature thus is the factor that more strongly affects the disappearance of PAH under in-vessel composting conditions. In addition, higher temperatures may facilitate PAHs volatilization [51,88] and result in a combination of various removal mechanisms, thus higher temperatures should be avoided to minimize the transfer of PAHs from the solid phase to the atmospheric phase. Moisture content is the second most important factor, and thus water amendments may be normally necessary during composing processes. It has been previously reported that the amendment of PAHs contaminated soils may enhance PAH removal [64]. For an elevated operational yield (i.e. higher amount of solid waste treated), a higher S:GW ratio will be advantageous, although a S:GW ratio that is too high may have inhibitory effects on the resident microbial population.

Kinetics of PAH Removal during in-Vessel Composting

Most of the PAH losses occurred within the first 45 days of treatment with a plateau forming by 98 days. These temporal profiles suggested typical biphasic kinetics indicating relatively rapid removal of PAHs during the initial phase of treatment followed by slower removal during the later phase. The rates of removal of PAHs were estimated by linear regression of the temporal profiles of PAH concentration over 98 days of continuous treatment. First order constants of ΣPAH reached a maximum of about 0.014 day^{-1} during optimal composting conditions at S:GW 0.8:1, MC 60% and T 38^{0}C. The removal rates reported in this investigation are in agreement with those reported by McFarland and Qiu [51] and Potter *et al.* [42] during composting of PAH-contaminated wastes. The reduction in degradation over time in the kinetic study can be explained by reduced bioavailability of PAHs due to immobilization in micropores or changes in binding forms [27,50,89].

C. Organic Matter Dynamics in an Aged Coal-Tar Contaminated Soil During in-Vessel Composting

The application of composting as a bioremediation technology of contaminated wastes has two goals: firstly, to maximize disappearance of the contaminant(s); secondly, to produce mature compost that could be used in land restoration for industrial, municipal or housing developments depending on final quality. By definition, composting accelerates the processes involved in the biological transformation of organic matter under controlled aerobic conditions to produce a stabilised product, compost, which implies the formation of humic-like substances (HS) [4]. The temporal dynamics of HS during composting may be used to determine the state of composting and the stability or maturity of the compost [4,90,91]. Although maturity of compost is an important factor, with agricultural and environmental implications [92-94], the majority of the reported studies on the use of composting approaches to bioremediate PAH contaminated wastes fail to present results on the assessment of maturity of the final compost [6]. This might be due to various reasons, but the lack of so-called "easy-to–perform" methodologies to characterize the maturity of composting/compost mixtures is certainly one of the most important. Among many available techniques [91,93,95,96], fluorescence spectroscopy has been applied to the characterization and differentiation of humic and fulvic acids of different origins and natures [97,98]. Further fluorescence spectroscopy in the excitation, emission, and synchronous scan modes has been recently applied to the characterization of whole compost from domestic solid wastes sampled at 1, 7, 15 and 30 days of windrow-composting treatment [91]. In this section, the dynamics of organic matter of a coal-tar contaminated soil during in-vessel composting-bioremediation will be assessed by using excitation-emission matrix (EEM) fluorescence spectroscopy.

Total Organic Matter Evolution During Composting

The TOM levels were high (approx. 62%) at the beginning of the composting process and then decreased, reaching a value of approx. 40% after 98 days treatment at 38^{0}C, thus indicating the occurrence of mineralization. At higher temperatures the TOM decrease was less, which may indicate that mineralization occurred to a lesser extent because higher temperatures constrained microbial growth [26]. That the TOC and ΣPAH concentration decreased during composting is consistent with this interpretation, and indicated that T, S:GW ratio and MC have an important influence on the behaviour of PAHs during composting.

To test whether bioavailability might be a limiting factor on the kinetics of in-vessel composting, relationships between the first order constant of losses and the total organic carbon content following 98 days of continuous treatment were sought. A linear correlation of k_1 (day^{-1}) *vs.* TOM (%) was found (R = 0.88), indicating that lower disappearance rates corresponded with higher organic content in the composting mixtures (Figure 2). Outliers occurred at operational conditions where MC was 40% and also at MC 60%, S:GW 0.6:1 and 70 ^{0}C where, in spite of the periodical water amendment, the reactors were probably under water stress conditions during long periods within the treatment process.

Correlations between microbial degradation of sequestered hydrophobic contaminants and the TOM of the soil in which the contaminant is present have been found [99,100], although TOM is not the sole contributor to the sequestration and desorption processes [72,100,101]. Table 2 summarizes the amount of ΣPAH bioavailable *vs.* the amount of ΣPAH lost after 21, 56 and 98 days of continuous composting treatment under the eighteen different operational conditions investigated. At 38 ^{0}C the majority of the ΣPAH bioavailable was lost. An increase in temperature treatment resulted in a decrease in the ΣPAH bioavailable and a decrease in the ΣPAH lost. At 70 ^{0}C a lower amount of ΣPAH was bioavailable and also a lower amount of ΣPAH was lost as compared to 55 ^{0}C for the identical MC and S:GW conditions. This indicates that different physico-chemical conditions in the composting mixtures due to different operational conditions may have changed the organic matter structure in a way that it has entrapped higher amounts of PAHs at higher temperatures, making them less available.

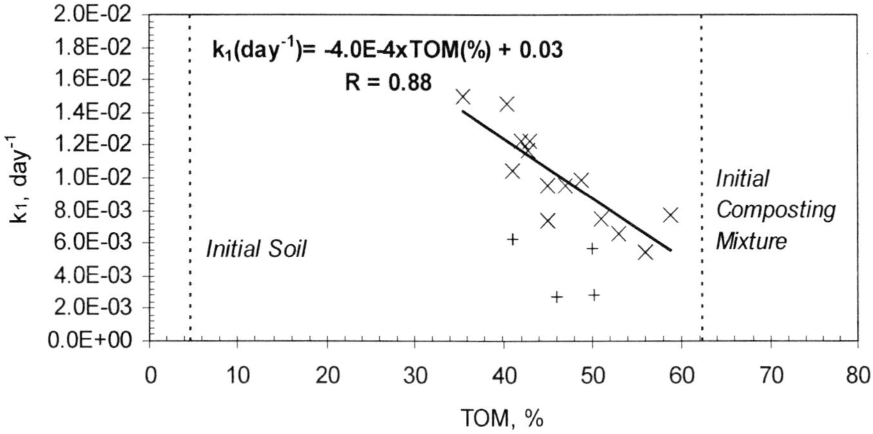

Figure 2. Kinetics of losses of ΣPAH vs. TOM after 98 days of continuous treatment for the eighteen experimental conditions investigated. (Legend: x refers to experimental conditions where MC was 60-80%; + refers to experimental conditions where MC was 40%, and the particular case where T 70 ^{0}C, S:GW 60:1 and MC 60%).

Contour EEM Fluorescence Spectra in Composting Mixtures of a Coal-Tar Contaminated Soil as Compared to other EEM Fluorescence Spectra

Contour EEM spectra of the composting mixtures after 21, 56 and 98 days treatment at three temperature levels (T; 38 ^{0}C, 55 ^{0}C and 70 ^{0}C), four soil to green waste ratios (S:GW; 0.6:1, 0.7:1, 0.8:1 and 0.9:1 soil to green waste mixture ratio on a dry weight basis) and three moisture contents (MC; 40%, 60% and 80%) did not exhibit major differences among them. Figure 3 illustrates the EEM spectra of the composting treatment at S:GW 0.8:1 and MC 60%. The spectra indicate the presence of two different fluorophores, each characterized by an Ex/Em wavelength pair, ~ 338/440 nm and ~ 239/430 nm, designated as primary peak (PP) and secondary peak (SP) respectively. These wavelength pairs have been computed from 54 wavelength pairs identified at eighteen different operational conditions.

The location of EEM peaks in this study was compared with the location of EEM peaks reported in the literature [102]. Excitation and emission boundaries were operationally defined into five regions based on supporting literature [103] to ease the interpretation of the results. Thus, the two different fluorophores observed in this study fell in two different regions defined by the boundaries of Chen *et al.* [103]. The PP fell in the region defined by longer excitation wavelengths (> 280 nm) and longer emission wavelengths (> 380 nm) related to humic acid-like organics. The SP fell in the region defined by shorter excitation wavelengths (< 250 nm) and longer emission wavelengths (> 350 nm) related to fulvic acid-like region.

The maximum Ex/Em wavelength pair ~338/440 nm coincided with peaks typical of humic-like organic substances. Thus, the presence of PP indicates the formation of humic acids during simulated in-vessel composting of an aged coal-tar contaminated soil. Peaks in this region have been previously reported.

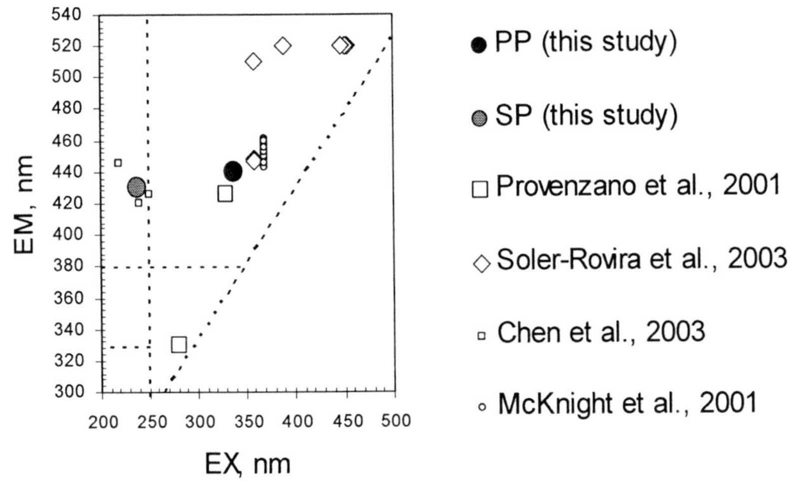

Figure 3. Location of EEM peaks based on literature reports and this study.

Provenzano *et al.* [91] reported peaks characterized by an Ex/Em wavelength pair 330/425 nm related to domestic solid waste following 30 days of composting treatment. Furthermore, Soler-Rovira *et al.* [97] reported peaks characterized by an Ex/Em wavelength pair 330/448 related to humic acids isolated from a composting mixture of a sewage sludge and wood chips and soil amended with this compost at a rate of 80 $Mg{\cdot}ha^{-1}$ in the presence and absence of barley crop after 180 days. McKnight *et al.* [98] reported peaks characterized by an Ex/Em wavelength pair 370/446 nm related to microbially derived organic matter.

Research in the literature indicated that fulvic acids (FAs) have only recently been investigated to monitor the maturity of composted wastes. Baddi *et al.* [104] reported a significant increase of nitrogen content, a high level of acidic functional groups and an increase of aromatic structures in FAs following 12 months of composting of olive mill wastes, which was related to a high degree of humification or maturation [94]. Molecular components which are potential contributors to the fluorescence of fulvic acids are hydroxycoumarins (Ex/Em wavelength pair 320-343/400-475 nm), chromone derivatives (Ex/Em wavelength pair 320-365/409-490 nm), flavones and isoflavones (Ex/Em wavelength pair 313-365/415-475 nm) and others [105]. These molecular components may derive from lignin and other degraded plant materials and subsequently possibly incorporated into the humic macromolecules during their formation (Senesi *et al.*, 1991). In this study a secondary peak characterized by an Ex/Em wavelength pair ~ 239/430 nm was observed, which additionally coincided with peaks reported by Chen *et al.* [103] typical of fulvic-acid like. Thus, the presence of SP indicates the formation of fulvic acids during simulated in-vessel composting of an aged coal-tar contaminated soil.

Impact of Temperature, Soil to Greenwaste Ratio and Moisture Content on Contour EEM Fluorescence Spectra

Following 98 days of continuous in-vessel composting treatment, the localization of the PP migrated to higher excitation wavelengths. When favourable operational conditions for degradation of PAHs were applied to the in-vessel composting reactors, the excitation wavelength increased from ~ 335 nm after 21 days of treatment to ~ 340 nm after 98 days of treatment. Similar shifts were noticed at MC 60% and S:GW 0.6:1-0.8:1 and MC 80% and S:GW 0.8:1. No shift of the emission peak position was observed at the mentioned operational conditions. No shift in the PP excitation wavelength was observed at MC 60% and S:GW 0.9:1 and MC 40% and S:GW 0.8:1. Differently, the localization of the SP did not migrate in the EEM fluorescence spectra at different operational conditions.

The optimal operational conditions for disappearance of PAHs from a contaminated aged coal-tar soil during in-vessel composting treatment occur at MC 60%, S:GW 0.8:1 and T 38 ^{0}C. By contrast, the poorest operational conditions for PAH disappearance was observed at MC 40%, S:GW 0.8:1 and MC 60%, S:GW 0.9:1. The EEM spectra in this investigation indicated that the treatment at optimal operational conditions for maximum degradation of PAHs coincide with the highest shift (~ 5 nm) in the excitation wavelength of the PP, while the worst operational conditions coincide with no shift in the excitation wavelength of the PP. The shift of PP toward higher excitation wavelengths may be indicative of the formation of increasing molecular size components, which is consistent with the formation of humic acid-like substances during composting treatment [106].

Fluorescence Intensity Evolution during in-Vessel Composting of an Aged Coal-Tar Contaminated Soil

An increase of fluorescence intensity may result from an increased molecular complexity associated with aromatic structures and conjugated double bonds. On the contrary, a decrease of fluorescence intensity suggests a lower molecular weight and lower degree of aromatic polycondensation [105], which may be related to mineralization of OM. In this study, an overall decrease of the fluorescence intensity of the PP and SP was observed under favourable operational conditions for loss of PAHs during simulated in-vessel composting ($p < 0.01$). Under these favourable conditions, the EEM spectra become simpler in appearance (less contour lines) with in-vessel composting time, which may be related to the mineralization or humification of organic matter that occur during composting [102]. However, the highest fluorescence intensities were observed at the worst operational conditions for biotic disappearance of PAHs, i.e. MC 40% - S:GW 0.8:1 and MC 60% - S:GW 0.9:1, regardless the temperature level, and the EEM spectra did not become simpler in appearance. The results from the multivariable statistical analysis indicated a significant influence of MC on the fluorescence intensity of PP ($p < 0.0001$) and the Tukey test indicated a significant higher fluorescence intensity of PP at MC 40% as compared to MC 60% and 80% ($p < 0.0001$). S:GW presented a less significant influence on the fluorescence intensity ($p < 0.05$) and T presented no significant influence on the fluorescence intensity. Complex EEM spectra may result due to the complex nature of the organic matter present in the coal-tar contaminated soil that under unfavourable conditions have not been metabolized and cause fluorescence in the composting mixture.

Fluorescence Index Evolution during in-Vessel Composting of an Aged Coal-Tar Contaminated Soil

McKnight *et al.* [98] suggested the use of a fluorescence index (ratio of fluorescence emission intensity at wavelength 450 nm to that at 500 nm, or E_{450}/E_{500}) to differentiate the microbially derived FAs (~ 1.9) from those of the terrestrially derived FAs (~ 1.4). In this study the fluorescence index measured at an excitation wavelength of 340 nm (instead of 370 nm used by McKnight *et al.* [98]) at the eighteen operational in-vessel composting conditions after 21, 56 and 98 days of continuous in-vessel composting treatment were recorded [102]. Fluorescence index ranged from ~ 1.5 to ~ 1.9 at 70 ^{0}C and 38 ^{0}C respectively, finding in general higher fluorescence indexes under favourable operational conditions for loss of PAHs during simulated in-vessel composting ($p < 0.01$). Additionally fluorescence indexes were correlated to TOC values of the composting mixture following 98days of continuous in-vessel treatment ($p < 0.01$). The results from the multivariable statistical analysis indicated a significant influence of temperature on the fluorescence index ($p < 0.01$). Fluorescence indexes at 38 ^{0}C were significantly different from those at 55 ^{0}C (Tukey-test, $p < 0.05$) and 70 ^{0}C (Tukey-test, $p < 0.005$). However, a dependence of fluorescence indexes on S:GW ratio or MC was not observed. The fluorescence index computed in this study during in-vessel composting of an aged coal-tar indicates the presence of microbially derived FAs, or microbially mediated mineralization. Additionally, these results indicate that although the fluorescence index was initially developed to elucidate the nature of fulvic acids in water media [98], this index may be also applied to elucidate the nature of humic substances in soil (composting) media.

D. Microbial Community Structure Changes in an Aged Coal-Tar Contaminated Soil During in-Vessel Composting

Active compost has a high level of microbial activity and a rapidly changing microbial community. Traditionally, the analysis of soil microbial communities has relied on culturing techniques using a variety of culture media designed to maximize the recovery of diverse microbial populations. However, only a small fraction (possibly < 0.1%) of the soil microbial community is amenable to investigation using this approach. To overcome these problems, other methods such as the analysis of phospholipid fatty acids (PLFA) have been used in efforts to study a greater proportion of the soil microbial community.

Analysis of Phospholipids Fatty Acids

In total, twenty three different microbial PLFA were identified during the composting process, although only eight were used as biomarkers in this study. The major shifts in the microbial community during the simulated in-vessel composting treatment could be ascertained using fungal to bacterial, and Gram-positive to Gram-negative bacterial PLFA ratios. The initial PAH-contaminated soil contained small amounts of PLFA that were indicative of fungi, giving a low value of the fungal to bacterial ratio (≈ 0.03); there were approximately equal proportions of Gram-positive to Gram-negative bacteria on the basis of signature PLFA content. When this PAH-contaminated soil was mixed with artificial green waste, the mixture contained PLFA indicative of fungi in a higher proportion than did the initial soil (fungal to bacterial PLFA ratio ≈ 1.29), with a dominance of Gram-positive bacteria (Gram-positive to Gram-negative ratio ≈ 1.97). Frostegard and Bääth [80] found that the fungal to bacterial PLFA ratio varied from 0.02-0.04 in different agricultural soils that were low in organic matter to a ratio of 0.3-0.5 in forest soils that were dominated by fungal biomass. Bolta *et al.* [1] found that the fungal to bacterial PLFA ratio in a composted household organic waste with shredded wood varied from 0.12 to 0.15. Andersson *et al.* [107]

reported a fungal to bacterial PLFA ratio of 0.5 in an aged PAH-contaminated soil mixed with birch wood. Fungal to bacterial PLFA ratios calculated in the present study for the initial PAH-contaminated soil and a mixture of the same soil with green waste correlate with the fungal to bacterial PLFA ratios reported in the literature. Thus, the initial composting mixture was dominated by fungal biomass.

Fungal to Bacterial Ratio Changes during Composting

Analysis of the structure of major microbial groups during composting revealed a high proportion of bacterial biomass over fungal biomass (fungal to bacterial PLFA ratio < 1) for the first three weeks that was maintained to the end of the experiment after 98 days (Figure 4). The relative proportion of fungi in the in-vessel composting reactors was lowest at the highest in-vessel composting temperature investigated, namely 70°C ($p < 0.05$).

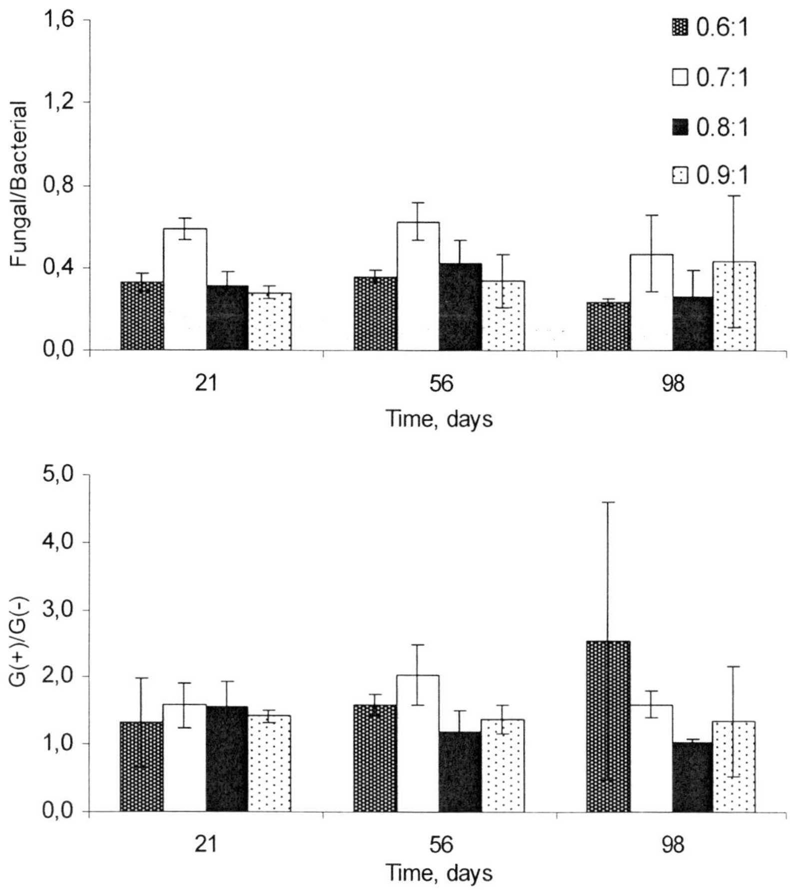

Figure 4. Temporal profiles of (a) fungal to bacterial PLFA ratio and (b) Gram-positive to Gram-negative bacterial biomass ratio at MC 60% and T 38^{0}C.

These results are comparable with the findings of Klamer and Bååth [83], who reported a rapid decrease in the fungal to bacterial PLFA ratio from 0.37 to 0.007 during the heating phase reaching 69^{0}C in the composting mixtures of straw materials. They observed a posterior increase in the fungal to bacterial PLFA ratio which never exceeded 0.1 after 100 days of treatment. However, Carpenter-Boggs *et al.* [84] reported that the PLFA markers for fungi

(18:2ω6c, 18:3ω6c) did not change significantly over 60 days following a conventional composting temperature profile that reached a maximum temperature of 60°C, although they suggested that the types of fungi that were present and active at any stage may have varied significantly. Bolta *et al.* [1] reported a temporal decrease in the relative proportion of fungi during the course of in-vessel composting of a mixture of shredded wood plus organic household waste, in which the average temperature during the first 17 days was close to 60^0C, decreasing to 15^0C after 30 days. No significant influence of the S:GW ratio or MC on the fungal to bacterial ratio was observed in the composting reactors.

Fungal to bacterial PLFA ratios were within the range 0.02 to 0.56 after 98 days of continuous composting treatment, showing that fungi are an important microbial group for in-vessel composting of a contaminated soil, and that drastic changes in microbial community structure occur during in-vessel composting at different operational conditions.

Gram-Positive to Gram-Negative Bacterial Ratio Changes during Composting

Temperature, S:GW ratio and MC all had a significant influence on the Gram-positive to Gram-negative bacterial ratios ($p < 0.05$) during the in-vessel composting treatment (Fig. 4). Thermophilic organic composting systems are largely comprised of bacilli and actinomycetes, and thus a higher relative proportion of Gram-positive bacterial PLFA would be expected at higher temperatures . This was observed at 38 and 55^0C, but not at 70^0C following 98 days of continuous in-vessel composting treatment ($p < 0.001$), probably because 70^0C is above the upper growth limits of such bacteria. The proportions of Gram-positive bacteria at 38^0C and 55^0C were similar, rising during the early stages of composting to a plateau after 21 days of treatment that was maintained to the end of the experiment after 98 days. A similar observation was made by Carpenter-Boggs *et al.* [84], who used an initial temperature of 60^0C decreasing to near 42^0C and 22^0C after an average 28 and 56 days, respectively, in a conventional composting treatment. A high ratio of Gram-positive (which include thermophiles) to Gram-negative bacteria, corresponded to the presence of a large amount of branched-chain fatty acids such as i15:0 and i17:0 that are common in species of *Bacillus*, a genus well known to be dominant in compost at high temperatures [108,109].

In general, the Gram-positive to Gram-negative bacterial PLFA ratio increased while PAH concentration in the composting mixtures decreased, which indicated that Gram-positive bacteria were inhibited in the presence of a high concentration of PAHs. Accordingly, the Gram-positive to Gram-negative bacterial ratio decreased from 21 to 98 days of in-vessel composting, but only in those bioreactors having the larger populations. These results are similar to the findings of Carpenter-Boggs *et al.* [84], who reported a decrease of indicators of general bacteria (15:0 and 17:0) and aerobic bacteria (16:1ω7c) over time. A different trend was observed in the composting reactors treated at 70^0C following 98 days of treatment, which showed a slight decrease of Gram-positive to Gram-negative bacterial ratio at lower PAH concentrations in the composting reactors. The effects of soil to green waste ratio and moisture content on the Gram-positive to Gram-negative bacterial ratio were also investigated. The relative proportion of PLFA indicative of Gram-negative bacteria increased with respect to Gram-positive bacteria at 60% moisture content and a soil to green waste ratio of 0.8:1 to 0.9:1 ($p < 0.01$). The Gram-positive to Gram-negative bacterial ratio changed significantly following the first 21 days of in-vessel composting treatment ($p < 0.05$) but thereafter no significant changes were observed when the moisture content varied from 40 to 80%.

E. Different in-Vessel Composting Approaches to Remove PAHs in an Aged Coal-Tar Contaminated

One important question yet to be answered for the in-vessel composting approach is how fresh green waste compost (FWGC) instead of green waste (GW), used as composting amendment will impact on the disappearance of the contaminant(s). Thus, the fate of 16 US EPA-listed PAHs during in-vessel composting of the same coal-tar contaminated soil amended with either green waste or fresh green waste composts collected from two landfill sites in the UK were compared during 56 days of continues treatment. Soil to amendment ratio was 0.8:1 on a dry weight basis, MC 60% and T 38^0C, 55^0C and 70^0C [110].

The TOM levels at the beginning of the composting process in the soil amended with the FGWC from Site 1 and Site 2 were 25.60±0.61% and 32.49±0.06% respectively. A decrease in TOM levels was observed in the soil/compost mixture of the soil amended with FGWC from Site 1 treated at all temperature regimes, to 21.48±0.16% within the first 21 days at 38^0C. The soil/compost mixture from Site 2 also presented a significant decrease in TOM level of 23.64±1.04% ($p < 0.005$) treated at 55^0C. Using FGWC instead of GW resulted in a lower disappearance of ΣPAHs during in-vessel composting of the same coal-tar contaminated soil (Figure 5). Both GW and FGWC sustain populations of microorganisms with the potential to degrade a variety of organic contaminants and they can improve the contaminated soil environment for indigenous or introduced microorganisms by changing the soil pH, nutrient status, aeration, and moisture retention characteristics [58]. Amendment with GW facilitated a higher oxygen transfer than FGWC amendment and this may explain why a higher disappearance of ΣPAHs was observed in the soil amended with GW than in the same soil amended with FGWC.

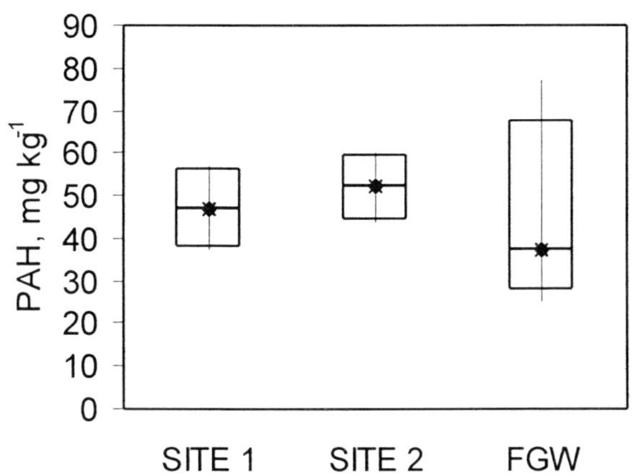

Figure 5. Concentration of PAHs at different temperature profiles after 56 days of in-vessel composting-bioremediation of a coal-tar contaminated soil amended with FGWC from Site 1, Site 2 and GW.

One major concern of using compost as a bioremediation approach is the problem of mixing non-contaminated material with contaminated soil, resulting in a greater quantity of contaminated material if the treatment does not succeed [6]. Thus, although the use of FGWC may offer important operational advantages during the application of the technology at a field scale (i.e. homogenization), this investigation indicated that in order to obtain a higher disappearance of ΣPAHs during in-vessel composting of an aged coal-tar contaminated soil, the use of GW as amendment resulted in a higher disappearance of ΣPAHs than using FGWC as amendment.

V. Regulatory and Sustainable Aspects of In-Vessel Composting as a Bioremediation Technology

Building a more sustainable society by using more sustainable waste management strategies to achieve recycling and landfill diversion, is not only concerned with technical aspects and optimization of physico-chemical and biological parameters, but is also very connected with regulatory and sustainable aspects.

A. Regulatory Aspects

In Europe, the Dutch List provides criteria relating to the clean up of PAH-contaminated soils, although regulatory objectives and priorities vary by country [6]. In the United Kingdom there is an action level of 500 mg $PAH{\cdot}kg^{-1}$ air-dried soil for land used for recreation and 10,000 mg $PAH{\cdot}kg^{-1}$ air dried soil for land with a hard covering. According to Italian regulatory guidelines, the maximum PAHs content in soils is 10 mg $PAH{\cdot}kg^{-1}$ for residential use and 100 mg $PAH{\cdot}kg^{-1}$ for industrial or commercial use. The United Kingdom and Italy take the view that land should be cleaned to make it fit for its intended use. By contrast, The Netherlands adopt a multifunctional approach whereby land must be fit for any use. Consequently, soil quality limits are much lower than in the United Kingdom, *i.e.* The Netherlands has an action level of 40 mg $PAH{\cdot}kg^{-1}$ air-dried soil [71,111].

Regulatory objectives and priorities relating to composting of biodegradable waste also vary by country. In the United Kingdom, composting is classified as a waste recovery operation under the Waste Framework Directive. In addition, composting of waste is a vital component of meeting the Waste Strategy 2000 [16] targets for recycling and composting set at 30% by 2010 and 33% by 2015. In Europe, the EC Landfill Directive 1999 [112] sets a target for reduction of biodegradable waste to landfill of 25% by 2010, 50% by 2013 and 65% by 2020. In relation to the longer-term requirements of the EC Landfill Directive, composting of kitchen waste, green waste, and general biodegradable organic waste, which might include contaminated soil, will probably have an important role to play [113]. Additionally, the EU Animal By-Products Regulation [15] requires that the composting of catering waste containing meat (or meat-derived products) must take place in a "closed composting reactor", operated at least $70^{0}C$ for 1 hour. This means that catering waste containing meat cannot be treated in an open windrow, except as a second stage after it has first been treated in a closed reactor.

Figure 6 summarizes the regulations to consider at the different stages of the application of composting-bioremediation to PAH-contaminated waste. The 'A' indicates regulations setting up an action concentration of PAHs present in soil, i.e. the New Dutch List, whilst 'B' indicates regulations dealing with organic wastes to compost, i.e. EC Landfill Directive.

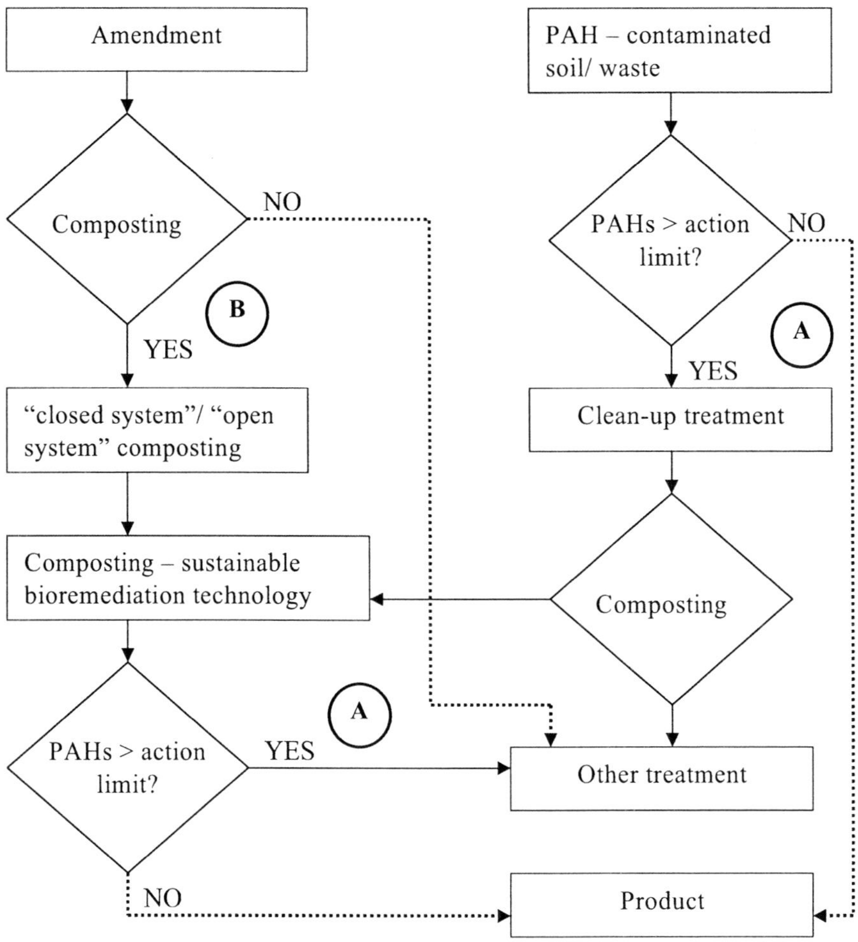

Figure 6. Regulatory considerations for bioremediation of PAH-contaminated soil using composting approaches.

B. Sustainable Aspects

Traditionally biodegradable waste has not been source-segregated and it has been buried in landfills, decomposed via a complex series of microbial and abiotic reactions, producing mainly methane (CH_4) and carbon dioxide (CO_2). The EC Landfill Directive 1999 [112] imposes strict engineering requirements on landfills, requiring landfill CH_4 to be captured and used. In the United Kingdom (UK), there are more than 1,000 landfill sites [114], only ca. 150 engineered landfills [115], thus the largest source of CH_4 emission is from landfill sites where CH_4 escapes through the landfill cover into the atmosphere. A low-cost approach such as increasing microbial activity to increase CH_4 oxidation using a suitable cover layer can help reducing CH_4 emissions from landfills.

This chapter has indicated that an aged PAH-contaminated coal-tar soil mixed with a biodegradable amendment can be treated by applying composting as a bioremediation technology, and accomplish a final concentration of PAHs below the regulatory limits. Thus, the final compost may be applied to land. It is here suggested that in order to abate CH_4

emission from landfill sites, compost obtained from the treatment of a PAH-contaminated soil/waste using composting as a bioremediation technology would be applied as a landfill cover. By doing so, global CH_4 emissions will be reduced since during composting there are no CH_4 emissions, and posterior application of compost as a landfill cover will reduce possible CH_4 emissions and consequently global warming.

Compost produced from municipal solid waste used as a fertilizer has been demonstrated to improve the quality of soil [116,117]. The application of compost produced following composting of a PAH-contaminated soil or waste may raise public concern if applied to food crops. However, the compost produced following composting of a PAH-contaminated soil or waste may be applied to energy crops. The use of compost as a fertilizer will not only reduce the energy requirements to synthesize inorganic fertilizers, but will also contribute to the production of biofuels. The use of biofuels for transport is becoming increasingly important for a number of reasons such as environmental concerns relating to climate change, depletion of fossil fuel reserves, and reduction of reliance on imports [118,119].

Therefore, the use of in-vessel composting as a sustainable bioremediation technology may contribute to reducing the amount of biodegradable waste buried in landfills and additionally lower CH_4 greenhouse gas emission from traditional waste management practice. Additionally, compost may contribute to energy production. This approach agrees with the recent European Sustainable Development Strategy, which addresses seven key challenges as priorities until 2010, being climate change and clean energy, better management and natural resources among them [108].

VI. Conclusions

Whilst more research on comparisons of composting/compost bioremediation technologies vs. conventional technologies remain limited, the investigations presented in this chapter have proved that composting, regardless the approach used, is a good sustainable environmental technology that may be used to remove PAH from contaminated soils and wastes. These bioremediation technologies have proved to be beneficial for the amelioration of contaminated soils by reducing the concentration of PAHs, which promotes soil sustainability and soil re-use in contrast to landfill or incineration approaches. Additionally, the final compost may be applied as a landfill cover to abate CH_4 emission from non-engineered landfills and contribute to sustainable development.

Acknowledgments

We are grateful to Cleanaway Ltd. and London Remade for providing support for this study through the Entrust scheme. The contribution of Dr Angus J. Beck and Dr Joe Lopez-Real to the realization of the research work summarized here and reported elsewhere [6, 26, 43, 102, 110] is specially acknowledged. We also thank Dr Jeremy Birnstingl for providing the coal-tar contaminated soil and Miss Katarina Spanova and Mr Mark Bennett for their assistance in the chemical analysis.

References

[1] S.V. Bolta, R. Mihelic, F. Lobnik, D. Lestan, Microbial community structure during composting with and without mass inocula, *Compost. Sci. Util.* 11 (2003) 6-15.

[2] C. Liang, K.C. Das, R.W. McClendon, The influence of temperature and moisture contents regimes on the aerobic microbial activity of a biosolids composting blend, *Bioresour. Technol.* 86 (2003) 131-137.

[3] J.S. VanderGheynst, F. Lei, *Microbial community structure dynamics during aerated and mixed composting,* 46 (2003) 577-584.

[4] E. Epstein, *The Science Of Composting,* Technomic Publishing Company, Lancaster, 1997.

[5] M.S. Finstein, F. Miller, P.F. Strom, Waste treatment composting as a controlled system. in: H. Rehm, G. Reed (Eds.), *Biotechnology,* Verlag Chemie, Weinheim, Federal Republic of Germany, 1986, pp. 363-398.

[6] B. Antizar-Ladislao, J.M. Lopez-Real, A.J. Beck, Bioremediation of polycyclic aromatic hydrocarbon (PAH)-contaminated waste using composting approaches, *Crit. Rev. Environ. Sci. Technol.* 34 (2004) 249-289.

[7] R.M. Seymour, D. Donahue, M. Bourdon, J.R. Evans, D. Wentworth, Intermittent aeration for in-vessel composting of crab processing waste, *Compost. Sci. Util.* 9 (2001) 98-106.

[8] B.S. Fraser, A.K. Lau, The effects of process control strategies on composting rate and odor emission, *Compost. Sci. Util.* 8 (2000) 274-292.

[9] K. Das, H.M. Keener, Numerical model for the dynamic simulation of a large scale composting system, *Transactions ASAE* 40 (1997) 1179-1189.

[10] V. Sasek, M. Bhatt, T. Cajthaml, K. Malachova, D. Lednicka, Compost-mediated removal of polycyclic aromatic hydrocarbons from contaminated soil, *Arc. Environ. Contam. Toxicol.* 44 (2003) 336-342.

[11] R.T. Haug, *The Practical Handbook of Compost Engineering*, Lewis Publishers, CRC Press, Boca Raton, Florida, 1993.

[12] R.T. Williams, K.R. Keehan, Hazardous and Industrial Waste Composting. in: H.A.J. Hoitink, H.M. Keener (Eds.), *Science and Engineering of Composting: Design, Environmental, Microbiological and Utilization Aspects*, Renaissance Publications, Worthington, Ohio, 1993, pp. 363-381.

[13] E.J. Walter, J.M. Lopez-Real, J. Wharfe, Composting of sewage sludge and straw: Laboratory scale simulation and evaluation of selected temperatures and effect on composting performance. in: C. Balis, M. de Bertoli, G.L. Ferrero, V. Manow, E. Kapetanios (Eds.), *ISHS Acta Horticulturae 302*, American Society for Horticultural Sciences, Athens, Greece, 1992, pp. 113-124.

[14] P. Bardos, J.M. Lopez-Real, The composting process and susceptible feedstocks, temperature, microbiology, sanitisation and decomposition. in: W. Bidlingmaier, P. L'Hermite (Eds.), *Compost Processes in Waste Management,* Guyot SA, Belgium, 1989, pp. 179-190.

[15] EC, EU Animal By-Products Regulations (2003/31/EEC), L273/1-95. *European Commission*, 2003.

[16] DETR, *Waste Strategy 2000: England and Wales*, London, Department of the Environment, Transport and the Regions: The Stationary Office, 2000.

[17] V. Sasek, T. Cajthaml, M. Bhatt, Use of fungal technology in soil remediation: a case study, *Water Air Soil Pollut.* 3 (2003) 5-14.

[18] M. Bhatt, T. Cajthaml, V. Šašek, Mycoremediation of PAH-contaminated soil, *Folia Microbiol.* 47 (2002) 255-258.

[19] T. Eggen, Application of fungal substrate from commercial mushroom production - *Pleurotus ostreatus* - for bioremediation of creosote contaminated soil, *Int. Biodet. Biodeg.* 44 (1999) 117-126.

[20] T. Eggen, V. Šašek, Use of edible and medicinal *Oyster Mushroom* [*Pleurotus ostreatus* (Jacq.:Fr.) Kumm.] spent compost in remediation of chemically polluted soils, *Int J Medicinal Mushrooms* 4 (2002) 255-261.

[21] R. Canet, J.G. Birnstingl, D.G. Malcolm, J.M. Lopez-Real, A.J. Beck, Biodegradation of polycyclic aromatic hydrocarbons (PAHs) by native microflora and combinations of white-rot fungi in a coal-tar contaminated soil, *Bioresour. Technol.* 76 (2001) 113-117.

[22] R. Canet, J.M. Lopez-Real, A.J. Beck, Overview of polycyclic aromatic hydrocarbon biodegradation by white-rot fungi, *Land Contam. Reclam.* 7 (1999) 191-197.

[23] S.H. Ferguson, P.D. Franzmann, I. Snape, A.T. Revill, M.G. Trefry, L.R. Zappia, Effects of temperature on mineralisation of petroleum in contaminated, *Antarctic terrestrial sediments*, 52 (2003) 975-987.

[24] R. Cavicchioli, *Extremophiles and the search for extraterrestrial life*, 2 (2002) 281-292.

[25] D.A. Cowan, N.J. Russell, A. Mamais, D.M. Sheppard, *Antarctic Dry Valley mineral soils contain unexpectedly high levels of microbial biomass*, 6 (2002) 431-436.

[26] B. Antizar-Ladislao, J. Lopez-Real, A.J. Beck, Laboratory studies of the remediation of polycyclic aromatic hydrocarbon contaminated soil by in-vessel composting, *Waste Manag.* 25 (2005) 281-289.

[27] A.J. Beck, D.L. Johnson, K.C. Jones, The form and bioavailability of non-ionic organic chemicals in sewage sludge-amended agricultural soils, *Sci. Total. Environ.* 185 (1996) 125-149.

[28] P.H. Dyke, C. Foan, H. Fiedler, PCB and PAH releases from power stations and waste incineration processes in the UK, *Chemosphere* 50 (2003) 469-480.

[29] P. Fernandez, J.O. Grimalt, R.M. Vilanova, Atmospheric gas-particle partitioning of polycyclic aromatic hydrocarbons in high mountain regions of Europe, *Environ. Sci. Technol.* 36 (2002) 1162-1168.

[30] C.A. Brandt, J.M. Becker, A. Porta, Distribution of polycyclic aromatic hydrocarbons in soils and terrestrial biota after a spill of crude oil in Trecate, Italy, *Environ. Toxicol. Chem.* 21 (2002) 1638-1643.

[31] R.C. Brenner, V.S. Magar, J.A. Ickes, J.E. Abbott, S.A. Stout, E.A. Crecelius, L.S. Bingler, Characterization and fate of PAH-contaminated sediments at the Wyckoff/Eagle Harbor superfund site, *Environ. Sci. Technol.* 36 (2002) 2605-2613.

[32] H.T. Yu, Environmental carcinogenic polycyclic aromatic hydrocarbons: photochemistry and phototoxicity, *J. Env. Sci. Health.* C 20 (2002) 149-183.

[33] F. Haeseler, D. Blanchet, V. Druelle, P. Werner, J.P. Vandecasteele, Ecotoxicological assessment of soils of former manufactured gas plant sites: Bioremediation potential and pollutant mobility, *Environ. Sci. Technol.* 33 (1999) 4379-4384.

[34] Z. Zheng, J.P. Obbard, Oxidation of polycyclic aromatic hydrocarbons (PAH) by the white rot fungus, *Phanerochaete chrysosporium*, *Enzyme Microbial Technol.* 31 (2002) 3-9.

[35] G. Maini, A.K. Sharman, C.J. Knowles, G. Sunderland, S.A. Jackman, Electrokinetic remediation of metals and organics from historically contaminated soil, *J. Chem. Technol. Biotechnol.* 75 (2000) 657-664.

[36] A.O. Thomas, J.N. Lester, The microbial remediation of former gasworks sites - A review, *Environ. Technol.* 14 (1993) 1-24.

[37] M. Eriksson, E. Sodersten, Z.T. Yu, G. Dalhammar, W.W. Mohn, Degradation of polycyclic aromatic hydrocarbons at low temperature under aerobic and nitrate-reducing conditions in enrichment cultures from northern soils, *Appl. Environ. Microbiol.* 69 (2003) 275-284.

[38] D. Dean-Ross, J. Moody, C.E. Cerniglia, Utilization of mixtures of polycyclic aromatic hydrocarbons by bacteria isolated from contaminated sediment, *FEMS Microbiol. Ecology* 41 (2002) 1-7.

[39] E.J. Joner, C. Leyval, Rhizosphere Gradients of Polycyclic Aromatic Hydrocarbon (PAH) Dissipation in Two Industrial Soils and the Impact of Arbuscular Mycorrhiza, *Environ. Sci. Technol.* 37 (2003) 2371-2375.

[40] C. Mougin, Bioremediation and phytoremediation of industrial PAH-polluted soils, *Polycyclic Aromatic Compounds* 22 (2002) 1011-1043.

[41] T.F. Guerin, The differential removal of aged polycyclic aromatic hydrocarbons from soil during bioremediation, *Environ. Sci. Pollut. Res.* 7 (2000) 19-26.

[42] C.L. Potter, J.A. Glaser, L.W. Chang, J.R. Meier, M.A. Dosani, R.F. Herrmann, Degradation of polynuclear aromatic hydrocarbons under bench- scale compost conditions, *Environ. Sci. Technol.* 33 (1999) 1717-1725.

[43] B. Antizar-Ladislao, J. Lopez-Real, A.J. Beck, Degradation of polycyclic aromatic hydrocarbons (PAHs) in an aged coal-tar contaminated soil under in-vessel composting conditions, *Environ. Pollut.* 141 (2006) 459-468.

[44] ISPAH, International Society for Polycyclic Aromatic Compounds, in: www.ispac.org/.

[45] S.L. Crawford, G.E. Johnson, F.E. Goetz, The potential for bioremediation of soils containing PAHs by composting, C*ompost. Sci. Util.* 1 (1993) 41-47.

[46] K.D. Racke, C.R. Frink, Fate of organic contaminants during sewage-sludge composting, *Bull. Environ. Contam. Toxicol.* 42 (1989) 526-533.

[47] A.O. Adenuga, J.H. Johnson, J.N. Cannon, L. Wan, Bioremediation of PAH-contaminated soil via in-vessel composting, Water Sci. Technol. 26 (1992) 2331-2334.

[48] M. Civilini, Fate of creosote compounds during composting, *Microbiol.* Europe 2 (1994) 16-24.

[49] X.J. Qiu, M.J. McFarland, Bound residue formation in PAH contaminated soil composting using *Phanerochaete-chrysosporium*, Hazard. *Waste Hazard. Mat.* 8 (1991) 115-126.

[50] M.J. McFarland, X.J. Qiu, J.L. Sims, Remediation of petroleum impacted soils in fungal compost bioreactors, *Water Sci. Technol.* 25 (1992) 197-206.

[51] M.J. McFarland, X.J. Qiu, Removal of benzo(a)pyrene in soil composting systems amended with the white-rot fungus *Phanerochaete-chrysosporium*, *J. Hazard. Mater.* 42 (1995) 61-70.

[52] J.F. Joyce, C. Sato, R. Cardenas, R.Y. Surampalli, Composting of polycyclic aromatic hydrocarbons in simulated municipal solid waste, *Water Environ. Res.* 70 (1998) 356-361.

[53] H. Kirchmann, W. Ewnetu, *Biodegradation of petroleum-based oil waste through composting*, 9 (1998) 151-156.

[54] C. Loser, H. Ulbricht, P. Hoffmann, H. Seidel, Composting of wood containing polycyclic aromatic hydrocarbons (PAHs), *Compost. Sci. Util.* 7 (1999) 16-32.

[55] J. Ahtiainen, R. Valo, M. Järvinen, A. Joutti, Microbial toxicity tests and chemical analysis as monitoring parameters at composting of creosote-contaminated soil, *Ecotoxicol. Environ. Safety* 53 (2002) 323-329.

[56] S. Amir, M. Hafidi, G. Merlina, H. Hamdi, J.C. Revel, Fate of polycyclic aromatic hydrocarbons during composting of lagooning sewage sludge, *Chemosphere* 58 (2005) 449-458.

[57] P. Oleszczuk, Influence of different bulking agents on the disappearance of polycyclic aromatic hydrocarbons (PAHs) during sewage sludge composting, *Water Air Soil Pollut.* 175 (2006) 15-32.

[58] B. Mahro, G. Schaefer, M. Kästner, Pathways of microbial degradation of polycyclic aromatic hydrocarbons in soil. in: A.R. Hinchee, L. E., I. Semprini, S.K. Ong (Eds.), *Bioremediation Of Chlorinated And Polycyclic Aromatic Hydrocarbons,* Lweis, Boca Raton, FL, 1994, pp. 203-217.

[59] B. Mahro, M. Kästner, Mechanisms of microbial degradation of polycyclic aromatic hydrocarbons (PAHs) in soil-compost mixtures. in: F. Arendt, G.J. Annokkee, R. Bosman, W.J. van den Brink (Eds.), *Contaminated Soil '93*, Kluwer Academic Publishers, The Netherlands, 1993, pp. 1249-1256.

[60] M. Kästner, S. Lotter, J. Heerenklage, M. Breuer-Jammali, R. Stegmann, B. Mahro, Fate of ^{14}C-labeled anthracene and hexadecane in compost-manure soil, *Appl. Microbiol. Biotechnol.* 43 (1995) 1128-1135.

[61] M. Kästner, B. Mahro, Microbial degradation of polycyclic aromatic hydrocarbons in soils affected by the organic matrix of compost, *Appl. Microbiol. Biotechnol.* 44 (1996) 668-675.

[62] M. Kästner, S. Streibich, M. Beyrer, H.H. Richnow, W. Fritsche, Formation of bound residues during microbial degradation of C-14 anthracene in soil, *Appl. Environ. Microbiol.* 65 (1999) 1834-1842.

[63] A. Haderlein, M.C.B. Aly-Hassan, R. Legros, B.A. Ramsay, The design and use of aerated microcosms in mineralization studies, *Biodegradation* 10 (1999) 437-442.

[64] A. Haderlein, R. Legros, B. Ramsay, Enhancing pyrene mineralization in contaminated soil by the addition of humic acids or composted contaminated soil, *Appl. Microbiol. Biotechnol.* 56 (2001) 555-559.

[65] Y. Laor, P.F. Strom, W.J. Farmer, Bioavailability of phenanthrene sorbed to mineral-associated humic acid, *Water Res.* 33 (1999) 1719-1729.

[66] H. Wischmann, H. Steinhart, The formation of PAH oxidation products in soils and soil/compost mixtures, *Chemosphere* 35 (1997) 1681-1698.

[67] C.J. Carlstrom, O.H. Tuovinen, Mineralization of phenanthrene and fluoranthene in yardwaste compost, *Environ. Pollut.* 124 (2003) 81-91.

[68] T. Cajthaml, M. Bhatt, V. Šašek, V. Mateju, Bioremediation of PAH-contaminated soil by composting: a case study, *Folia Microbiol.* 47 (2002) 696-700.

[69] K.L. Lau, Y.Y. Tsang, S.W. Chiu, Use of spent mushroom compost to bioremediate PAH-contaminated samples, *Chemosphere* 52 (2003) 1539-1546.

[70] H.I. Atagana, Co-composting of PAH-contaminated soil with poultry manure, *Lett. Appl. Microbiol.* 39 (2004) 163-168.

[71] L.M. Moretto, S. Silvestri, P. Ugo, G. Zorzi, F. Abbondanzi, C. Baiocchi, A. Iacondini, Polycyclic aromatic hydrocarbons degradation by composting in a soot-contaminated alkaline soil, *J. Hazard. Mat.* B126 (2005) 141-148.

[72] B.W. Bogan, Sullivan, W.R., Physicochemical soil parameters affecting sequestration and mycobacterial biodegradation of polycyclic aromatic hydrocarbons in soil., *Chemosphere* 52 (2003) 1717-1726.

[73] P. Conte, A. Zena, G. Pilidis, A. Piccolo, *Increased retention of polycyclic aromatic hydrocarbons in soils induced by soil treatment with humic substances*, 112 (2001) 27-31.

[74] M.L. Brusseau, R.E. Jessup, P.S. Rao, Modelling the transport of solutes induced by multiprocess non-equilibrium, *Water Resourc. Res.* 25 (1989) 1971-1988.

[75] J.G.A. Birnstingl, An Investigation Into The Bioremediation Of Polycyclic Aromatic Hydrocarbons In A Manufactured Gas Plant Soil. *Ph.D thesis*, University of Lancaster, 1997.

[76] T. Richard, Cornell Composting, Cornell Waste Management Institute, in: compost.css.cornell.edu/Composting_homepage.html.

[77] N.T. Faithful, *Methods In Agricultural Chemical Analysis: A Practical Handbook*, ed., Institute of Rural Studies. University of Wales, Aberystwyth, UK, 2002.

[78] L.S. Clesceri, A.E. Greenberg, A.D. Eaton, *Standard Methods for the Examination of Water and Wastewater*, 20th ed., American Public Health Association, Washington D. C., 1998.

[79] M. Kates, *Techniques in Lipidology*, 2nd ed., Elsevier, Amsterdam, The Netherlands, 1985.

[80] A. Frostegard, E. Bääth, The use of phospholipid fatty acid analysis to estimate bacterial and fungal biomass in soil, *Biol. Fertil. Soils* 22 (1996) 59-65.

[81] L. Zelles, Fatty acid patterns of phospholipids and lipopolysaccharides in the characterisation of microbial communities in soil: a review, *Biol. Fertil. Soils* 29 (1999) 111-129.

[82] J.S. Buyer, D.P. Roberts, E. Russek-Cohen, Microbial community structure and function in the spermosphere as affected by soil and seed type, *Can. J. Microbiol.* 45 (1999) 138-144.

[83] M. Klamer, E. Bääth, Microbial community dynamics during composting of straw material studied using phospholipid fatty acid analysis, *FEMS Microbiol. Ecol.* 27 (1998) 9-20.

[84] L. Carpenter-Boggs, A.C. Kennedy, J.P. Reganold, Use of phospholipid fatty acids and carbon source utilization patterns to track microbial community succession in developing compost, *Appl. Environ. Microbiol.* 64 (1998) 4062-4064.

[85] D.C. Erickson, R.C. Loehr, E.F. Neuhauser, PAH loss during bioremediation of manufactured gas plant site soil, *Water Res.* 27 (1993) 911-919.

[86] H. Feitkenhauer, M. R., H. Markl, Degradation of polycyclic aromatic hydrocarbons and long chain alkanes at 60-70^{0}C by *Thermus* and *Bacillus* spp., *Biodegradation* 14 (2003) 367-372.

[87] K.T. Steffen, A. Hatakka, M. Hofrichter, Degradation of benzo[a]pyrene by the litter-decomposing basidiomycete *Stropharia coronilla*: role of manganese peroxidase, *Appl. Environ. Microbiol.* 69 (2003) 3957-3964.

[88] L. Lazzari, L. Sperni, M. Salizzato, B. Pavoni, Gas chromatographic determination of organic micropollutants in samples of sewage sludge and compost: behaviour of PCB and PAH during composting, *Chemosphere* 38 (1999) 1925-1935.

[89] L.J. Ehlers, R.G. Luthy, *Contaminant bioavailability in soil and sediment*, 37 (2003) 295A-302A.

[90] M. Domeizel, A. Khalil, P. Prudent, UV spectroscopy: a tool for monitoring humification and for proposing an index of the maturity of compost, *Bioresour. Technol.* 94 (2004) 177-184.

[91] M.R. Provenzano, S.C. de Oliveira, M.R. Santiago-Silva, N. Senesi, Assessment of maturity degree of composts from domestic solid wastes by fluorescence and fourier transform infrared spectroscopies, *J. Agric. Food Chem.* 49 (2001) 5874-5879.

[92] P. Wang, C.M. Changa, M.E. Watson, W.A. Dick, Y. Chen, H.A.J. Hoitink, Maturity indices for composted dairy and pig manures, *Soil Biol. Biochem.* 36 (2004) 767-776.

[93] M.P. Bernal, C. Paredes, M.A. Sánchez-Monedero, J. Cegarra, Maturity and stability parameters of compost prepared with a wide range of organic wastes, *Bioresour. Technol.* 63 (1998) 91-99.

[94] N. Senesi, G. Brunetti, in: M. de Bertoldi, P. Sequi, B. Lemmes, T. Papi (Eds.), *The Science of Composting*, Chapman and Hall, London, UK, 1996, pp. 195-212.

[95] S. Perez, M. Guillamon, D. Barcelo, Quantitative analysis of polycyclic aromatic hydrocarbons in sewage sludge from wastewater treatment plants, *J. Chrom.* A 938 (2001) 57-65.

[96] Y. Laor, Y. Avnimelech, Fractionation of compost-derived dissolved organic matter by flocculation process, *Org. Geochem.* 33 (2002) 257-263.

[97] P.A. Soler-Rovira, G. Brunetti, A. Polo, N. Senesi, Effects of amendment with composted sludge on soil humic acid properties, *Compost. Sci. Util.* 11 (2003) 176-184.

[98] D.M. McKnight, E.W. Boyer, P.K. Westerhoff, P.T. Doran, T. Kulbe, D.T. Anderson, Spectrofluorometric characterization of dissolved organic matter for indication of precursor organic material and aromaticity, Limnol. *Oceanograph.* 46 (2001) 38-48.

[99] N. Chung, M. Alexander, Differences in sequestration and bioavailability of organic compounds aged in dissimilar soils, *Environ. Sci. Technol.* 32 (1998) 855-860.

[100] N. Chung, M. Alexander, Effect of soil properties on bioavailability and extractability of phenanthrene and atrazine sequestered in soil, *Chemosphere* 48 (2002) 109-115.

[101] S.C. Hwang, T.J. Cutright, Preliminary exploration of the relationships between soil characteristics and PAH desorption and biodegradation, *Environ. Int.* 29 (2004) 887-894.

[102] B. Antizar-Ladislao, J. Lopez-Real, A.J. Beck, Investigation of organic matter dynamics during in-vessel composting of an aged coal-tar contaminated soil using fluorescence excitation-emission spectroscopy, *Chemosphere* 64 (2006) 839-847.

[103] W. Chen, P. Westerhoff, J.A. Leenheer, K. Booksh, Fluorescence excitation-emission matrix regional integration to quantify spectra for dissolved organic matter, *Environ. Sci. Technol.* 37 (2003) 5701-5710.

[104] G.A. Baddi, M. Hafidi, J. Cegarra, J.A. Alburquerque, J. Gonzalvez, V. Gilard, J.C. Revel, Characterization of fulvic acids by elemental and spectroscopic (FTIR and C-13-

NMR) analyses during composting of olive mill wastes plus straw, *Bioresour. Technol.* 93 (2004) 285-290.

[105] N. Senesi, T.M. Miano, M.R. Provenzano, G. Brunetti, Characterization, differentiation and classification of humic substances by fluorescence spectroscopy, *Soil Sci.* 152 (1991) 259-271.

[106] F.J. Stevenson, *Humus Chemistry. Genesis, Composition, Reactions*, John Wiley and Sons, New York, 1982.

[107] B.E. Andersson, L. Welinder, P.A. Olsson, S. Olsson, T. Henrysson, Growth of inoculated white-rot fungi and their interactions with the bacterial community in soil contaminated with polycyclic aromatic hydrocarbons, as measured by phospholipid fatty acids, *Bioresour. Technol.* 73 (2000) 29-36.

[108] T. Beffa, M. Blanc, P.F. Lyon, G. Vogt, M. Marchiani, J.L. Fischer, M. Aragno, Isolation of Thermus strains from hot composts (60 to 80 degrees C), *Appl. Environ. Microbiol.* 62 (1996) 1723-1727.

[109] E.A. Paul, F.E. Clark, *Soil Microbiology and Biochemistry*, ed., Academic Press, San Diego, California, 1996.

[110] B. Antizar-Ladislao, J.M. Lopez-Real, A.J. Beck, Bioremediation of a PAH in an aged coal-tar contaminated soil using different in-vessel composting approaches, *J. Hazard. Materials* (In press)

[111] S.C. Wilson, K.C. Jones, Bioremediation of soil contaminated with polynuclear aromatic hydrocarbons (PAHs): a review, *Environ. Poll.* 81 (1992) 229-249.

[112] EC, *Council Directive On The Landfill of Waste* (1999/31/EEC), L182/1. European Commission, 1999.

[113] S. Burnley, The impact of the European landfill directive on waste management in the United Kingdom, *Resour. Conserv. Recy.* 32 (2001) 349-358.

[114] DEFRA, Operational Waste Facilities in England and Wales, London, Department of Environment, Food and Rural Affairs, 2006.

[115] H. Willumsen, Landfill gas plants worldwide: number and types, in: *Sardinia '03 Waste Management and Landfill Symposium, CISA*, University of Cagliari, Sardinia, Italy, 2003, pp.

[116] M. Negre, S. Zancolo, E. Malusa, G. Piccone, Fertilisation of an urban park soil with municipal solid waste compost. *Effects on soil properties and plant growth*, 15 (2006) 200-206.

[117] G.M. Zinati, Y.C. Li, H.H. Bryan, Utilization of compost increases organic carbon and its humin, humic and fulvic acid fractions in calcareous soil, *Compost. Sci. Util.* 9 (2001) 156-162.

[118] EC, *The European Parliament and of the Council of 8 May 2003 on the promotion of the use of biofuels or other renewable fuels for transport* (2003/30/EC), L 123/42-46. 2003, pp.

[119] A. Wingren, M. Galbe, G. Zacchi, Techno-economic evaluation of producing ethanol from softwood: Comparison of SSF and SHF and identification of bottlenecks, *Biotechnol. Prog.* 19 (2003) 1109-1117.

In: Environmental Research Trends
Editor: Marvin A. Cato, pp. 63-103

ISBN: 978-1-60021-556-8

Chapter 2

DISTRIBUTION, SOURCE AND STATUS OF HEAVY METALS AND ORGANOCHLORINE PESTICIDES IN SEDIMENTS OF SUNDERBAN MANGROVE WETLAND, INDIA

***S. K. Sarkar*[*], *C. Maggi*[**], *M. Chatterjee*[*], *J. Bianchi*[**], *A. Bhattacharya*[*], *L. Manfra*[**], *B. D. Bhattacharya*[*] and *K. K. Satpathy*[***]**

[*]Department of Marine Science, Calcutta University,
35 Ballygunge Circular Road, Calcutta-35, India
[**]ICRAM-Central Institute for Applied Marine Research,
Via di Casalotti, 300-00166 Roma, Italy.
[***]Indira Gandhi Center for Atomic Research,
Environmental and Industrial Safety Section, Safety Group,
Kalpakkam 603102, Tamil Nadu, India

ABSTRACT

The paper elucidates a comprehensive account of heavy metals (Ba, Cr, Cu, Fe, Mn, Ni, Pb, Zn, Cd, As and Hg) and organochlorine pesticide residues in sediments from Sunderban mangrove wetland, West Bengal, India, one of the luxuriant mangrove ecosystems and biosphere reserves in the world. Sedimentary organic carbon, ph and textural properties were worked out in order to assess their role on metal distribution and pesticide residues. The sampling sites have been selected covering various geomorphological areas of intertidal zones that are under variable stresses both from natural and anthropogenic sources. The coastal processes of wave-tide-current climate together with riverine hydrodynamics act as the primary control of sedimentation in these areas. Sandflats (2-4 φ), mixed flats (sand+mud) and mudflats (silt+clay; φ <4) characterize the intertidal zone depending on the energy regimes and availability of the material. The heavy metallic pollutants as well as the organochlorine pesticides, as measured with the help of ICPAES and GC equipped with electron capture detector respectively, have accumulated in the sediment matrices from multifarious sources.

Distribution of metals in the sediment cores is characteristically different from each other with elevated values either in surface or deeper layers. Regarding organochlorine pesticides, an elevated level of ΣHCH, ΣDDT and metabolites of HCH and DDT, β-HCH, pp'-DDT and pp'-DDE were found to be dominant. The sources of contaminants are closely related to human activities, such as domestic and industrial discharges, agricultural chemical applications and soil erosion due to deforestation. The resulting compositional dataset was subjected to principal component analysis, cluster analysis, analysis of variance, correlation matrix and regression analysis as and when required. Pollution Load Index (PLI) values range from 0.35 to 0.76 and index of geoaccumulation (I_{geo}) exhibited higher values in case of Zn and Cd revealing moderately contaminated status. The study is compared to other estuarine environment in India and abroad. The present data will serve as a baseline against which future anthropogenic effects may be assessed.

INTRODUCTION

Sediment is a matrix of materials, which is comprised of detrital, inorganic, or organic particles, and is relatively heterogeneous in terms of its physical, chemical and biological characteristics (Hakanson, 1992). It is often stated that sediments have a marked ability for converting inputs of metals from various sources into sparingly soluble forms, either through precipitation as oxides or carbonates, or through formation of solid solutions with other minerals (Salomons and Förstner, 1984). Thus aquatic sediments constitute the most important reservoir or sink of metals and other pollutants. However, due to various diagenetic processes, the sediment-bound metals and other pollutants may remobilize and be released back to overlying waters, and in turn impose adverse effects on aquatic organisms (Li et al., 2000).

Different sources of heavy metal contaminants in estuary regimes, including atmospheric deposition, waste water discharge, leachate and surface runoff. In estuarine ecosystems, sediments are increasingly recognized as the most important pollution sink and as a carrier and possible future source of contaminants. Heavy metals are not necessarily fixed permanently to sediments, but they may be recycled via chemical, biological and hydrodynamic processes, within sedimentary compartments and back to the water column (Förstner, 1983; Tessier and Campbell, 1988). The accumulation of contaminants in sediments can thus pose serious environmental problems in surrounding areas (Santschi et al., 1984; Tiller et al., 1989; Millward and Glegg, 1997). Moreover, those metals have strong affinities for sediments in the receiving water. Therefore the distribution of heavy metals in sediments can serve as an indication of time history and extent of pollutant discharge in specific areas (Chow et al. 1973; Bruland et al., 1974; de Groot et al., 1976). In this study the vertical distribution of trace elements within cores collected from various location along the estuary has been determined. Since our aim was to provide comparitive data, the sampling sites were chosen to represent zones of sediments with high inputs of municipal and/or industrial wastes and zones at greater distances from the potential sources of pollution.

The study of sediment cores can provide historical records of various influences on the aquatic system by indicating both natural background levels and the man-induced accumulation of elements over an extended period of time (Goldberg et al., 1977; Tiller et al., 1989; Winkels et al., 1993; Buckley et al., 1995). In order to assess the long-term environmental impacts of contaminated sediments, the chemical forms and, potential and toxicity of heavy metals need detailed studies (Förstner, 1983; Tessier and Campbell, 1988; Stamoulis et al., 1996).

Organochlorine compounds (OC) such as hexachlorocyclohexanes (HCHs), DDT and its metabolites, endosulfan and endosulfan sulphate are ubiquitous, persistent, toxic and bioaccumulate in nature. The unique properties of such compounds make them environmentally persistent with a global distribution and they are, thus, exerting chronic-toxic effects on wildlife and humans (Loganathan and Kannan, 1994). As a result, organochlorines in various environmental media have received much attention (Tanabe et al., 1994; Hong et al., 1995, 1999). These are usually regarded as the ultimate sink for persistent pollutants discharged into the environment (Gibbs, 1973), and are one of the best media for the long-term monitoring of many contaminants (Larsen and Jensen, 1989). Some developing countries are still using these compounds because of their low cost and versatility in industry, agriculture and public health (Tanabe et al., 1994). Consequently, environmental problems associated with toxic contamination in these countries are of great concern (Guruge and Tanabe, 2001).

Rivers and estuaries can serve as important sources of metal and organic contaminants to coastal marine environments. Organochlorines enter into the marine environments through different pathways. Estuaries input is, in particular, important because estuaries are a major interface between the land and the sea (Hong et al., 1999). The Hugli estuary is one of the important estuaries in northeast coast of India and can be considered an important pathway for pollutant transport into the Bay of Bengal. Over 50% of the metals discharged into the Ganges settle out in the estuarine and river mouth regions and only 15% reach the Bay of Bengal (Darsan Kumar and Hadar, 1999). Inadequate treatment of wastes has manifested itself as increasing contamination of this aquatic environment. Despite the Hugli river's importance, information on organochlorine levels in sediment has not been studied in detail. However, many reports are available describing the distribution of the conservative pesticide residues in the biota and sediment in the marine environment of other parts of India (Shailaja and Nair, 1997; Kannan and Sen Gupta, 1987; Shailaja and Sen Gupta, 1989, 1990; Shailaja and Singbal, 1994; Ramesh et al., 1990; Rajendran et al., 1992). Hence the present paper attempts to address this deficit and presents the results of a comprehensive survey of HCHs, DDTs, endosulfan and endosulfan sulphate in surficial sediments of the Hugli estuary, and compared with other estuaries and seas to carry out an investigation for understanding the current status of their contamination to prevent biological hazards. There is no doubt that the distribution of organochlorines and their metabolites in the environment of India are of great significance in the context of global distribution and accumulation of these compounds.

A cursory survey of literature reveals that there exists a wealth of information with respect to accumulation of heavy metals by the biota of different trophic levels in the lower stretch of Hugli estuary (Sarkar et al., 1999, 2002, 2003, 2004). The present work attempts to study the distribution of heavy metals in sediment profiles and organochlorine pesticide residues in the surface sediments of Sunderban mangrove wetland, northeast coast of Bay of Bengal, India. The objective of this research is to assess the accumulation and environmental impacts of metal and pesticide contaminants in the sediments which would help to assess the environmental impacts of the contamination, to find suitable treatment techniques for polluted materials and propose measures for strategic environmental control and management in the

Sunderban delta region for the next century. The results would serve as a baseline against which future anthropogenic effects can be assessed.

MATERIALS AND METHOD

Study Area

The Hugli estuary (88^0 00′-89^0 28′E and 21^0 00′- $22^0$30′N), which supports the world's largest magnificent mangrove block, the Sunderbans, is a typical and unique ecosystem of the Indian subcontinent and has been recognized as a World Heritage Site, covering approximately an area of 4267 km^2 (Figure 1). It is mostly populated., low-lying and highly vulnerable coastal environment where silting of rivers has become a common problem. The deltaic tidal stretch receives heavy discharge of industrial effluents to the tune of 430 million liter per day from various industries situated on both the banks of Hugli River (Sen et al., 1994). Tide is semi-diurnal with distinct diurnal variability and belongs to macrotidal regime (tidal range up to 6 m) (Sarkar et al., 2002). The predominant wind direction reverses from southwest in the rainy season to northeast in the dry season. These two factors produce strong flushing of the coastal areas, effectively dispersing pollutants.

This macrotidal estuary together with the luxuriant Sunderban mangrove complex forms the genetically diverse ecosystems. This river system feeds the major rural, agricultural, urban and industrial areas and flows into the Bay of Bengal. The coastal environment has been undergoing the rapid urban expansion and subsequent socio-economic development and has put tremendous pressure on the local environment. A significant ecological change is pronounced in Hugli estuarine environment due to huge discharges of domestic and industrial wastes along with the rapid emergence of Haldia port complex, a major oil disembarkment terminal in eastern India. The estuary has become vulnerable to chemical pollutants such as heavy metals, organochlorine pesticides, petroleum-derived hydrocarbons, chlorinated hydrocarbons etc., which may have changed the estuary's geochemistry and affected the quality of the local coastal environment (Sarkar et al., 1994, 1999, 2002, 2003, 2004, Saha et al., 2006).

Coastal sediments in the vicinity of the mouth of the Hugli estuary is a complex mixture of various components originating from both the Hugli river and tidal and longshore sediments from the Bay of Bengal. A substantial portion of the sediments is transported landward by waves and tides and is deposited along the river course and tidal flats of the estuary. The river borne sediments carried downstream mix with the tidal sediments to form the tidal bars and tidal islands at the funnel mouth of the estuary. Hugli River collects and transports natural weathering products as well as potentially harmful elements derived from anthropogenic sources. Abbas and Subramanian (1984) calculated that at Kolkata (former Calcutta) the Ganges annually supplies 411 X 10^6 t (i.e., 328 X 10^6 t sediment + 83 X 10^6 t solute load) of total load of the Hugli estuary.

Tropical cyclones also affect the coastal zone 3-4 times per year. The climate is humid (up to 96%) and the seasons are pronounced with four-month duration. Premonsoon (March - June) is the dry season, the monsoon (July – October) is accompanied by heavy rainfall (annual average precipitation 1600 mm) and the winter season (November – February) is characterized by low temperature and lower precipitation. The studied stations belong to different tidal environments; have different wave energy fluxes and distances from the sea (Bay of Bengal) and diverse human interference with a variable degree of exposure to heavy metallic contamination.

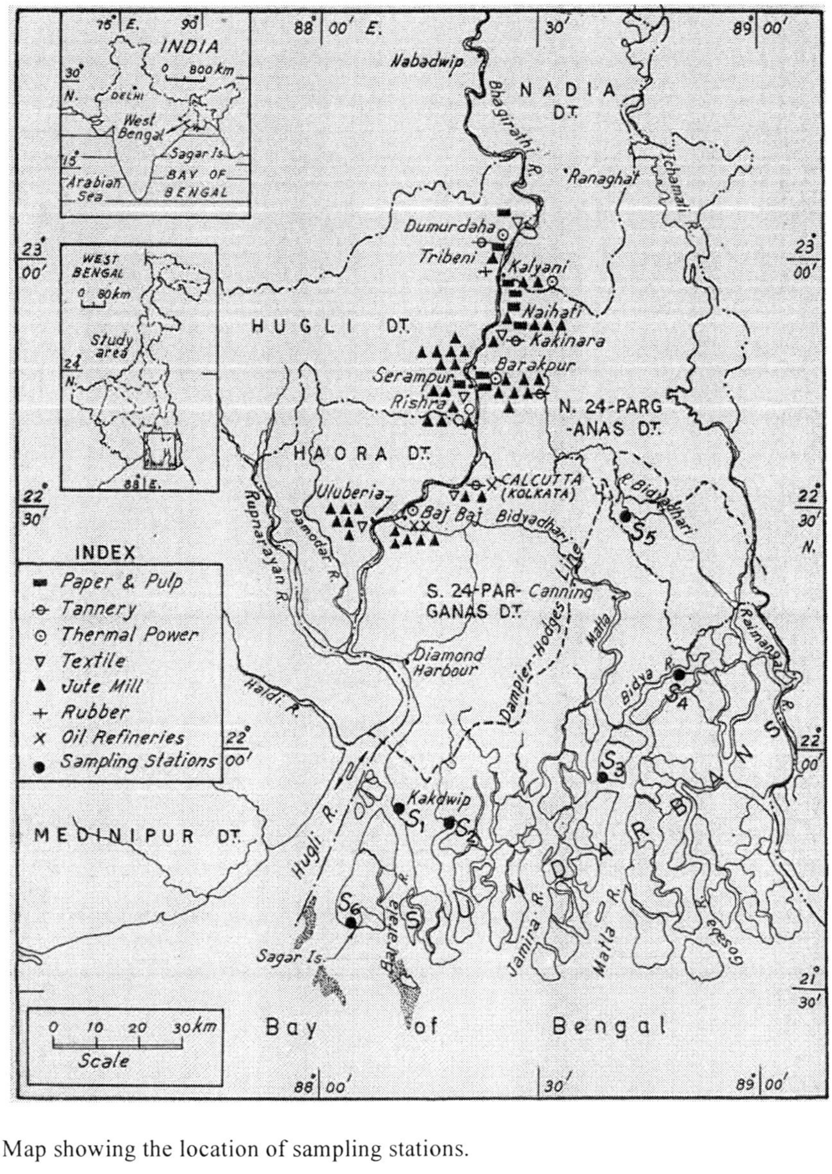

Figure 1. Map showing the location of sampling stations.

The estuarine stretch, from Diamond Harbor to Gangasagar is a wide divergent one with eroding banks, submerged flats and unstable mid channel bars. The main estuary is identified as positive estuary in the mixohaline region (Pantalu, 1966).

Five sampling sites (Kakdwip S_1; Jharkhali S_2 Gangadharpur S_3, Gosaba S_4 and Ghusighata S_5) have been chosen for analyses of heavy metals in the Sundarban wetland. Both upper and lower littoral zones were considered for sampling station Kakdwip while the lower littoral zone was considered for rest of the stations. The widths of the mudflats range from 200 – 300 m under typical tropical macrotidal conditions. In addition, a solitary sampling station at the southern tip of the Sagar Island (Gangasagar S_6) was selected for organochlorine pesticide analysis in the surface sediments (Figure 1).

Table 1. Analysis of Certified Reference Material (PACS-2)

Metal	PACS - 2	
	certified concentration (mg/Kg)	measured concentration *(mg/Kg)*
Cd	2.11 ± 0.15	2.125±0.058
Cr	90.7 ± 4.6	73.87±1.52
Mn	440 ± 19	420.87±6.58
Ni	39.5 ± 2.3	37.78±1.47
As	26.2 ± 1.5	21.53±0.75
Cu	310 ± 12	328.03±11.69
Pb	183±8	168.85±2.51
Zn	364±23	320.11±10.65
Fe(%)	4.09±0.06	3.21±0.06
Hg	3.04±0.20	2.991±0.139

Morphodynamic Description of the Stations

Station S_1 (Kakdwip) Sediment samples from this station has been collected from the mudflat of east-west trending Moynapara creek about 1.5 Km east of Muriganga River, the eastern arm of the Hugli river. The mud flat is dominantly silty clay with organic carbon ranging between 0.57 - 0.73% with the average value of 0.61% .pH is generally alkaline (Table 2). The mud flat, in its lower part, contains marshy vegetation with a few saplings of mangroves, and is bioturbated by the burrows of *Sesarma* sp (crab), *Nereis* sp and *Dendroneries* sp (Polychaete). Deep burrows (3-4 m) with surface mounds at and around the burrow openings are produced by *Thalassina* sp (crab) in the upper intertidal to supratidal mud flat areas.

The mud flat is infected by the sewage discharged by the Central Inland Brackish water Aquaculture farm (CIBA) situated on the east of the mudflat. Marshy vegetation areas are profusely grazed by domestic animals. Bioturbational and tidal activities play a key role in the vertical and lateral transport of bed load and suspended load materials.

Station S_2 (Gangadharpur): The mudflat of Gangadharpur sampling sites is situated on the western bank of the wide Saptamukhi river, a major tidal inlet in the Hugli-Matla delta complex. The sediments are chiefly silty clay with insignificant contribution of sand sized particles. The organic carbon content ranges from 0.65- 0.89 % with the average value of 0.84%. The sediments are mostly neutral to slightly acidic in nature. Sediment recycling and water circulation are mostly restricted by the moribund nature of the associated creeks and thus, there is little opportunity of sediment resuspension. Resuspension of sediments by biogenic activities is extremely insignificant because of extremely low energy situation. The high concentration of a large number of metals (e.g., Mn, Ni, Pb, Zn) can be ascribed to be the result of accumulation in the mudflat behaving like a total sediment sink.

Station S_3 (Jharkhali): The mudflat of the Jharkhali area occurs in a cove at the juncture of the eastern bank of Matla river and northern bank of the Bidya river, a major distributory of the Matla. The sediments are silty clay with an admixture of 7-10% of sand. Organic Carbon ranges from 0.66- 0.87% with the average value of 0.78%. The sediments are slightly acidic to slightly alkaline (Table 2). Metals like Pb and Zn show a higher concentration at this station as its geomorphic situation favors accumulation of the heavy metals. Vertical sediment movement by bioturbating macro infauna is limited and hence lateral transport of sediment by physical process (as like, wave and tidal action) is inhibitated to a great extent.

Table 2. Sediment quality characteristics of the sediment core samples at 5 stations in Sunderban delta

Sample No.	pH	Organic Carbon (%)	Sand (%)	Silt (%)	Clay (%)	Texture
S1//0-4/LL	8.5	0.59	10.44	25.86	63.7	Clayey very fine
S1/4-8/LL	8.4	0.59	6.28	24.1	69.62	Clayey very fine
S1/8-12/LL	8.1	0.59	7.2	21.2	71.6	Clayey very fine
S1/12-16/LL	7.9	0.57	13.92	73.16	12.92	Coarse silty
S1/0-4/UL	7.9	0.73	5.08	40.44	54.48	Clayey
S1/4-8/UL	7.9	0.67	1.6	68.2	30.2	Fine silty
S1/8-12/UL	7.9	0.61	1.52	33.62	64.86	Clayey very fine
S1/12-16/UL	7.9	0.57	2.68	46.02	51.3	Silty clay
S2/0-4/ LL	6.5	0.85	0.48	40.1	59.42	Silty clay
S2/4-8/LL	6.6	0.65	0.16	43.3	56.54	Silty clay
S2/8-12/LL	6.7	0.83	0.28	33.96	65.76	Clayey very fine
S2/12-16/LL	7.1	0.90	0.32	37.62	62.06	Clayey very fine
S2/16-20/LL	7.1	0.89	1.20	37.14	61.66	Clayey very fine
S2/20-24/LL	7.4	0.87	0.24	43.24	56.52	Silty clay
S2/24-28/LL	7.1	0.87	0.28	31.58	68.14	Clayey very fine
S2/28-32/LL	7.3	0.89	0.28	85.98	13.74	Coarse silty
S3/0-4/ LL	6.5	0.81	13.04	68.36	18.6	Fine silty
S3/4-8/LL	6.6	0.87	7.72	27.56	64.72	Clayey very fine
S3/8-12/ LL	7.2	0.81	9.36	22.18	68.46	Clayey very fine
S3/12-16/ LL	7.5	0.73	9.96	37.68	52.36	Clayey fine
S3/16-20/LL	7.5	0.66	13.84	67.3	18.86	Fine silty
S3/20-24/LL	7.7	0.83	16.12	32.42	51.46	Clayey fine
S3/24-28/LL	7.9	0.79	10.76	18.06	71.18	Clayey very fine
S4/0-4/ LL	6.5	0.73	0.4	44.9	54.7	Silty Clay
S4/4-8/ LL	7.9	0.73	2.4	66.56	31.04	Fine silty
S4/8-12/LL	8.1	0.61	0.6	51.7	47.7	Silty clay
S4/12-16/LL	7.7	0.83	0.4	35.94	63.66	Clayey very fine
S4/16-20/LL	7.7	0.63	1.28	38.16	60.56	Clayey very fine
S4/20-24/LL	7.8	0.63	0.52	61.92	37.56	Silty clay loam
S5/0-4/LL	6.9	0.91	10.76	73.46	15.78	Fine silty
S5/4-8/ LL	6.9	1.20	7	53.24	39.76	Silty clay loam
S5/8-12/ LL	7.0	1.20	10.08	56.88	33.04	Silty clay loam
S5/12-16/LL	7.2	1.30	11	68.18	20.82	Fine silty
S5/16-20/LL	7.3	1.40	11.72	64.42	23.86	Fine silty
S5/20-24/ LL	7.4	1.40	10.56	50.82	38.62	Silty clay loam

LL = Lower Littoral Zone: UL = Upper Littoral Zone.

Station S_4 (Gosaba): The mudflat of this sampling station occurs in the meander bend of the Bidya river, a major distributory of the Matla river. The Mudflat is cresentic in outline and forms the point bar on the southern flank of Bidya River. The mudflat sediments are composed of sub-equal proportions of silt and clay with insignificant admixture of sand (<2%). The organic carbon content ranges from 0.61- 0.83% with the average value of 0.69%. The sediments are generally slightly alkaline. Because of regular tidal flushing and erosion-accretion process as in the point bar, the heavy metallic contaminants get dispersed and so the levels of metal concentrations are relatively low in this station. Bioturbational activities of macrobenthic organisms, particularly by crabs enhance sediment dynamics together with the influence of physical processes.

Station S5 (Ghusighata): The mudflat of the Ghusighata station occurs by the side of the north-south trending Karati river, a tidal channel upstream of the Matla river in its south and Bidyadhari river in its northwest. The sediments are dominantly silty (average 62%) with some admixture of silt (average 28%) and sand (average 10%). The organic carbon content is highest in this station ranging from 0.91 to 1.40% with the average value of 1.23%. The sediments are generally neutral to slightly acidic. Being fed by the huge discharge of the sewage sludge through the sewage disposal canal of the Kolkata Metropolitan city, the sediments of the tidal mudflats of this station are heavily infected by pollutants like Cr, Cu and Zn which show an appreciably higher values (Table 2). Some selective macro benthic in-fauna of the polychaete group viz, *Mastrobranchus* sp and *Dendronereis* sp are the commonest species as they can inhabit the pollution inflected substrates and help clustering the bluffs and sediment grains while forming their tubes.

Station S_6 (Gangasagar) occurs at the southwestern edge of Sagar Island and faces the eastern margin of the funnel- mouthed Hugli estuary. The station faces the severity of the wave climate of the coastal Bay of Bengal and macrotidal amplitude almost throughout the year. This station is thus prone to erosion specially, during the episodic phases of tropical cyclones- the severe types of which have a recurrence interval of 3.28 years within 100 Km of island (Bandyopadhyay, 1998). The brick pavements over the high embankments made surrounding this area are destroyed almost regularly. Breaching of these embankments create pathways for the ingressive tidal waters to reach up to the abandoned old marginal embankments. The sandy beach of this station often shows admixture of coarse to fine particles that are derived from breaking down of the pavements. The upper intertidal zone materials are generally more homogeneous in terms of their mud contents and texture compared with that of the lower intertidal zone. The lower intertidal sediments are subject to constant reworking by waves and tidal currents and thus are composed of more or less clear sand and silt, the clay fraction being absent. The sand: mud ratio in the upper intertidal zone of this station is about 20:80, whereas, in the lower intertidal flat this ratio becomes 95:5 suggesting that the lower flat at Light House is a sandy beach environment with an admixture of cleaned sand and silt (Table 2). There is no contribution of clay fraction in the mud ratio as these are dispersed to some distant zone beyond wave breaking.

Sampling and Analytical Protocol

During winter months (January-February 2006) , cores of 30 cm in length were collected by steel corers (45 cm length and 5 cm diameter) at the sampling sites as discussed above. At

Kakdwip station, cores were taken from two transects of the littoral zones (upper and lower) and the upper 40 cm of each core was arbitrarily divided at 4 cm fractions (sub samples) with the help of PVC spatula. All visible marine organisms and shell fragments were removed by hand.

The slice size differs between the two transects and the stations due to variations in nature of the sediments. Core fractions were stored in labeled polyethylene bags stored in iceboxes and transported to the laboratory where they were frozen to -20 ^{0} C. Within two days, a portion of each sample was placed in a ventilated oven at a very mild temperature (max. 40 ^{0}C). Dried samples were fractioned into two parts: one part was used for analyses organic carbon, pH and texture. Organic carbon (C_{org}) content of the soil was determined following a rapid titration method (Walkey and Black, 1934). The textural analyses of the samples were performed by sieving-cum-pipetting method. Sieving was done in a Ro-Tap Shaker (Krumbein and Pettijohn, 1938), and statistical computation of textural parameters was done by using formulae of Folk and Ward (1957). The other dried fraction was dislodged and disaggregated very gently with utmost care using agate mortar and pestle and sieved through 63μ sieve and stored in hematic plastic bags until analyses for heavy metals. In the field, soil temperature was recorded in degree centigrade by using a thermometer of 0.1^{0}C sensitivity. The water salinity and pH were analyzed by Mohr-Knudsen method and with the help of a pH meter respectively.

Metal dissolution was conducted using microwave-assisted digestion (Milestone MLS Ethos TC high performance microwave digestion unit). 0.5 g of oven-dried (35°C, 48 h) sediment was digested in Teflon bombs with 3 ml of HNO_3 and 9 ml of HCl. After cooling, the solution was diluted into a 50 ml volumetric flask to the mark with distilled water. Cd and As were determined in digests and extracts by a graphite furnace atomic absorption with Zeeman background correction technique (SpectrAA –220Z, Varian). Chemical modifier used for cadmium analysis is palladium solution (500□g/mL) plus a reducing agent such as ascorbic acid (1%), while chemical modifier used for arsenic analysis is $Ni(NO_3)_2$ (500 mg/ml). Cr, Fe, Al, Mn, Ba, Zn, Cu, Pb and Ni were measured with coupled emission plasma ICP-OES (Liberty AX, Varian). For Hg analysis a Direct Mercury Analyzer (DMA-80, FKV) instrument was used. The DMA 80 is a direct atomic absorption with gold amalgamation for direct mercury determination in solid and liquid samples without the need of sample chemical pre-treatment (EPA Method 7473). The Hg analysis was performed using dried sediment samples (35°C, 48h) without any further sample preparation.

In order to guarantee QA/QC, quality parameters such as accuracy, quantification limit and repeatability were estimated . The accuracy for total content of metals was evaluated using certified reference materials (CRM) as shown in Table 1. Quantification limits were been: 0.01 μg/g for mercury, 0.01 μg/g for cadmium, 0.05 μg/g for lead and 1μg/g for the other metals. All samples (and reference materials) were run in triplicate. All reagents used were analytical grade and all glassware was acid-washed and deionized water rinsed prior to use.

For pesticides analyses, samples of surface sediments were collected with the help of a grab sampler from the extreme fringe of Sagar Island, northeastern part of the Bay of Bengal. Ten gram of sediment samples was hot extracted with hexane (50 ml) for 3 hours. The samples were filtered using vacuum and transferred quantitatively to 1000 ml separating funnel. This was added with 100 ml of acetonitrile, mixed well and shaken for 1 minute. The layers were allowed to separate. The acetonitrile layer was collected in conical flasks separately. Similarly, 2 x 100 ml of acetonitrile was partitioned with the sample and pooled in respective flasks. The collected samples were then transferred to the 1000 ml separator funnel containing 200 ml of distilled water and 2 ml of saturated sodium chloride. 100 ml of hexane was added to the sample and shaken for 1 minute. The hexane layer was collected in a 500 ml round bottom flask after passing through a bed of sodium sulfate. Similarly, 2x100 ml of hexane was used for the sample and pooled the hexane fraction. Then the collected samples were evaporated to dryness on a rotary evaporator-using vacuum. Quantification of organochlorines was performed by injecting an appropriate aliquot of final extract into a gas chromatograph (Hewlett Packard 5890 Series II) equipped with a ^{63}Ni electron capture detector (ECD) and a moving needle type injection system. The column consisted of 30 QC5/BPX.50, 30 m long, 0.53 mm i.d. wide bore capillary. The column temperature was programmed from 60 ^{0}C (1 min hold) at a rate of 20 ^{0}C/ min to 160 ^{0}C (10 min hold) and then at a rate of 2^{0}C/ min to 270 ^{0}C (15 min hold). Injector and detector temperatures were maintained at 250^{0}C and 300^{0}C respectively. Argon was used as the carrier gas while nitrogen was the make-up gas. Pesticides were quantified by comparing individual peak area of sample with the corresponding peak area of the standard. Matrix blanks and method blanks were analyzed with every batch of 10 samples. The analytical procedure was calibrated against a standard reference material (IAEA-383) supplied by the International Atomic Energy Agency. All the results were expressed in μg g^{-1} on dry weight basis.

Sediment Quality Test

Sediment pollution in the present study was assessed using enrichment factor (EF), geoaccumulation Index values (Igeo) and Pollution Load Index (PLI). The trace metal concentrations were normalized taking Aluminium as a grain size proxy (Rubio et al., 2000; Angelidis and Aloupi, 1995; Balls et al., 1997) employing the expression given by Rubio et al., (2000) and the world average shale concentration as background value (Turekian and Wedepohl, 1961). Aluminium is generally preferred to other abundant elements, because of its conservative nature, analytical precision and major constituent of clay minerals.

Possible sediment enrichment of metals was evaluated in terms of the I_{geo} of Müller (1979). The formula used for the calculation of I_{geo} is $\log_2 (C_n/1.5\ B_n)$, where C_n is the measure content of element "n", and B_n the element's content in "average shale" (Turekian and Wedepohl, 1961). The geoaccumulation Index (I_{geo}), originally defined by Müller (1979) for a quantitative measure of the metal pollution in aquatic sediments, has adopted in the present study.

Based on reported results, we calculated the Tomlinson Pollution Index (PLI) for the upper most samples of the cores to assess the relative heavy-metal toxicity of the surficial sediments. The PLI index (Angulo, 1996) is obtained as a concentration factor (CF) of each heavy metal with respect to the background value in the sediment. The equations applied are the following:

$$CF_{metal} = C_{metal}/C_{background}$$

$$PLI = \sqrt[n]{(CF_1 X CF_2 X \ldots\ldots\ldots\ldots\ldots X CF_n)}$$

The PLI index represents the number of times by which the heavy-metal concentrations in the sediment exceeds the background concentration, and gives a cumulative indication of the overall level of heavy-metal toxicity in a particular sample. In this study, the lowest concentration of a heavy metal detected among all the samples was used as the background for that heavy metal.

Statistical Analyses

Two-way analysis of variance (ANOVA) was carried out on the log- transformed data in order to eliminate the dependence of variance on the mean (Brix and Lyngby, 1984). Mulvariate techniques were used to assess the regional distribution pattern of the element assemblages in the study areas. Principal component analysis was used to characterize the metal composition in sediments, and cluster analysis was used for grouping 6 sampling stations. The raw data were classified objectively using factor analysis. The purpose of factor analysis is to reduce the complexity within the similarity matrix of a multivariate data collection, transform it into a simpler and more easily interpreted factor matrix. Regression analyses and Pearson product correlation coefficients between trace metals and organic carbon and mud (silt and clay%) in sediment were also worked out. Statistical software packages (Statistica 6.0, Minitab, S plus, etc) were used in analyzing data sets.

RESULT AND DISCUSSION

Physicochemical Parameters

The sediment quality characteristics such as pH, organic carbon (%), and percentages of sand, silt and clay obtained in different profiles of the five sampling sites has been shown in Table 2 and Figure 2 (a-e). Mean values obtained for organic carbon (0.53 to 1.40 %) are comparatively low in comparison to other coastal regions of India, e.g., Gulf of Mannar, southeast coast of India described by Jonathan and RamMohan (2003) and Cochin, southwest coast of India worked out by Sunil Kumar (1996). The low organic carbon values obtained might be the result of marine sedimentation and mixing processes at the sediment water interface where the rate of delivery as well as rates of degradation by microbial-mediated processes can be high (Canuel and Martens, 1993). This is also related with the poor absorbability of organics on negatively charged quartz grains, which predominate in sediments in this estuarine environment (Sarkar et al., 2004). A relatively lower percentage of organic carbon in the top sediment surface (0- 4 cm) at the study sites could be attributed to the constant flushing activity by tides along with the impact of waves.

The sediments of the studied stations have shown varying pH from slightly acidic to basic (6.5 to 8.5) where the acidic nature is mainly pronounced at Jharkhali and Gangadharpur and this was partly due to the oxidation of FeS_2 and FeS to H_2SO_4 and partly resulted from the decomposition of mangrove litter and hydrolysis of tannin in mangrove plants releasing various kinds of organic acids (Liao, 1990). The sediment samples show a variable admixture of sand, silt and clay. Clay fractions dominate in low energy areas of suspensional deposits.

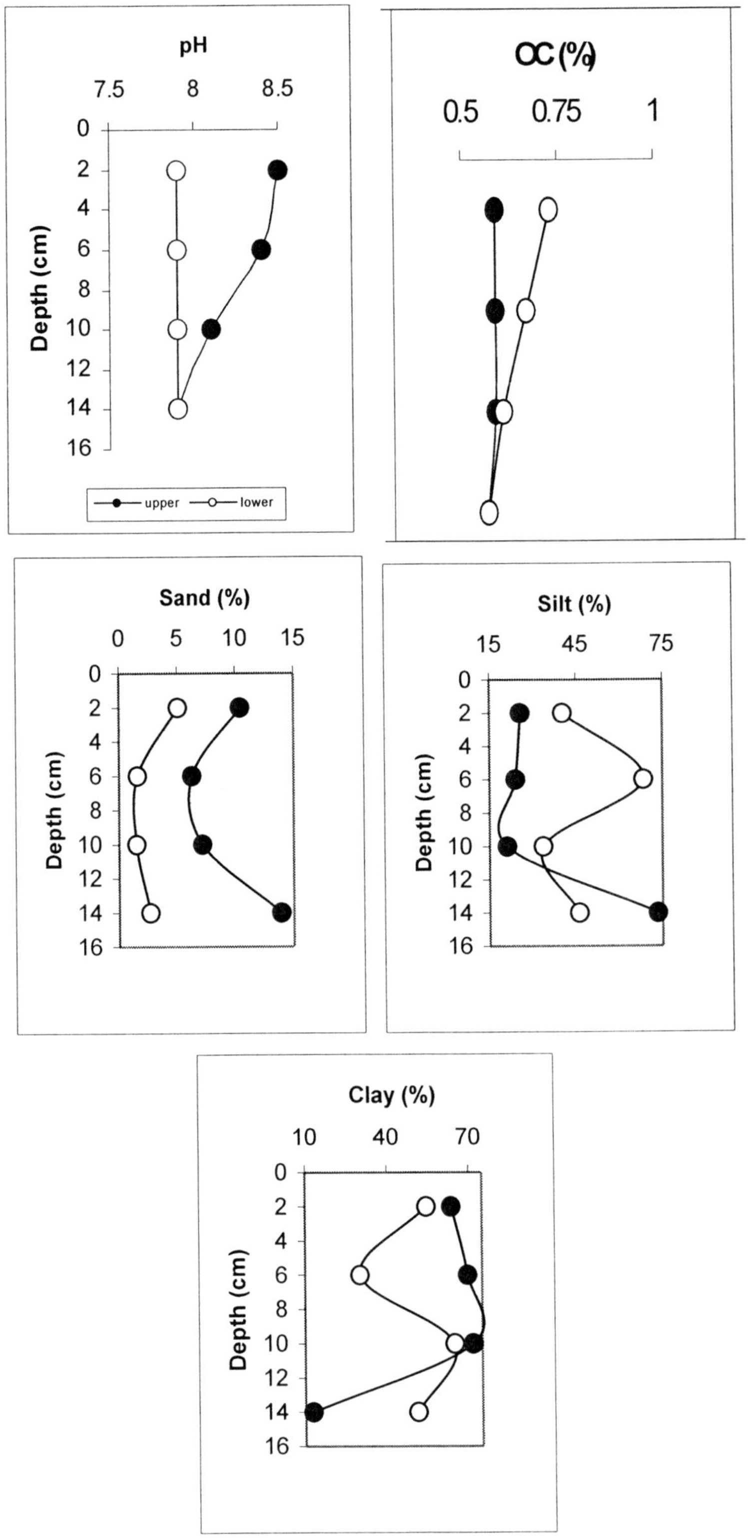

Figure 2a. Sediment quality Characteristics in sediment profiles at Kakdwip (S1).

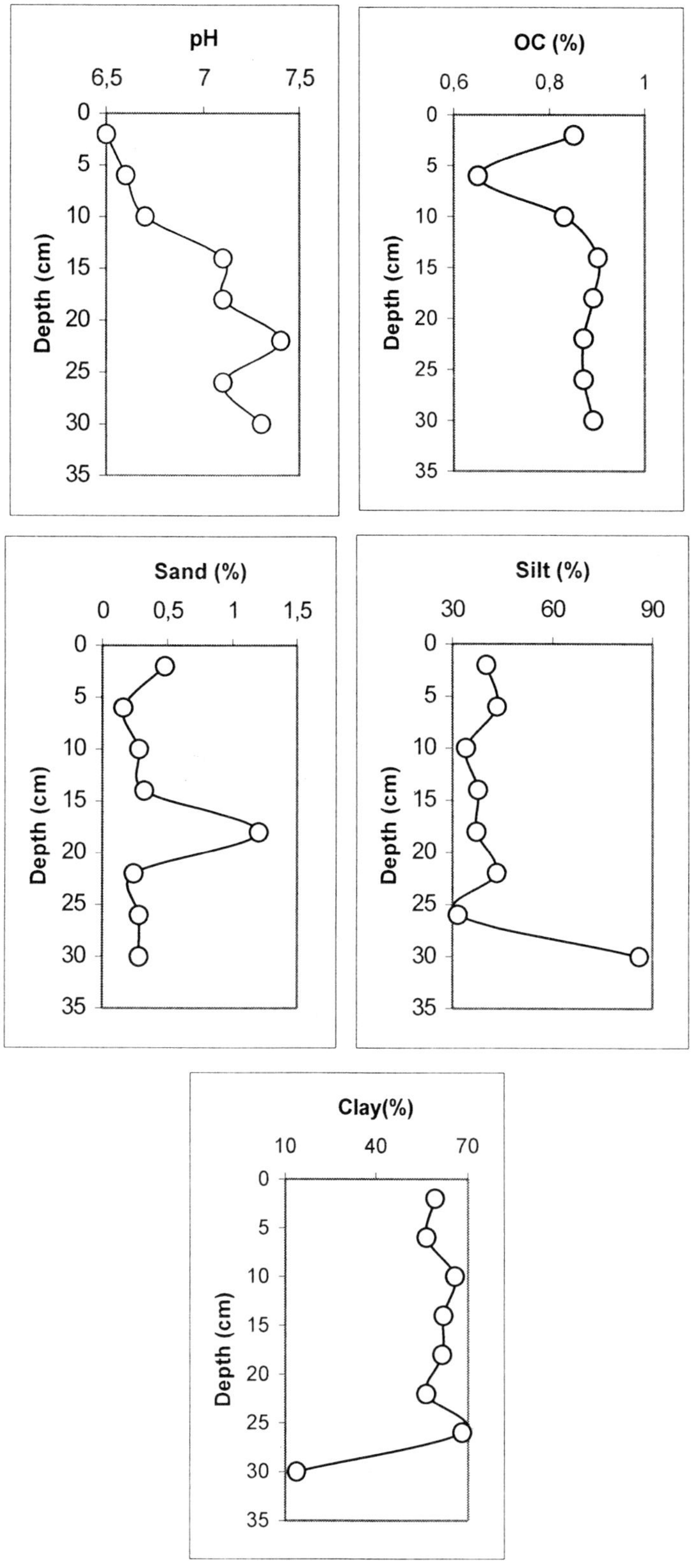

Figure 2b. Sediment quality Characteristics in sediment profiles at Gangadharpur (S2).

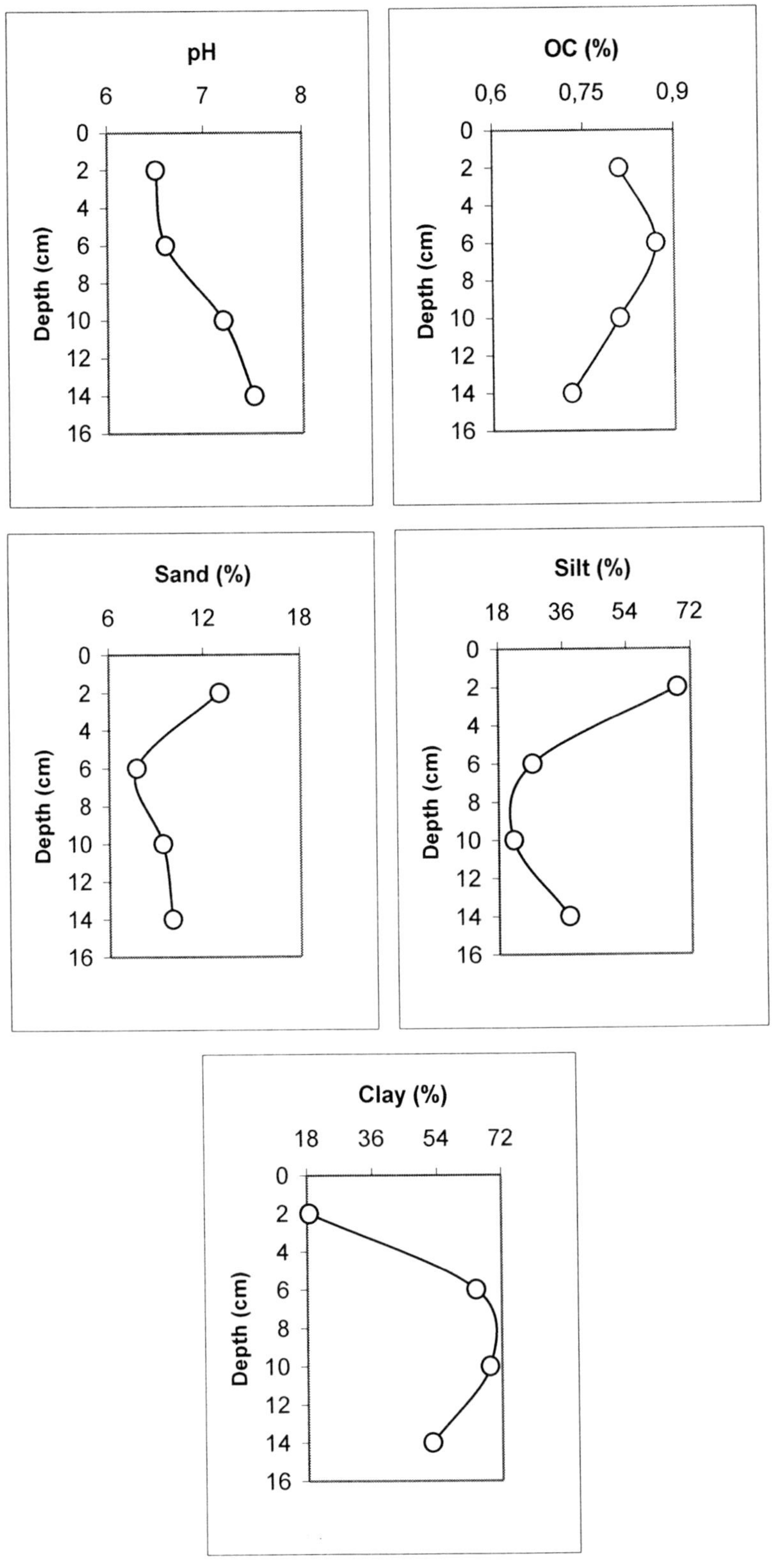

Figure 2c. Sediment quality Characteristics in sediment profiles at Jharkhali (S3).

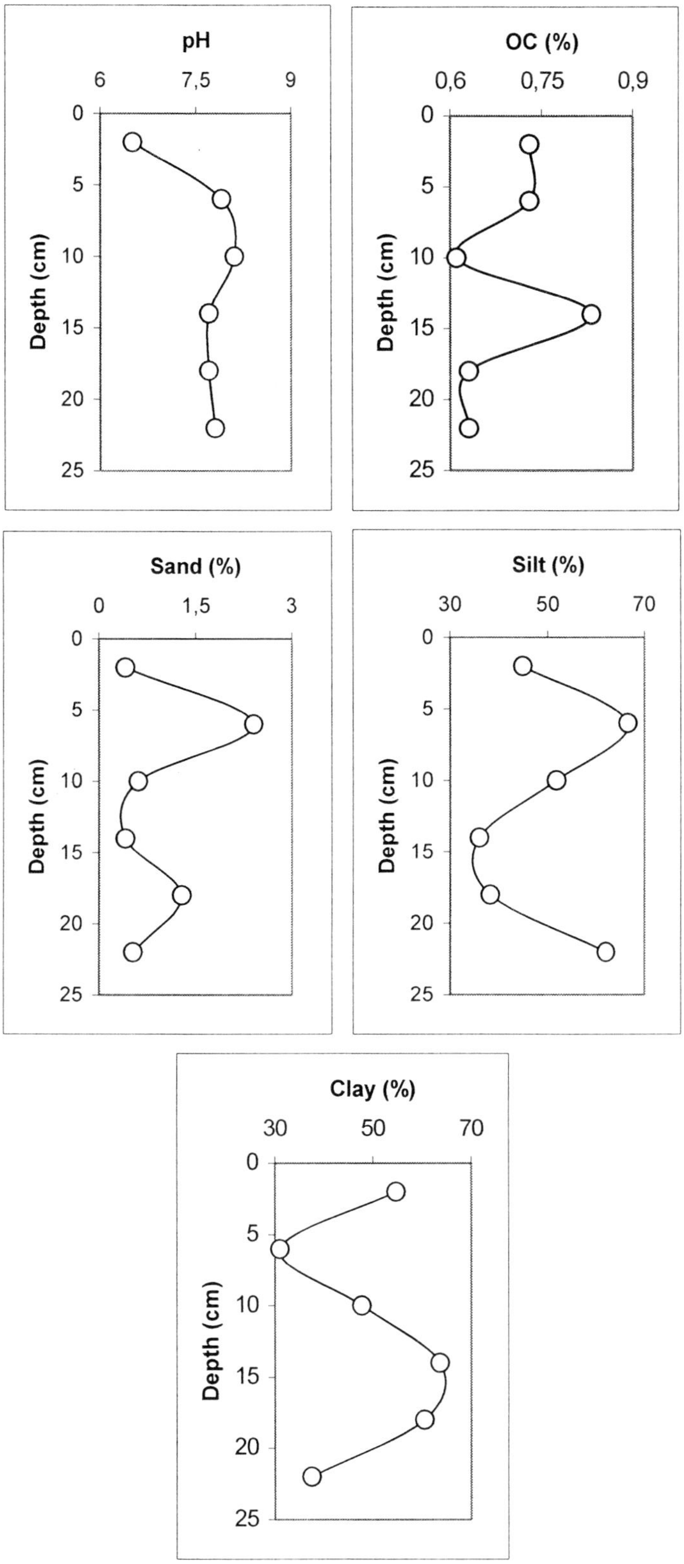

Figure 2d. Sediment quality Characteristics in sediment profiles at Gosaba (S4).

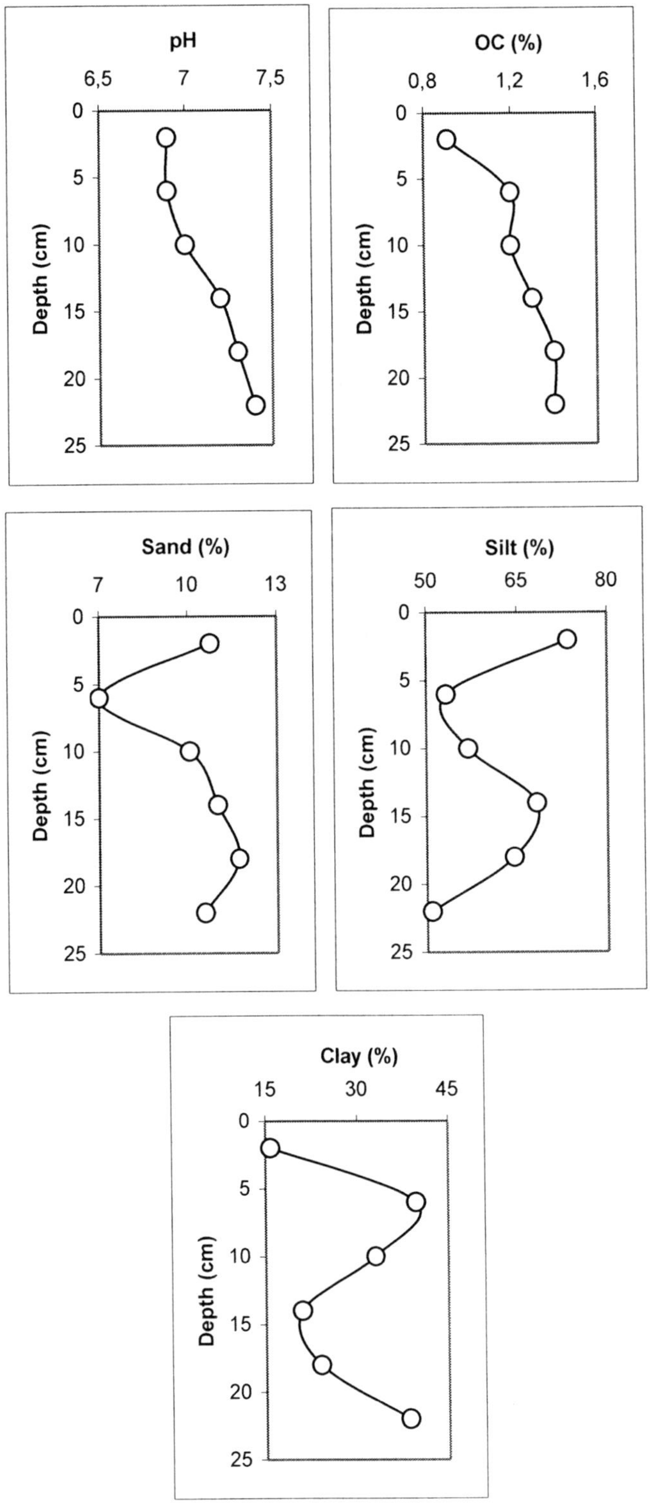

Figure 2e. Sediment quality Characteristics in sediment profiles at Gushighata (S5).

Where the energy level is high in the estuary, more and more clean sand and silt dominate in the lower flats giving the depositional area a sandy beach environment. A variable amount of erosion and depositions in the study stations reflected from variable admixture of sand and clay fractions in individual sample.

Variations of Elements in the Sediment Cores

There exists a sharp difference between the concentration of sediments in the upper and lower littoral zone at Kakdwip (S 1) with three to four fold increase in the upper zone. The enrichment in the upper zone might be due to the lixiviation processes caused by extremely high tidal regime, easy infiltration of polluted soil from adjacent coastal areas and sediment resuspension processes. Intensive fishing and ferry services, sewage drainage from the mainland and other commercial activities are supposed to be potential sources for the enrichment of these elements at this region.

From all elements studied Al showed the higher mean concentration in the five sampling stations (Table 3), followed by Fe. The maximum mean concentration values for Fe (54600 mg kg^{-1}), Cu (151.84 mg kg^{-1}); Zn (1202.22 mg kg^{-1}); Mn (1072.61 mg kg^{-1}); Pb (35.11 mg kg^{-1}) and As (12.77 mg kg^{-1}) were obtained at Gangadharpur showing no uniform behavior in the sediment core. The hydraulic conditions (e.g., circulation currents, tidal action etc.,) and the estuarine zone dynamics (local mixing, resuspension and erosion) might be responsible in facilitating deposition of metal contaminants at this site. The synchronous elevated values of the two redox-sensitive metals Cu and Zn are interesting. These two metals are classified as having very high pollution potential (Pringle et al., 1968) and are also known to act synergistically (Perkins, 1974). Copper concentrations in clean, non-geochemically enriched sediments are n the order of 10 mg/Kg or less, while values in excess of 2000 mg/Kg have been reported for copper polluted sediments (Legorburu and Canton, 1991; Bryan and Langston, 1992). In the current study, elevated of Cu was encountered at S2 Gangadharpur (151.841 mg/Kg) and S5 Ghusighata (110.599 mg/Kg) and might be classified as "moderately polluted" according to USEPA criteria (Giesy and Hoke, 1990) indicating the anthropogenic influence caused by the input of untreated domestic sewage, as this element has a preferential association with organic matter (Hirner et al., 1990), through industrial effluent discharge as well as extensive use of antifouling paints by shipping activities (Zingde and Desai, 1981). Differences in grain size between the sampling sites might also related for Cu contamination. Uncontaminated sediments typically contain zinc concentrations of 5-50 mg/Kg depending upon local geology. Zinc concentrations found during the present study spanned two orders of magnitude-" moderately contaminated sites (Kakdwip, Gosaba and Gushighata) as well as "heavily polluted" sites (Jharkhali and Gangadharpur) as per USEPA's criteria (Griesy and Hoke, 1990).

An erratic behavior of the distribution pattern of metals and metalloid in sediment cores was pronounced in all the stations, as shown in Table 3 , which might be ascribed to the metal deposition in coastal sediments through natural processes as well as human activities (Förstner and Wittman, 1983). The prevalent enrichment pattern of the metals Fe, Al, Mn, Ni, Zn and Hg at surface/subsurface in the core was recorded in few cases. The metal enrichment at the surface of the profile may reflect the impacts of contaminants input from recent urbanization and industrial developments around the Sunderban wetland.

Table 3. Concentrations of trace metals in sediment profiles at five stations of Sudarban (concentration expressed in mg/Kg, only Fe and Al as %)

Sample code	Ba mg/kg	Cr mg/kg	Cu mg/kg	Fe %	Al %	Mn mg/kg	Ni mg/kg	Pb mg/kg	Zn mg/kg	Cd mg/kg	As mg/kg	Hg mg/kg
S1/0-4/LL	89.501	38.223	21.945	2.96	3.06	695.641	25.881	11.328	49.536	0.046	5.534	0.019
S1/4-8/LL	109.182	46.010	26.598	3.40	3.81	749.528	31.390	14.014	55.835	0.051	5.814	0.016
S1/8-12/LL	105.531	42.277	23.232	3.04	3.35	694.313	28.559	11.757	52.320	0.046	5.779	0.013
S1/12-16/LL	81.212	46.062	21.952	3.00	2.89	699.107	26.672	11.365	49.502	0.052	4.699	0.015
S1/0-4/UL	148.787	61.326	37.802	4.15	5.06	829.105	41.239	19.618	75.620	0.081	8.127	0.025
S1/4-8/UL	139.992	64.870	37.428	4.49	5.26	856.510	43.829	18.535	76.225	0.069	8.072	0.025
S1/8-12/UL	189.883	72.528	38.679	4.56	5.99	864.522	44.902	18.003	75.723	0.073	8.694	0.023
S1/12-16/UL	161.551	61.359	34.609	4.04	5.11	837.103	38.951	17.208	66.903	0.068	7.654	0.022
S2/0-4/LL	189.162	92.940	51.804	5.20	7.33	1000.83	51.169	21.949	1202.21	0.045	11.866	0.026
S2/4-8/LL	158.365	73.774	151.841	4.59	6.13	912.800	47.659	25.067	1004.55	0.116	9.563	0.039
S2/8-12/LL	177.462	87.275	53.892	5.09	6.52	1072.61	53.163	22.043	726.152	0.049	12.773	0.027
S2/12-16/LL	212.224	95.811	80.247	5.40	7.04	941.097	56.881	35.107	144.980	0.049	10.285	0.029
S2/16-20/LL	311.646	104.796	68.676	5.46	8.26	964.053	57.636	27.038	123.654	0.042	11.002	0.027
S2/20-24/LL	274.861	98.117	69.496	5.31	7.75	970.715	56.608	27.959	118.020	0.047	11.320	0.027
S2/24-28/LL	176.578	91.380	69.895	5.14	6.91	947.248	52.300	24.105	107.920	0.051	8.785	0.028
S2/28-32/LL	249.605	100.302	70.269	5.34	7.62	899.536	58.128	31.535	127.590	0.052	9.410	0.030
S3/0-4/LL	238.811	70.963	47.757	4.15	6.39	703.757	45.551	23.331	1260.63	0.051	9.352	0.027
S3/4-8/LL	239.085	69.468	66.057	4.04	6.51	684.166	44.515	21.220	1065.85	0.059	8.257	0.025
S3/8-12/LL	245.819	68.227	44.107	4.06	6.26	670.407	41.150	22.437	202.583	0.053	9.052	0.021
S3/12-16/LL	205.733	61.503	35.709	3.75	5.51	634.098	39.377	18.883	88.240	0.042	7.866	0.018
S3/16-20/LL	194.590	57.499	34.221	3.68	5.29	644.636	38.344	18.415	95.168	0.051	10.652	0.019
S3/20-24/LL	280.410	60.924	28.385	3.96	5.07	764.403	33.068	16.292	71.245	0.045	12.270	0.017
S3/24-28/LL	131.254	49.578	31.279	3.65	4.06	718.207	35.229	16.289	72.996	0.046	8.724	0.018
S4/0-4/LL	302.602	88.192	47.296	4.73	7.96	769.966	47.864	23.945	1011.54	0.036	7.117	0.030
S4/4-8/LL	351.719	89.996	43.021	4.62	8.46	687.181	46.126	20.730	273.498	0.040	9.914	0.022
S4/8-12/LL	403.173	87.874	44.630	4.59	9.24	628.684	49.305	24.381	93.263	0.042	8.570	0.025
S4/12-16/LL	298.123	74.870	44.023	4.58	7.65	731.816	47.708	22.910	87.213	0.036	9.049	0.026
S4/16-20/LL	320.968	82.669	43.975	4.85	7.88	716.705	48.293	25.167	87.930	0.038	8.345	0.022
S4/20-24/LL	213.698	75.696	45.997	4.26	6.57	557.779	44.656	24.719	84.270	0.034	9.931	0.024
S5/0-4/LL	135.103	80.640	66.805	2.65	3.31	538.337	24.972	14.933	106.297	0.626	6.099	0.078
S5/4-8/LL	216.230	104.221	62.557	2.85	4.33	513.148	27.000	14.883	83.266	0.926	6.487	0.030
S5/8-12/LL	193.551	99.878	50.517	2.97	4.20	505.767	29.839	17.866	83.520	1.609	7.756	0.036
S5/12-16/LL	209.128	158.583	110.599	2.92	4.57	513.309	28.614	24.963	139.981	2.256	5.717	0.062
S5/16-20/LL	213.955	147.474	79.506	2.87	4.55	501.915	30.384	21.132	102.831	1.446	5.681	0.049
S5/20-24/LL	239.130	201.874	57.226	3.48	4.85	537.972	32.848	21.723	97.151	2.407	7.318	0.063

LL = Lower Littoral Zone : UL = Upper Littoral Zone.

The presence of distinct concentration maximum in the top of recently deposited surface sediments is a common pattern of distribution usually observed in areas receiving anthropogenic inputs of trace metals (Young et al., 1973; Crecelius and Piper, 1973; Bruland et al., 1974; Thomson et al., 1974; Kemp et al., 1976; Knauer, 1977; Eisler et al., 1977; Hershelman et al., 1981; El-Sayed, 1982).

Barium showed irregular distribution in the sediment profiles with the elevated values of 403.107 mg Kg^{-1} at Gangadharpur. Variations in the concentration of Ba have been related to the use of barite ($BaSO_4$) in drilling muds (Sharma et al., 1999). A good correlation between location of oil and gas wells and high Ba concentration has been reported in the south Texas Bays (Holmes et al., 1974). Thus the present results are consistent with the use of Ba as possible markers for exploitation activities.

The chromium levels in sediments indicate increasing trend from west to east part of Sundarban reaching maximum value of 201.874 mg/kg at Gushighata at 20 – 24 cm depth. The highly elevated Cr level in sediments at this site is related to the mixing of tannery effluents in the sewage canal from the adjacent tanneries containing highly toxic chromium compounds along with high organic matter and salt content. Elevated level of Cr in the body tissues of the capitellid worm Masobranchus indicus (Family Capitellidae) was also recorded at this site (Sarkar et. al., 2005), which is in conformity with the present findings. Abrupt elevated values of Cd (maximum 2.407 mg/kg at 20-24 cm) were also observed at this site which could be attributed to the combined contribution from sewage effluents as endorsed by Abu-Hilan and Badran (1990) in sediment cores from Gulf of Aquaba (Red Sea).

Mercury concentration showed lack of any spatial variability with an almost uniform decreasing trend from surface / subsurface layer to deep cores reaching maximum value (0.078 mg / kg) at Gushighata at 0 – 4 cm. Concentrations of Hg in surface sediments, therefore, may be the result of post – depositional diagenetic processes that remobilize the metal from deeper sediments and cause upward migration in the sediment column (Rasmussen, 1994). Canãrio et al., (2005) also found high Hg in surface sediments of the Tagus estuary (Portugal) and interpreted that Hg may be mobilized upward under subtoxic conditions and retained in the upper sediment layers. The correlation between the Hg and organic carbon content was negative (-0.104) (Table 4), also recorded by Canãrio et al., (2005). They emphasized the hypothesis of active processes of methylation/demethylation in surface sediments influencing the accumulation of MMHg rather than total content of Hg. Redox conditions can also restrict mobility of Hg from surface sediments, for example, Fe and Mn oxides and hydroxides are more stable in surface sediments because of higher redox potential and enrichment of these constituents is likely to reduce mobility of Hg due to the tendency of Hg and Hg – bound organic complexes to adsorb on these surfaces (Rasmussen, 1994). The most important anthropogenic sources of mercury pollution in Sunderban environment are atmospheric deposition, erosion, urban discharge, agricultural materials, mining and combustion along with the untreated discharge from industries. It is reported that on the global scale, Asia (especially China and India) accounts for about 50% of total anthropogenic mercury emissions (Pacyna and Pacyna, 2002). Combustion of coal as an energy source is the major source of mercury emissions in Asia (Wang et al., 2000; Pacyna and Pacyna, 2002).

Concentrations of Pb revealed erratic distribution pattern in individual cores as well as between sites. The trend of decreasing Pb concentration down the profile is mostly evident at Gangadharpur reaching maximum level of 35.107 mg Kg^{-1} at the depth of 12-16 cm. This may reflect some historical contamination with Pb either from power generation plants and other industrial activities in the region resulting from the atmospheric inputs. The site is located in the high tidal current area of the estuary and is subject to strong mixing by ocean currents. The high salinity marine waters (27 – 31 o/oo) can also facilitate element diffusion through the sediment profiles (Salmons et al., 1988). As evidenced in Table 3 the prevalent Pb values showed comparative high trend at Gangadharpur, which might be ascribed to river borne sources (Förstner, 1983) as well as intensive human activities (Alagarsamy, 2006) including agriculture in the drainage basin (Monbet, 2006), auto-exhaust emission together with atmospheric deposition (Adriano, 1986). Emissions from automobiles consuming leaded petrol allover India are the major source of atmospheric Pb (Huntzicker et al., 1975). A high proportion of the Pb from vehicle emissions may also be transferred to the estuary by way of road runoff through storm water drains. In addition, a substantial contribution from the factories located in the upstream of the Hugli river dealing with lead producing lead ingots and lead alloys plays a vital role as mentioned by Sarkar et al., (2006). Lead concentrations in clean coastal sediments are around 25-μg g^{-1} or less (UNEP, 1985; Bryan and Langston, 1992) and the average Pb levels in Indian river sediments is about 14 mg Kg^{-1} (Dekov et al., 1999). However the present values are below the USEPA, 1996 prescribed maximum values of 90 mg Kg^{-1} for non-polluted sediments, (Giesy and Hoke, 1990).

Table 4. Correlation (Pearson) coefficient matrices between trace metal concentrations and sediment properties at 5 stations (n=35). Significant values are expressed in bold

	Ba	Cr	Cu	Fe	Al	Mn	Ni	Pb	Zn	Cd	As	Hg	OC%	Mud%
Cr	0.370													
Cu	0.129	**0.518**												
Fe	**0.488**	0.015	0.183											
Al	**0.824**	**0.177**	0.213	**0.859**										
Mn	-0.033	-0.233	0.107	**0.798**	**0.434**									
Ni	**0.517**	0.057	0.260	**0.978**	**0.886**	**0.732**								
Pb	**0.577**	0.450	**0.570**	**0.742**	**0.764**	0.376	**0.809**							
Zn	0.108	-0.004	0.317	0.294	0.311	0.291	0.333	0.226						
Cd	-0.013	**0.811**	0.351	-0.515	-0.349	-0.583	-0.477	-0.018	-0.171					
As	0.423	-0.020	0.109	**0.757**	**0.631**	**0.607**	**0.722**	**0.516**	0.301	-0.392				
Hg	0.018	**0.719**	**0.581**	-0.293	-0.162	-0.360	-0.234	0.176	0.026	**0.741**	-0.254			
O.C%	0.270	0.069	-0.022	0.150	0.194	0.079	0.107	0.046	-0.300	0.012	0.138	-0.104		
Mud%	0.314	0.037	0.253	**0.776**	**0.700**	**0.621**	**0.773**	**0.567**	0.151	-0.376	0.376	-0.141	0.078	

A typical example that demonstrates sharp synchronous elevated levels of all the metals at Kakdwip in 8– 12 cm sediment profile in the upper littoral zone (Table 3). The overall elevated concentrations of the elements at this zone may be attributed to multiple factors like fouling of sediments by frequent dredging activities to maintain the navigation channel along the international shipping line, use of mechanized boat for fishing, erosion due to prawn seed catch by the local inhabitants. Coagulations due to mixing of salt and freshwater leading to greater sedimentation in the upper littoral zones might be one of the potential factors.

Table 5. Summary of Principal Component Analysis of metal concentration in sediments

Eigenanalysis of the Correlation Matrix		
Eigenvalue	Proportion	Cumulative
5.4186	0.417	0.417
3.2435	0.250	0.666
1.5194	0.117	0.783
0.9294	0.071	0.855
0.6276	0.048	0.903
0.4809	0.037	0.940
0.3261	0.025	0.965
0.2326	0.018	0.983
0.1305	0.010	0.993
0.0463	0.004	0.997
0.0272	0.002	0.999
0.0103	0.001	0.999
0.0076	0.001	1.000

Variable	PC1	PC2	PC3
Ba	0.256	0.200	0.398
Cr	0.017	0.516	0.087
Cu	0.109	0.391	-0.307
Fe	0.419	-0.036	-0.002
Al	0.392	0.086	0.166
Mn	0.309	-0.186	-0.226
Ni	0.420	0.007	-0.030
Pb	0.336	0.270	-0.022
Zn	0.165	0.038	-0.533
Cd	-0.212	0.447	0.064
As	0.344	-0.047	0.001
Hg	-0.114	0.470	-0.151
O.C%	0.068	0.013	0.591

On the other hand, Fe and Mn have also fairly close distribution patterns in sediment cores which indicate the strong association to the geochemical matrix between the two elements and their variations between stations could be associated with different facts such as, differences in hydrodynamics, churning, erosion, bioturbation, periodic dredging activities etc., This association is not unusual and has been previously recognized by several authors. (Abu- Hilal and Badran, 1990, Krauskopf, 1965; Friedman, 1968; Nohara and Yokota, 1978; El-Sayed, 1982). Noteworthy are the highest levels of Fe in the study area, which could be mainly attributed to the presence of floating old rusty and stranded barges. These barges are major sources for particulate Fe, which settle down and mix with the bottom sediments in these regions. The precipitated Fe in the form of oxyhydroxide has the affinity to scavenge other metals such as Cu, Zn and Pb as they pass through the water en route to the sediments (Waldichuk, 1985). The range in sedimentary Fe pooled mean values between 2.2 and 49.7% are somewhat higher than that reported from other Indian estuaries (Subramanian et al., 1988).

In the present study significant correlations of mud with most of the elements (Fe, Al, Mn, Ni and Pb) are observed as shown in Table 4, which endorse that, the sediment mud is acting as metal carrier and, also their complexation with mud playing an important role in their distribution patterns. Intermetallic relationship revealed that high degree of correlation existed indicating their identical behavior during transport in estuarine environment. The negative or insignificant relationships of Mn with majority of the metals suggest that Mn-oxide may be only a minor host phase for these elements in the estuarine environments.

Similar poor associations of Mn with other trace metals in sediments were also noticed by Alagarsamy (2006) from Mandovi estuary, west coast of India. Non-significant correlations of Zn, Cu and Cd with all the elements may be possibly due to the different processes like biological effects and external inputs operating in estuarine sediments.

The study results show that the concentration factor values of most of the elements like Cu, Fe, Mn, Zn, Cr, Co, Ni and Ba are low (<1). However CF for elements like Pb and B were found to be higher (>1) at stations 2 and 3 whereas similar higher CF values for Zn and Pb were encountered at station 1. A consistent and steady increase of Pb levels (maximum value 44.47 µg g -1 at 24-28cm) was recorded in the upper littoral zone at St. 3.

Statistical Analyses of the Data Set

A cluster analysis was carried out to identify any analogous behavior pattern between the six sampling sites including differential groups of moderately polluted and extremely polluted areas as shown in Figure 3. The Euclidean distances of standardized data were chosen as dissimilarity measurement. The lower and upper littoral sites of Kakdwip as well as Gosaba and Gangadharpur form two well-differentiated groups experiences different anthropogenic stresses. Rest two stations viz Gushighata and Jharkhali respectively are associated with the two groups distantly.

PCA of the present metal data set indicates that three PCs explain 10.16% of 9 metals variability. The influence of anthropogenic inputs is clearly shown by high positive loadings for Ba, Cu, Fe, Al, Mn and Ni in PC1. PC2 distinguishes between high positive loadings for Ba, Cr, Cu, Cd and Hg and high negative loadings for Mn.

The parameter loading for the fourteen components from the principal component analyses of the dataset are given in Table 6. It can be seen from this table that most of the variables associated with each factor are well defined and contribute very little to the other factors, which helps in the interpretation of results. The exceptions for this are the variables Ba, Cu, Zn and organic carbon. Further, it can be inferred that the first factor, which explains 37.96% of the total variance, is related to the association of Fe, Al, Mn, Ni, Pb and Hg. While the parameters Cd and Hg are negatively loaded with the factor, the metals Fe, Al, Mn, Ni and Hg are positively loaded with this Factor. The association of mud and trace metals indicates mutual affinities among those elements. This association may lead to conclude that these elements may act in a synergetic manner. Factor 2, on the other hand, explains 26.03%

of the total variance and is positively loaded with Cr, Cd and Hg. These three toxic metals are loaded separately to other metals forming one unit by themselves and exhibiting their maximum values at Gushighata. An inverse association is also revealed for Fe, Al, Mn, Ni, As and mud. Factor 3 explains 21.39% of the total variance and is loaded with Ba and Al. Mn, Cd and Hg are negatively loaded with this factor. This factor may be termed as heavy metal removal factor. Factor 4 explains 13.8% of the total variance and loaded with the metalloid As-an environmentally important trace element and has origins from anthropogenic sources. (e.g., from fertilizers used in the adjacent agricultural land of Sundarban wetland).

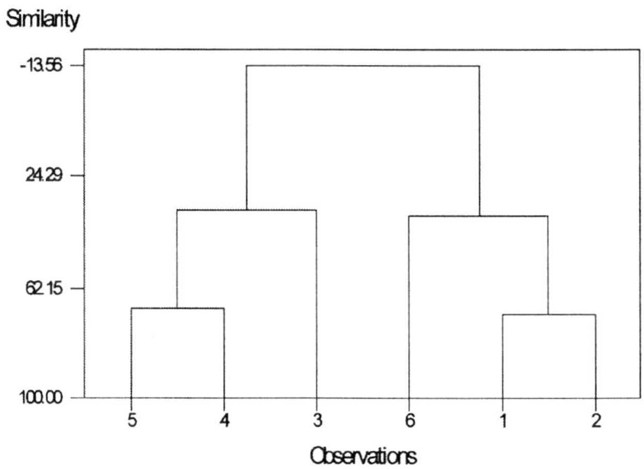

Figure 3. Dendogram showing the relationship between sediment samples from 6 cores collected from 5 different sampling stations using Ward's method.

Table 6. Results of the factor analysis for trace metals and sediment quality parameters at 5 stations in Sundarban

Principal Component Factor Analysis of the Correlation Matrix Rotated Factor Loadings and Communalities Varimax Rotation				
Variable	Factor1	Factor2	Factor3	Factor4
Ba	0.163	0.146	0.955	0.135
Cr	0.059	0.951	0.225	0.007
Cu	0.186	0.382	0.031	0.034
Fe	0.789	-0.145	0.329	0.387
Al	0.608	-0.046	0.719	0.214
Mn	0.723	-0.272	-0.174	0.367
Ni	0.773	-0.122	0.368	0.338
Pb	0.541	0.240	0.428	0.228
Zn	0.128	-0.041	0.080	0.118
Cd	-0.351	0.894	-0.073	-0.151
As	0.348	-0.122	0.264	0.874
Hg	-0.134	0.729	-0.042	-0.109
O.C%	0.052	0.012	0.134	0.051
Mud%	0.957	-0.067	0.172	-0.006
Varience	3.6673	2.5918	2.0692	1.3312
% Var	0.2620.146	0.185	0.148	0.095

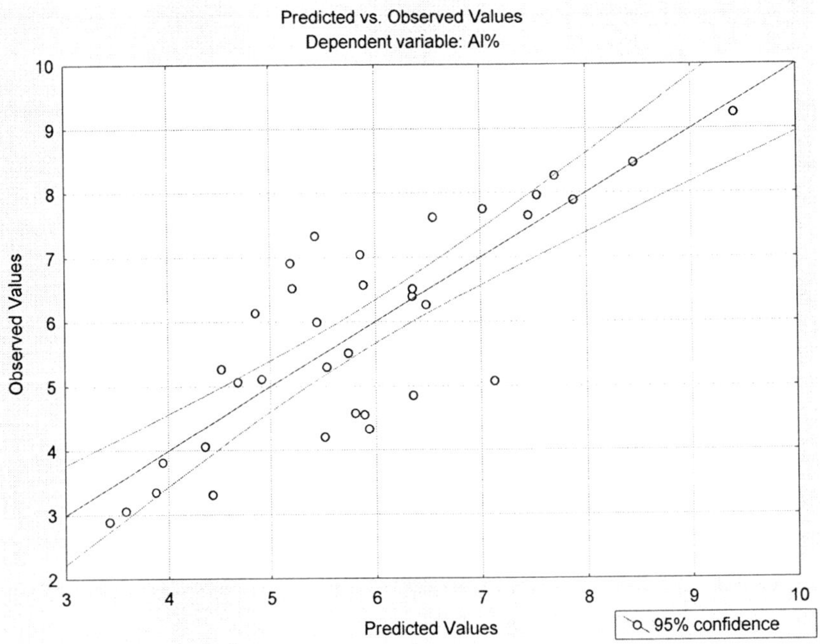

Figure 4. a) Al % vs. Ba.

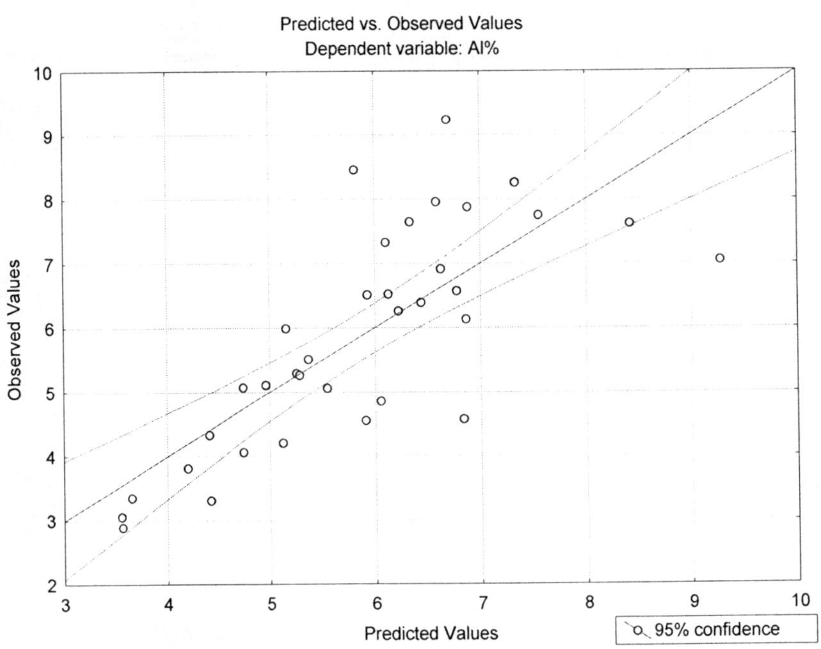

Figure 4. b) Al% Vs Pb.

Figure 4. (Continued on next page)

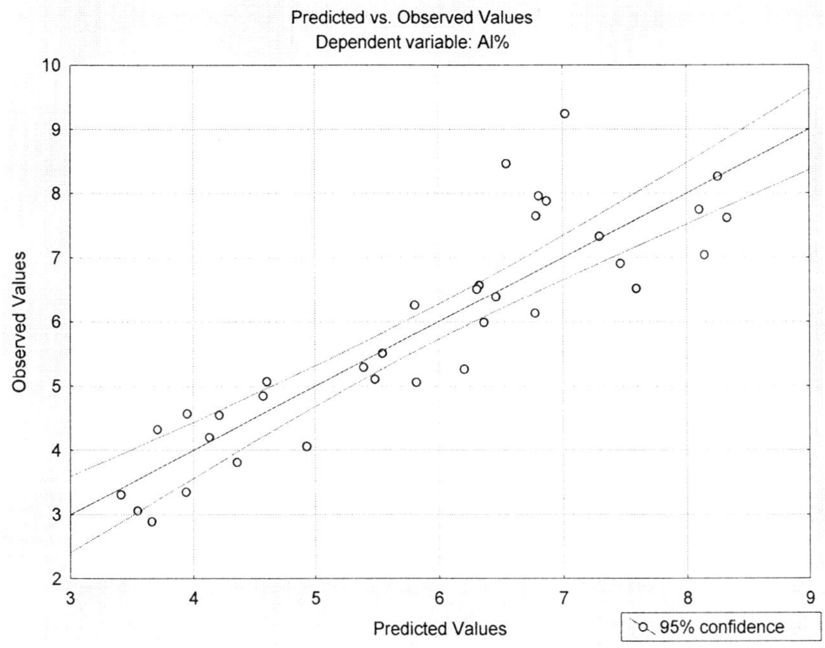

Figure 4. c) Al% Vs Ni.

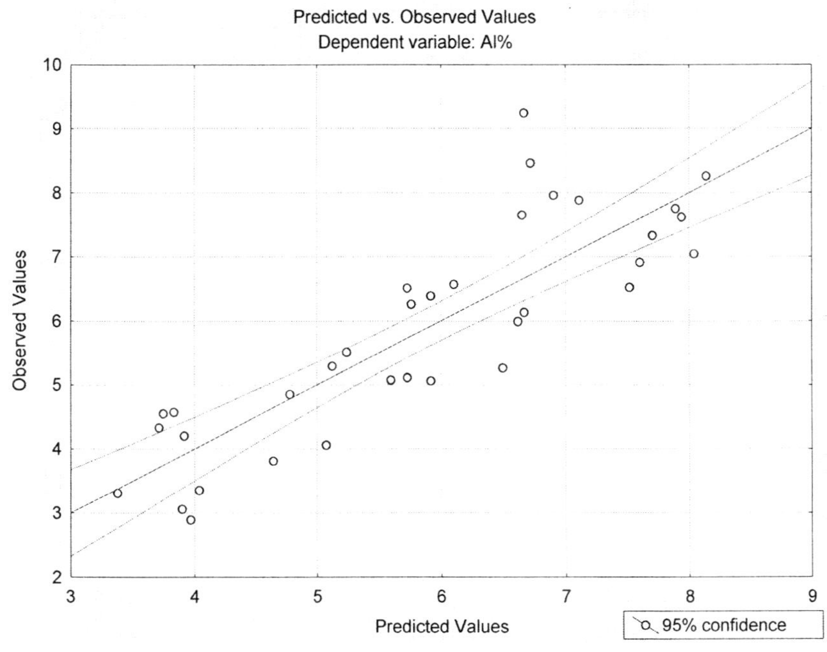

Figure 4. d) Al% Vs Fe%.

Figure 4. (Continued on next page)

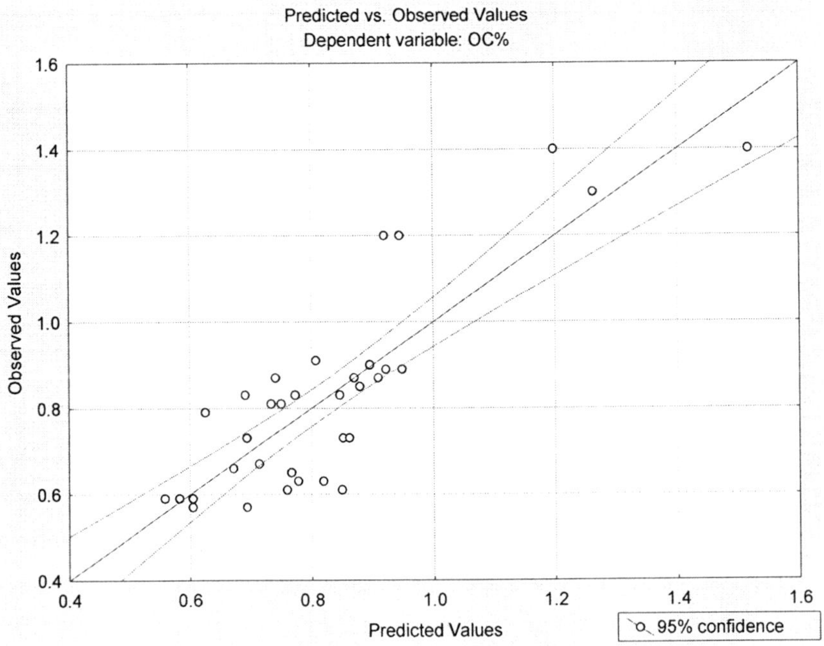

Figure 4. e) OC% vs Cr.

Figure 4. Regression lines for correlation of Al% with (a) Ba, (b) Pb, (c) Ni, (d) Fe% respectively and (e) OC% with Cr. The broken lines represent the 95% confidence interval for all data.

Regression line of Al concentration with Ba, Pb, Ni and Fe concentrations and Cr with OC% are shown in Figure 4a, b, c, d and e respectively. The broken lines represent the 95% confidence interval for all the data. There is a good fit between Al and the dependent variables i.e., a good linear relationship can be shown. The graph shows that the individual points lay within the 95% confidence level excepting few cases. Two-way ANOVA comparisons, using location and sediment cores as factors, made to test the Null hypothesis showed that there was no difference in sediment trace metal concentrations at 3 sampling sites (F-statistics and p values are 0.04, 1.000 and 0.72, 0.489 respectively).

Sediment Quality Values and Interpretation

To estimate the possible environmental consequences of metals analyzed, our results were also compared to US NOAA's sediment quality guidelines. In this study the effects range-low (ER - L) and Effects range-median (ER - M) concentrations are considered. The ER - L represents chemical concentrations below which adverse biological effects were rarely observed, while the ER - M represents concentrations above which effects were more frequently observed (Long et al., 1995,1997). Computing the mean values of the elements in each studied station, majority of the elements revealed lower values excepting Zn (150 mg/Kg) and Cu (34 mg/Kg) in some cases.

Differentiating the elements originating from human activity and those from natural weathering is an essential part of geochemical studies. One such technique largely applied is 'normalization' where metal concentrations were normalized to a textural or compositional characteristic of sediments. There is usually a close relationship between metal concentration and grain size in estuarine and marine sediments since metals tend to associate with finer-grained particles (Livens and Baxter, 1988; Cundy et. al. 1997). In order to compare samples it is necessary to compensate for this by applying a correlation. Otherwise any finer-grained samples may show relatively high metal concentrations. Whilst rationing metal concentrations to grain size for each sample can compensate for the grain size effect, this has the disadvantage of being laborious and so many authors have normalized metal concentration to some element found dominantly in the clay-size fraction. Aluminium has been frequently used as a grain size proxy in both marine and estuarine sediments (Zwolsman et. al., 1993; Cundy and Crodace, 1995). Normalizing elements relative to Al is widely used to compensate for variations in both grain size and composition, since it represents the quantity of aluminosilicates, which is the predominant carrier phase for adsorbed elements in coastal sediments. According to Nolting et al. (1999), this method is also a powerful tool for the regional comparison of trace metal content in sediments and can also be applied to determine enrichment factors (EFs).

The values of EFs can be obtained using the equation $EF=(metal/Al)_{sediment}/(metal/Al)_{shale}$. EFs close to unity point indicate crustal origin while those greater than 10 are considered to be non-crustal source (Nolting et al., 1999). The minimum EFs obtained for many elements (Ni, Ba, Cd and Hg) are less than unity in majority of the cases implying that these elements are depleted in some of the phases relative to crustal abundance in the study area. However, it is evident that elements with a mean EF value greater than unity reveal sediment contamination, for example the overall higher EF values for Pb, Zn, Cu and Cr for all the stations. An overall higher EF values for Pb and Zn in all the three stations considering both upper and lower littoral zones suggests the presence of contaminated sediments derived from multifarious sources like domestic sewage, power-plant operation, major storm events or dumping of sediments dredging along the international shipping zones as endorsed by Batley and Brockbank, 1990. An overall higher enrichment factor values (>1) for Pb at all the stations can be ascribed to the influence of external discrete sources like sewage and agricultural runoff or can result from intense fishing and boating activities across the estuary throughout the year as endorsed by Abu-Hilal (1987). The values of Pollution Load Index (PLI) were found to be very low in two transects of three stations. The present abrupt higher EF values for Zn (6.25 at Jharkhali) and Pb (9.58 at Gushighata) are a matter of great concern.

Table 7. Mean Enrichment Factors (EFs) of metals in sediments from 6 cores collected from 5 different stations

Station	Ba	Cr	Cu	Fe	Mn	Ni	Pb	Zn	Cd	As	Hg
Kakdwip lower	0.41	1.08	1.23	1.36	1.49	0.93	1.37	1.16	0.36	0.94	0.09
Kakdwip UP	0.40	1.17	1.27	1.60	2.03	1.01	1.48	1.33	0.39	1.02	0.09
Gangadharpur	0.42	1.15	1.90	1.22	1.26	0.89	1.49	4.82	0.21	0.91	0.08
Jharkhali	0.55	1.01	1.33	1.19	1.18	0.85	1.42	6.25	0.24	1.06	0.08
Gosaba	0.54	0.92	1.00	0.98	0.81	0.69	1.19	2.88	0.13	0.68	0.06
Gushighata	0.65	2.73	2.94	1.17	1.14	0.79	1.79	2.00	9.58	0.93	0.25

Mean concentrations of total trace elements for the present work with those obtained from the Bay of Bengal including estuarine and tidal zones of southeast India, The Gulf of Mannar, crustal average (Taylor, 1964) and the upper crust (Taylor and MacLennan, 1985) have been presented in Table 9. In general the mean concentrations of total Mn, Cr, Cu, Ni and Zn are lower than the concentrations of these metals in sediments of the Bay, River Palar and creek and tidal zones of southeast coast of India. Lead has a higher mean concentration than sediments from estuaries and coastal zones of southeast coast of India (Hema Achyuthan et al., 2002) and the crustal average suggesting lead contamination. The geoaccumulation index values (I_{geo}) showed low values in majority of the cases excepting Cu (0.19 at Gangadharpur) and Zn (0.94, 1.52 and 1.53 at Gosaba, Jharkhali and Gangadharpur respectively) indicating sediments are uncontaminated to moderately contaminated (Muller, 1979) as a result of anthropogenic activities according to the I_{geo} classification (Muller, 1979) (Table 8).

Table 8. Description of sediment quality I_{geo} classification (Muller, 1979)

I_{geo}	I_{geo} class	Description of sediment quality
>5	6	Extremely contaminated
4-5	5	strongly to extremely strongly contaminated
3-4	4	Strongly contaminated
2-3	3	Moderately to strongly contaminated
1-2	2	Moderately contaminated
0-1	1	Uncontaminated to moderately contaminated
<0	0	Uncontaminated

Table 9. Mean concentrations of total trace metals in sediments off Hugli estuary, compared with concentrations of total trace metals from other areas of SE coast of India, Kalpakkam, the Gulf of Mannar, crustal average and upper crust

	Mn	Cr	Cu	Ni	Co	Pb	Zn
Present Study							
Min. conc.	228.3	12.91	4.30	6.86	4.02	nd	22.96
Max. conc.	716.9	84.74	45.29	52.45	23.32	44.47	204.99
Mean conc.							
Std. Dev. (N=49)	76.79	12.66	6.83	9.89	2.66	3.29	41.19
Kalpakkam, Bay of Bengal[a]	356	57	20	30	9	16	71
Shallow cores sediment,							
Bay of Bengal[b]	529	84	26	64	-	-	-
Bed sediments river Palar[c]	1131	338	64	-	-	45	-
Intertidal sediments,							
SE coast of India[d]	51-2005	11-394	0-105	-	-	5-130	13-144
Core sediments, 0-10.5 cm interval	345	318	-	426	10	2	168
Adyar Estuary, SE coast of India[e]							
Core sediments, Tidal zones, SE coast of India[e]							
1.0-5 cm interval, Muttukadu	409	48	-	49	8	1	68
2.0-1 cm interval, Mamallapuram	777	87	-	53	-	13	106
3.0-1 cm interval, Marakkanam	442	58	-	12	7	11	87
Surface sediments,							
the Gulf of Mannar[f]	296	167	-	24	7	16	73
Core sediments, 0-2.5 cm interval	391	643	-	49	11	17	105
Enore Creek, SE coast of India							
Crustal average[g]	950	100	55	75	25	12.5	70
Upper crust[h]	600	35	25	20	10	20	71

[a] Selvaraj et al., 2004; [b]Sarin et al., 1979; [c]Mohanachandran, 1988; [d]Mohanachandran and Subramanian (1990);[e]Hema Achyuthan et al.,2002; [f]Jonathan and Ram Mohan (2003); [h]Taylor and McLennan (1985).

Table 10. Levels of organochlorine pesticide residues in the surface sediments collected from the mouth of the Hugli estuary during March 1998 - August 2000 (values express in μg/g dry wt.)

Month	α-HCH	β-HCH	χ-HCH	δ-HCH	op′ DDT	pp′ DDT	op′ DDE	pp′ DDE	pp′ TDE	α-endo sulfan	Endo sulfan sulphate
Mar 98	BDL	0.12	BDL	BDL	BDL	0.10	BDL	0.003	BDL	0.002	0.10
April	BDL	0.17	0.16	BDL	BDL	0.04	0.001	0.003	0.002	BDL	0.09
May	BDL	0.25	0.002	BDL	BDL	0.06	BDL	0.002	0.001	BDL	0.11
June	BDL	0.20	0.003	BDL	BDL	0.11	0.005	0.003	0.001	BDL	0.12
July	BDL	0.11	0.001	BDL	BDL	0.03	BDL	0.001	0.002	BDL	0.11
August	BDL	BDL	0.003	BDL	BDL	0.03	0.002	0.008	0.006	BDL	0.08
Septem	BDL	0.12	0.002	0.002	BDL	0.002	BDL	0.003	BDL	0.008	0.08
Octob	0.003	0.01	0.004	0.002	BDL	0.002	0.01	BDL	BDL	BDL	BDL
Novem	BDL	0.08	0.001	BDL	BDL	0.04	BDL	0.001	0.003	BDL	0.08
Decem	BDL	0.10	0.001	BDL	BDL	0.07	BDL	0.001	0.003	BDL	0.13
Jan 99	BDL	0.11	0.002	0.002	BDL	0.02	BDL	0.003	0.001	0.003	0.13
Feb	BDL	0.04	0.007	0.002	0.03	BDL	0.001	0.001	BDL	BDL	BDL
March	BDL	0.12	BDL	0.004	BDL	BDL	BDL	0.004	0.002	0.001	0.18
April	BDL	0.17	BDL	0.003	0.04	0.05	BDL	0.003	BDL	0.002	0.19
May	BDL	0.20	0.001	0.006	BDL	0.08	0.001	0.006	0.002	0.003	0.40
June	BDL	0.10	0.002	0.004	BDL	0.07	BDL	0.003	0.001	0.002	0.28
July	BDL	0.10	BDL	0.004	BDL	BDL	BDL	0.003	BDL	BDL	0.17
August	BDL	0.10	0.001	0.004	BDL	0.04	BDL	0.003	BDL	0.002	0.17
Septem	BDL	0.10	BDL	0.003	0.03	BDL	BDL	0.004	BDL	BDL	0.28
Octob	BDL	0.13	0.001	0.004	BDL	0.07	BDL	0.005	0.003	0.001	0.29
Novem	BDL	0.15	0.003	BDL	BDL	0.07	0.001	0.002	0.03	BDL	0.07
Decem	0.003	0.02	BDL	0.004	BDL	0.005	0.02	BDL	BDL	BDL	0.06
Jan 00	BDL	0.17	0.001	0.003	BDL	0.07	0.003	0.005	0.012	BDL	0.13
Feb	BDL	0.21	0.007	0.003	BDL	0.05	0.008	0.003	0.01	BDL	0.13
March	BDL	0.18	0.01	0.002	BDL	0.06	0.008	0.002	0.009	BDL	0.12
April	BDL	0.15	0.002	0.003	BDL	0.06	0.003	0.002	0.006	BDL	0.10
May	BDL	0.18	BDL	0.003	BDL	0.06	BDL	0.002	0.02	0.001	0.13
June	BDL	0.15	BDL	0.002	BDL	0.05	BDL	0.01	0.002	BDL	0.09
July	BDL	0.14	0.001	0.003	BDL	0.08	BDL	0.002	0.003	0.001	0.09
August	BDL	0.11	0.001	0.001	BDL	0.04	BDL	0.001	0.003	BDL	0.06

BDL = below detection limit

Table 11. Comparison of organochlorine residue levels ($ng\ g^{-1}$) in surface sediment collected from different estuaries and seas

Location	HCHs		DDTs		References
	range	mean	range	mean	
Victoria Harbour HongKong	BDL-2.3	0.58	1.38-25.4	10.2	Hong et al., 1995
Pearl River estuary PRC	0.28-1.23	0.68	1.36-8.99	2.84	Hong et al., 1999
Xiamen Western Bay, PRC	0.14-1.12	0.45	4.45-3.11	42.8	Hong et al., 1995
Juilong River Estuary, PRC	3.7-13	9.1	8.7-69	26	Chen et al., 1986
Macao estuary, PRC	---	---	BDL-79.0	---	Zhang et al., 1999
West coast of Sri Lanka	0.086-0.33	---	0.09-1.6	---	Gurge & Tanabe, 2001
Eastern coast of India	0.008-0.02	0.10-0.97	---		Sarkar, 1994
Mediterranean	---	0.18	---	---	Burns & Villeneuve, 1987

BDL = below detection level.

Interpretative Problems

In coastal ocean and estuarine systems, bioturbation and physical mixing of the sediments can occur (Benninger et al., 1979; Nittrouer et al., 1983; 1984) and these are very much pronounced at Kakdwip and Ghusighata. As discussed earlier, the mudflats of these two sites are populated with various macrozoobenthos, which stabilize the bottom against erosion and also enhance sediment accumulation. The processes may obliterate the information contained in the top part of the sedimentary column. Another typical problem is concerned with the role of diagenesis, particularly for trace metal remobilization as pointed out by Valette-Silver (1993). The vertical distribution of some metals (Fe, Mn etc.,) can be affected after sediment deposition by diagenetic processes that solubilize the elements in the anoxic portions of the core and redeposit them in the upper oxic layers. Based on studies conducted off the California coast, Shaw et al., (1990) described a variety of chemical behavior for the transition metals depending on the form under which they are transported to the sediments (e.g., scavenged by the Mn-Fe oxides, biogenic material, detrital material etc.,). Transport, burial and diagenesis play a key role in the preservation of historical records for metal contamination (Valette-Silver, 1993; Bhattacharya et al., 2000). Moreover, the phenomena such as tidal inundation, salinity changes as well as wind and waves allow the process of bioturbation, resuspension and erosion that are known to affect the metal concentrations in surface sediments (Bellucci et al., 2002). Hence for reliable results an excellent knowledge of the environment and of the problems specific to each area are to be taken into consideration.

DISTRBUTION OF PESTICIDES IN SURFACE SEDIMENTS

The concentration of four important metabolites of HCHs in the Hugli river basin is shown in Table 1, which reveals a heterogenic nature of distribution. The ΣHCH ranges from 0.003 to 0.33 with a mean value of 0.134 ± 0.073 whereas the ΣDDT ranges from 0.003 to 0.119 with a mean value of 0.058 ± 0.032. A high concentration of β-HCH was observed sharing a major portion of HCH followed by α - and γ- HCH, which were present in almost, uniform concentration. Regarding the HCH isomers, β-HCH was the highest (100 %) followed by δ-HCH (commercial name lindane) (0.90 to 48.4%) (Table 2). Conversely, the level of α- HCH was below the detection in all the samples excepting a solitary case. It could be noted that the β– HCH is always higher than γ -HCH confirming the characteristic feature of more stability and resistance to microbial degradation (Rajendran and Subramanian, 1997). This may suggests recent exposure to technical HCHs. In west coast of Sri Lanka, Guruge and Tanabe (2001) observed the HCH isomers in the following decreasing order: α - HCH > β - HCH > γ -HCH. HCH contamination might have occurred through atmospheric transport from other parts of India, where large amount of technical HCHs are still being used and high concentrations were found in biota (Ramesh et al., 1992). The occurrence of DDT isomers are predominant in the following order: pp′ DDT > pp′ DDE > pp′ TDE > op′ DDE > op′ DDT which is in conformity with the findings of James and Ramesh (2002) from Brahmaputra river basin, India. Among the DDT metabolites, pp′- DDT and pp′- DDE accounted for the greatest abundance with approximately 70 – 90% of the total DDT in sediments as the sum of these two forms as shown in Table 1. These distributions indicate an active degradation of DDT in the sediments and /or inputs of already degraded DDT to the area. DDT can be degraded to DDT or to DDE via dehydrochlorination produced by biotic or abiotic decomposition reaction (Fries, 1972; Gerlach, 1981). The dominance of either pp′ DDE or both pp′ DDT and pp′ DDE in sediments was also recorded by Guruge and Tanabe (2001) from west coast of India and by Booij et al., (2001) from north-west coast of Java, Indonesia. A comparison of organochlorine residue levels in sediments from different estuaries and seas is given in Table 2.

Data of OC concentrations in sediments are scarce for Asian developing countries. While comparing with the existing data it is revealed that the present data on OC concentrations in sediments were lower than those concentrations reported in polluted riverine and coastal sediments from the Phillipines (Lee et al., 1997), South India (Kannan et al., 1995a), Korea (Khim et al., 1999) and Sri Lanka (Guruge and Tanabe, 2001). A reason for this could be that the location in the present study was rather remote or less urbanized compared to those in other Asian locations. Nevertheless, the accumulation pattern of OC in sediment from the present observation do not follow any unique pattern though an elevated level was marked in the premonsoon season characterized by high salinity (20.75 to 28.66 o/oo) and pH values (8.2 to 8.6). The finding was also in agreement with the observation of Agnihotri et al. (1996) from the upper stretch of the Ganga river. Guruge and Tanabe (2001) observed that OCs concentrations in sediments were a few orders of magnitude less than those concentrations in biological samples. Organochlorine concentrations in sediments from tropical regions have been observed to be low due to rapid evaporation (Iwata et al., 1994). However, to understand the different OC accumulation pattern in sediments, further work with more samples is needed.

The OCs distribution pattern in the sediment, as revealed in the present study, was mainly related to tributary runoff, coastal sewage input and some localized anthropogenic input. The concentration of these pesticides may be attributed to the influx of huge amount of pesticides in a number of major rivers situated on the east coast of India into the Bay of Bengal as endorsed by Sarkar (1994). A huge amount of consumption of organochlorine pesticides mainly in agricultural activities contributes a considerable amount of their residues in the marine environment. Amount of isomers and metabolites of HCHs and DDT in the sediment, β-HCH and γ- HCH and pp′ DDT and pp′ DDE were found to be highly dominated over the others.

Various opinions exist in connection with the organochlorine accumulation in sediments. Belzuce et al., (2001) observed that the concentrations of OCs in surface sediments along the Nervion estuary (Northern Spain) were higher from the middle part of the estuary, in rivers and in areas of high deposition. They asserted that the fine texture of the sediments, their anoxic properties and the presence of organic matter are all directly related to the enrichment by trace organic compounds. Again, in Mississippi river delta, Santschi et al. (2001) interpreted that a combination of erosion processes of riverine and other sedimentary deposits might control the presence of DDTs in the sediments during floods.

To find out the correlation of organochlorines with other physicochemical parameters, Pearson correlation matrix was calculated. It has been revealed that an overall negative and significant relations exist between organic carbon in sediment with ∑HCHs ($r=-0.38$; $p < 0.05\%$)), ∑DDTs ($r= -0.40$; $p<0.05$ %) and endosulfan sulphate ($r=-0.38$; $p<0.05$ %). Water salinity is positively correlated with both ∑HCHs ($r=0.41$; $p<0.05$ %) and ∑DDTs ($r=0.38$; $p<0.05$ %) but negatively related with endosulfan sulphate ($r = -0.09$). On the contrary, the correlation values between soil pH and other OCs are insignificant in all the cases. One-way analysis of variance (ANOVA) (sediment organochlorines as factors) shows that the variations of concentrations of ∑HCHs, ∑DDTs and endosulfan sulphate over the months are highly significant ($F= 11.94$; $p < 0.0001$).

In India, an estimated 380000 tones of pesticides and other halogenated hydrocarbons are used each year, of which 55000 tones are used in agriculture. The total annual consumption of DDT and its isomers is 107000 tones per year (Glasby and Roonwal, 1995). Available data on consumption of pesticide in the neighboring countries in Indian Ocean region reveal the following features: 11000 tones in Pakistan, 3000 tones in Bangladesh and 28000 tones in Sri Lanka (Qasim, 2001). These chemicals are used as pesticides, herbicides and fungicides in agriculture and for controlling vector-borne diseases as malaria. It is recognized that a large portion of the pesticides used on land reaches the oceans through atmospheric transport and river discharge in India (GESAMP, 1989). The applied pesticide can be transported through surface run-off, leaching and vapor phase and ultimately accumulates and settles in the bottom sediment.

CONCLUSION

The work has assessed the status of the concentration of heavy metals and organochlorine pesticide residues in sediments in a heavily impacted estuarine system of northeast India. In general the pattern of metal concentrations reflects differential heavy metal loading due to human activities and habitats use pattern and ecology of the estuarine biota. The need for enacting suitable pollution control measures and regular monitoring programme has emerged out of this study. Agricultural run-off, sewage outfall and the redistribution of contaminated sediments from multifarious industries are all potential contemporary sources of contamination, although more detailed survey of surficial sediments is required to determine other current point sources of contamination. The coastal areas of West Bengal are in a stage of rapid development and new mining; dredging and chemical activities are being incorporated. Therefore, changes in pH, salinity, and organic carbon content of the sediment and degradation of organic complexes are expected in near future. These processes could mobilize these metals (Gil et al., 1999) and expose the biota to a chronic contamination, impacting the marine environment in this location, as well as causing public health and economical hazards. Leaching tests under different conditions should clarify the remobilization of heavy metals. Monitoring of nearby wells is necessary to observe heavy metal impacts on the groundwater especially arsenic as several cases of arsenic poisoning have already been detected in many areas of West Bengal including the lower Gangetic plain (Chakraborti et al., 2003) .The study also provides evidence that the Hugli estuary is not grossly polluted by organochlorines. However, as further expansion of agricultural activities occurs it becomes more critical to understand the factors influencing the transport of sediment and sediment-bound contaminants. Unfortunately, no substantial data about the organochlorine contamination in this environment is available and thus the present data can provide as a baseline to monitor the success of strategies for the conservation of the estuary. Recently, Kumar et al., (2006) and Lahiri et al., (2002) has detected a significant bioconcentration of DDT and HCH residues in breast milk in the north central and eastern part of India respectively and the newborn is a recipient of this bio-concentrated form of pesticides. DDT residues were also detected in the blubber of the endangered species, gangetic dolphin (*Platanista gangetica*) by Sinha and Sinha (2001) in the upstream of River Ganges, eastern part of India. These findings confirm the alarming stage of deterioration and hence intensive monitoring of persistent organic pollutants (POPs) in both biotic and abiotic compartments is to be undertaken by the environment managers, public health officials and persons responsible for enforcing policy standards.

ACKNOWLEDGEMENT

One of the authors, Mousumi Chatterjee, is thankful to Council of Scientific and Industrial Research (CSIR), New Delhi, India for extending a fellowship. Part of the present work is reproduced with some enhancement from publication "Environment International": volume no .29, Authors: Badal Bhattacharya, Santosh Kumar Sarkar, Nilanjana Mukherjee. Title of the article: "Organochlorine pesticide residues in sediments of tropical mangrove estuary, India: implications for monitoring". Page no.: 587-592. Copyright (2003) with permission from Elsevier.

REFERENCES

Abbas, N., Subramanian, V., (1984). Erosion and sediment transport in the Ganges river basin (India). *Journal of Hydrology,* 69: 173- 182.

Abu-Hilal A H. (1987). Distribution of trace elements in near shore surface sediments from the Jordan Gulf Of Aqaba (Red Sea). *Marine Pollution Bulletin,* 18, 190-193.

Abu-Hilal A. H. and Badran M. M. (1990). Effect of pollution sources on metal concentration in sediment cores from the Gulf of Aqaba (Red Sea). *Marine Pollution Bulletin*, Vol. 21, No. 4, 190-197.

Adriano D. C. (1986). *Trace Elements in Terrestrial Environments*. Springer-Verlag, New York.

Agnihotri, N. P; Kulshrestha, G; Gajbhiye, V. T; Mahopatra, S. P.; and Sing, S. B.(1996). Organochlorine insecticide residues in agricultural soils of the Indo-Gangetic plain. *Environ. Monit. Assess*. 40:279-288.

Bami, H.L (1992). Pesticides's risk – an overview of Indian scene. *Pesticides information,* vol. xviii, 4-13.

Alagarsamy R. (2006). Distribution and seasonal variation of trace metals in surface sediments of the

Mandovi estuary, west coast of India. Estuarine, *Coastal and Shelf Science*, 67, 333-339.

Angelidis, M. O. and Aloupi, M. (1995), Metals in sediments of Rhodes Harbour. Greece. *Marine Pollution Bulletin,* 31; 273-276.

Angulo, E., (1996). The Tomilson Pollution Load Index applied to heavy metal 'Mussel-Watch' data: a useful index to assess coastal pollution. *Science of the Total Environment.* 187; 19-56.

Balls, P.W; Hull, S.; Miller, B.S.; Pirie, J.M.; and Proctar, W., (1997). Trace metal in Scottish estuarine and coastal sediments. *Marine Pollution Bulletin*, 34, 42-50.

Bandopadhay, S., (1998). Coastal erosion and its management in Sagar Island, South 24 Parganas, West Bengal. *Indian Journal of Earth Science*, Vol. 24, Nos. 3-4, 51-69 p.

Batley, G. E. and Brockbank, C. I. (1990). Impact of ocean disposal of dredged sediments from the RTA Glebe Island bridge site- A coastal sediment survey. *CSIRO report, Center for Advanced Analytical Chemistry- Division of Coal and Energy Technology, Report FTAR 050.*

Bellucci, L. G., Frignani, M., Paolucci, D. and Ravanelli, M. (2002). Distribution of Heavy metals in sediments of the Venice Lagoon: The role of the industrial area. *The Science of the Total Environment*, 295, 35 – 49.

Benninger, L. K., Aller R.C. , Cochran, J.K. and Turekian K.K..(1979). Effects of biological sediment mixing on the ^{210}Pb chronology and trace metal distribution in a Long Island Sound sediment core. *Earth and Planet Science Letters* 43, 241-259.

Bhattacharya, A., Sarkar, S. K., Das, G. K., Giri, S., Bose, G., Mitra, K. and Dey, D. (2000). Holocene tidal flat sedimentation in the Sundarban Biosphere Reserve, northeast India. *S.D.M.C.E.T; Dharwad and IGCP*-367, Spl. Pub. V.2, 51-55.

Belzunce, M. J., Solaun, O. , Franco, J., Valencia, V., and Borja, A., (2001) Accumulation of organic matter, heavy metals and organic compounds in surface sediments along the Nervion estuary (northern Spain). *Marine.Pollution Bulletin*.42: 1407-1411.

Booij, K.; Hillerbrand, M.T.J.; Nolting, R.F.; and Ooijen, J.V.(2001) Nutrients, trace metals, organic contaminants in Banten bay, Indonesia. *Marine Pollution. Bulletin*. 42:1187-1190.

Brix, H and Lyngby, J.E. (1984). A survey of metallic composition of *Zostera marina* (L.) in the Linfjord, Denmark. *Archives of Hydrobiology*, 99, 347-359.

Burns, K., and Villeneuve, J. P., (1987) Chlorinated hydrocarbons in the open Mediterranean ecosystem and implications for the mass balance calculations. *Marine Chemistry*, 20:337-359.

Bruland, K. W., Bertine, K., Koide, M. and Goldberg, E. D. (1974). History of metal pollution in Southern California coastal zone. *Environmental Science and. Technology*, 8: 425-432.

Bryan, G. W. and Langston, W. J. (1992). Bioavailability, accumulation and effects of heavy metals in sediments with special reference to United Kingdom estuaries: A review. *Environmental Pollution*, 76, 89-131.

Buckley, D. E., Smith, J. N., and Winters, G. V., (1995). Accumulation of contaminant metals in marine sediments of Halifax Harbour, Nova Scotia: environmental factors and historical trends. *Applied. Geochemistry*. 10: 175-195.

Canãrio, J., Vale, C., and Caetano, M. (2005). Distribution of methymercury and mercury in surface sediments of the Tagus estuary. Marine Pollution Bulletin 50, 1142-1145.

Canuel, E. A. and Martens, C. S. (1993). Seasonal variability in the sources and alteration of organic matter associated with recently deposited sediments. *Organic Geochemistry*, 20(5); 563-77.

Chakraborti, D., Mukherjee, S. C., Pati, S., Sengupta, M. K., Rahman, M. M., Chowdhury, U. K., Lodh, D., Chanda, C. R., Chakraborti, A. K. and Basu, G. K.. (2003). Arsenic groundwater contamination in middle Ganga plain, Bihar, India: A future danger. *Environment Health Perspectives*.; III (9): 1194- 1201.

Chen, S, Lin Z, and Lin M.,(1986) Distributions of BHC and DDT concentration in the surface sediments of Xiamen Harbour. *Taiwan Strait* 5:32-37.

Chow, T.J., Bruland, K., Bertine, K., Soutar, A., Koide, M. and Goldberg, E.D. (1973). Lead pollution record in Southern California coastal sediments. *Science* 181, 551-552.

Cundy, A. B., Croudace, I. W., Thomson, J., and Lewis, J. T., 1997. Reliability of salt marshes as 'Geochemical Recorders 'of pollution input: A case study from contrasting estuaries in Southern England. *Environmental Science and Technology* 31, 1093-1101.

Cundy, A. B. and Croudace, I. W., 1995. Sedimentary and geochemical variations in a salt-marsh/ mud-flat environment from the mesotidal Hamble estuary, southern England. *Marine Chemistry* 51, 115-132.

Crecelus, E. A. and Piper, D. Z. (1973). Particulate lead contamination record in sedimentary cores from Lake Washington, Seattle. *Environment Science and Technology.* 7, 1053 – 1055.

Darshankumar, H.; and Hader, D., (1999) *Global aquatic and atmospheric environment. Springer-Verlag*; Berlin.

de Groot, A. J., Salomons, W. and Allersma, E. (1976). Process affecting heavy metals in estuarine sediments. In: *Estuarine Chemistry* (J. D. Burton and P.S. Liss, eds.), pp. 131-157. Academic Press, London.

Dekov, V. M., Subramanian, V. and Van Grieken, R. (1999). Chemical composition of riverine suspended matter and sediments from the Indian sub-continent. In: Ittekot. V., Subramanian. V., Annadurai. S. (Eds.). *Biogeochemistry of Rivers in tropical South and Southeast Asia.* Heft 82. SCOPE Sonderband. Mitteilugen aus dem Geologisch-Paläontolgischen Institut der Universität. Hamburg, pp. 99-109.

Eisler, R. Lapan, R. L., Telek, G., Davey, E. W., Soper, A. E. and Barry, M. (1977). Survey of metals in sediments near Quonset Point, Rhode Island. *Marine Pollution Bulletin.* 6, 85 – 87.

El-Sayed, M Kh. (1982). Effect of sewage effluent on the sediment of Nordasvatnet (a land-locked fjord), Norway. *Marine Pollution Bulletin,* 13, 85-88.

Folk, R. L. and Ward, W. C. (1957) Brazos River bar, a study of the significance of grain size parameters. *Journal of Sedimentary Petrology* 27: 3-26.

Förstner, U., (1983). Assessment of metal pollution in rivers and estuaries. In: Thornton, I. (Ed.), *Applied Environmental Geochemistry. Academic Press*, London, pp. 395-423.

Friedman G M. (1968). Geology and geochemistry of reefs, carbonate sediments and waters, Gulf of Aqaba, Red Sea. *Journal of Sedimentary Petrology*, 38, 895-919.

Fries, G.F. (1972) Degradation of chlorinated hydrocarbons under anaerobic conditions, pp. 256-270. In Gould, R.F. (ed.), Fate of organic pesticides in the aquatic environment. *Advances in Chemistry, Series III.* American Chemical Society, Washington, D.C.

Förstner, U. and Wittman, G.T (1983). *Metal Pollution in the aquatic environment.* Springer-Verlag, Berlin.

Gil MN, Harvey MA, Estenes J L. (1999). Heavy metals in intertidal surface sediments from the Patagonium coast, Argentina. *Bulletin of Environmental Contamination and Toxicology*; 63: 52- 58.

Gerlach, S.A.(1981), Marine Pollution : diagenesis and therapy. *Springer-Verlag*, New York.

GESAMP (1989). The atmospheric input of trace species to the world ocean. *GESAMP Reports and Studies* No. 38.

Gibbs, R.J. (1973). Mechanisms of trace metal transport in rivers. *Science* 180: 71 – 72.

Giesy, J.P. and Hoke, R. (1990). Freshwater sediment quality criteria: Toxicity bioassessment. In: Baudo, R., Giesy, J., Muntau, H. (Eds.), sediments: Chemistry and Toxicity of In-Place Pollutants. Lewis Publishers, Inc., pp. 265-348.

Gil, M.N., Hurvery, M.A. and Estenes, J.L. (1999). Heavy metals in intertidal surface sediments from Potagonium coast, Argentina. Bulletin of Environmental Contamination and Toxicology, 63, 52-58.

Glasby, G.P. and Roonwal, G.S.(1995), Marine pollution in India : an emerging problem. *Current. Science.* 68: 495 – 497.

Goldberg, E. D., Gamble, E., Griffin, J. J. and Koide, M., (1977). Pollution history of Narragansett Bay as recorded in its sediments. *Estuarine Coastal and Marine Science.* 5, 549-561.

Guruge KS, and Tanabe S.(2001) Contamination by persistent organochlorines and butylin compounds in the west coast of Sri Lanka. *Marine Pollution Bulletin.* 42(3): 179-186.

Guruge, K.S., Tanabe, S., Fukuda, M., Yamagishi, S. and Tatsukawa, R..(1997) Accumulation pattern of persistent organochlorine residues in common cormorants (*Pharacrocorax carbo*) from Japan. *Marine Pollution Bulletin.* 34: 186 – 193.

Hakanson, L. (1992). Sediment variability. In: Burton, Jr. G.A. (ed.) Sediment toxicity Assessment, *Lewis Publishers*, *Boca Raton*, FL, pp. 19-36.

Hema Achyuthan, Richardmohan, D., Srinivasalu, S. and Selvaraj, K. (2002). Trace metals in the sediment cores of estuary and tidal zones from northern part of Southeast coast of India. *Indian Journal of Marine Science*, 31, 141-149.

Hershelman, G. P., Schafer, H. A., Jan, T. K. and Young, D. R. (1981). Metals in marine sediments near a large California municipal outfall. *Marine Pollution Bulletin.* 12, 131 – 134.

Hirner, A.V., Kristsotakis, K., and Tobschall, H.J. (1990). Metal-organic association in sediments- I. Comparison of unpolluted recent and ancient sediments and sediments affected by anthropogenic pollution. *Applied Geochemistry*, 5; 491-505.

Holmes, C. W., Slade, E. A. and Mickerran, C. J. (1974). Migration and redistribution of zinc and cadmium in marine estuarine system. *Environmental Science and Technology*, 8: 254-259.

Hong H, Chen W, Xu L, Wang X, and Zhang L. (1999) Distribution and fate of organochlorine pollutants in the Pearl River estuary. *Marine Pollution Bulletin.* 39:376-382.

Hong, H., Xu, L., Zhang, L., and Chen, J.C.,(1995) Wong YS, Wan TSM. Environmental fate and chemistry of organic pollutants in the sediment of Xiamen and Victoria harbors. *Marine Pollution Bulletin.* 31:229-236.

Huntzicker, J. J., Friedlander, S. K. and Davidson, C. I. (1975). Material balance for automobile-emitted lead in Los Angels Basin. *Environmental Science and Technology*, 9, 448-457.

Iwata, H.; Tanabe, S.; Aramoto, M.; Sakai, N.; and Tatsukawa, R. (1994) Persistent organochlorine residues in sediments from the Chukchi Sea, Bering Sea and Gulf of Alaska. *Marine Pollution Bulletin.* 28:746-753.

James, R. A. and Ramesh, R. (2002). Distribution of pesticides in the river sediments. In Environmental hazards in south Asia, V. Subramanian (Editor), *Capital Publishing Company*, New Delhi, India, pp. 161 – 165.

Jonathan MP and Ram Mohan V. (2003). Heavy metals in sediments of the inner shelf off the Gulf of Mannar, Southeast coast of India. *Marine Pollution Bulletin* 46, 263-268.

Kannan, K.; Tanabe, S.; and Tatsukawa, R.(1995) Geographical distribution and accumulation features of organochlorine residues in fish in tropical Asia and Oceania. *Environmental. Science and Technology*. 29; :2673-2683.

Kannan, S. T.; and Sengupta, R. (1987) Organochlorine residues in zooplankton off Saurashtra coast, India. *Marine Pollution Bulletin*. 18: 92-94.

Kemp. A. L.W., Thomas, R. L., Dell, F. I. and Jaqet, J. M. (1976). Cultural impact on the geochemistry of sediments in Lake Erie. *Journal.of Fisheries Research Board of Canada*. 33, 440 – 462.

Khim, J. S.; Kannan, K.; Villeneuve, D. L.; Koh, C. H.; and Geisy, J. P.(1999) Characterization and distribution of trace organic contaminants in sediment from Masan Bay, Korea. I. Instrumental analysis. *Environmental. Science.and Technology*. 33:4199-4205.

Knauer, G. A. (1977). Immediate industrial effects on sediments metals in a clean coastal environment. *Marine Pollution Bulletin* 8, 249-254.

Krauskopf KB. (1965). Factors affecting the concentration of thirteen rare metals in seawater. *Geochemica. Cosmochima Acta* 9, 1-32.

Krumbein, W. C. and Pettijohn, F. J, (1938). *Manual of Sedimentary Petrology*. New York: Plenum; 1938. p.549.

Kumar, A., Dayal, P., Shukla, G. S. and Josepth, P. E. (2006). DDT and HCH residue load in mother's breast milk: a survey of lactating mother's from remote villages in Agra region. *Environment International*. 32: 248-251.

Lahiri, P., Pratihar, S., Chakroborti, J., and Mitra, M. (2002). Organochlorine pesticide residues in human milk in Kolkata. *Science and Culture*, 68(78), 182 – 185.

Larsen, B. and Jensen, A. (1989). Evaluation of the sensitivity of sediment stations in pollution monitoring. *Marine Pollution Bulletin*. 20: 556 – 560.

Lee, J. S., Tanabe, S., Takemoto, N., and Kubodera, T., (1997) Organochlorine residues in deep-sea organisms from Suruga Bay, Japan. *Marine Pollution Bulletin*. 34:250-258.

Legorburu, I. and Canton, L. (1991). Heavy metal concentrations in littoral sediments from Pasajes harbour, Spain. Marine Pollution Bulletin, 24, 462-464.

Li, X., Shen, Z., Wai, O.W.H., and Li, Y. (2000) Chemical partitioning of heavy metal contaminants in sediment of the Pearl River estuary. *Chemical Speciation and Bioavailability*. 12 (1): 17-25.

Logananthan, B. G., and Kannan, K.,(1994) Global organochlorine contamination trends: an overview. *Ambio* 23:187-191.

Liao, J. F. (1990). The chemical properties of the mangrove Solonchak in the northeast part of Hainan Island. *The Acta Scientiarum Naturalium Universities Sunyatseni. (Supp.)*. 9(4) 67-72.

Livens F. R., and Baxter, M. S., (1988). Particle size and radionuclide levels in some west Cumbrian soils. *The Science of The Total Environment*. 70, 1-17.

Long, E. R., Field, L. J. and MacDonald, D. D. (1997). Predicting toxicity in marine sediments with numerical sediment quality guidelines. *Environmental Toxicology and Chemistry*, 17 (4) : 714-727.

Long, E. R., MacDonald, D. D., Smith, S. C. and Calder, F. D. (1995). Incidence of adverse biological effects within ranges of chemical concentrations in marine and estuarine sediments. *Environmental Management* 19 (1), 81-97.

Millward, G. E., and Glegg, G. A., (1997). Fluxes and retention of trace metals in the Humber estuary. *Estuarine Coastal and Shelf Science*. 44, 97-105.

Mohanachandran G (1988). Heavy metal distribution in deltaic and coastal sediments in between Palar and Cauvery Rivers. *M.Phil. Dissertation. Unpublished.* Jawaharlal Nehru University, New Delhi, India.

Mohanachandran G, and Subramanium V (1990). Texture, mineralogy and elemental composition of sediments along the southeast coast of India. Indian J Mar Sci 19 1990, 128-132.

Monbet Ph. (2006). Mass balance of lead through a small macrotidal estuary: The Morlaix River estuary (Brittany, France). *Marine Chemistry*, 98: 59-80.

Muller, G. (1979). Schwermetalle in den sedimenten des Rhems Veranderungen seit 1971. *Umschan* 79, 778-783.

Nittrouer, C. A., D. J. Demaster, B. A. McKee, B. A. Cutshall and I. L. Larsen. (1983/1984). The effect of sediment mixing on ^{210}Pb accumulation rates for the Washington continental shelf. *Marine Geology*, 54, 201-221.

Nohara, M. and Yokota, S. (1978). The geochemistry of trace elements in pelagic sediments from the central Pacific Basin. *Journal of Geological Society of Japan* 84(4) , 165-175.

Nolting, R. F., Ramkema, A. and Everaarts, J. M. (1999). The geochemistry of Cu, Cd, Zn, Ni and Pb in sediment cores from the continental slope of Banc d' Arquin (Mauritani). *Continental Shelf Research*, 19, 665-691.

Pacyna, E.G., and Pacyna, J.M. (2002). Global emission of mercury from anthropogenic sources in 1995. *Water, Air and Soil Pollution*, 137: 149 – 165.

Pantalu V. R. (1966) *Contribution to the study of biology and fishery of some estuarine fishes.* Ph. D thesis, Calcutta University.

Perkins, E.J. (1974). *The biology of estuaries and coastal waters.* Academic Press, London.

Pringle, B.H., Hissong, D.E., Katz, E.L. and Mullkwarka S.T. (1968). Trace metal accumulation by estuarine mollusks. *Journal of Sanitary Engineering Division, American Society of Civil Engineers*, 94 (F.A. #), 455-475.

Qasim, S. Z,(2001). Some major pollutants of Indian Ocean. *Journal of Indian Ocean Studies.* 9 (3): 417 – 425.

Rajendran, R. B., and Subramanian, A. N. (1997) Pesticide residues in water from Kaveri, south India. *Chemistry. and Ecology.* 13:57-70.

Rajendran, R. B.; Karunagaran, V. M., Babu, S., and Subramanian, A.N. (1992). *Levels of chlorinated insecticides in fishes from the Bay of Bengal. Marine Pollution Bulletin, 24; 567-570.*

Ramesh, A., Tanabe, S., Iwata, H., Tatsukawa, R., Subramanian, A. N., Mohan, D., and Venugopalan, V. K.,(1990) Seasonal variation of persistent organochlorine insecticide residues in Vellar river waters in Tamil Nadu, South India. *Environmental. Pollution.* 67:289-304.

Ramesh, A., Tanabe, S., Kannan, K, Subramanian, A. N., Kumaran, P. L., andTatsukawa, R.,(1992) Seasonal variation of persistent organochlorine insecticide residues in Vellar River waters in Tamil Nadu, South India. *Arch. Environmental Contamination. And Toxicology.* 23:23-36.

Rasmussen, P. E. (1994) Current methods of estimating atmospheric mercury fluxes in remote areas. *Environmental Science and Technology*, 28, 2233 – 2241.

Rubio, B. and Nombela, M. A., Vilas, F. (2000), Geochemistry of major and trace elements in sediments of the Ria de Vigo (NW Spain): an assessment of metal pollution. *Marine Pollution Bulletin*, 40; 968-980.

Saha, M., Sarkar, S. K. and Bhattacharya, B.,(2006) Interspecific variation in heavy metal body concentrations in biota of Sunderban mangrove wetland, northeast India. *Environment International.* 32, 203-207.

Salomons, W/, Kerdlik, H., van Pagee, H., Klomp, R., and Schreur, A., (1988). Behaviour and impact assessment of heavy metals in estuary and coastal zone. In: Seeliger, U., de Lacerda, L. D., Patchineelam, S. R. (Eds.), *Metals in coastal environments of Latin America.* Springer-Verlag, New York, pp. 159-198.

Salomons, W. and Förstner, U. (1984). *Metals in the Hydrocycle*, Springer, Berlin.

Santschi, P. H., Presley, B. J., Wade, T. L., Garcia-Romero, B., and Baskaran, M., (2001) Historical contamination of PAHs, PCBs, DDTs and heavy metals in Mississippi River Delta, Galveston Bay and Tampa Bays sediment cores. *Marine Environmental* Research 52:51-79.

Santschi, P.H., Nixon, s., Pilson, M. and Hunt, C., (1984). Accumulation of sediments, trace metals (Pb, Cu) and total hydrocarbons in Narragansett Bay, Rhode Island. *Estuarine, Coastal and Shelf Science.* 19, 427-449.

Sarin MM, Borole DV, and Krishnaswami S (1979). Geochemistry and geochronology of sediments from the Bay of Bengal and equatorial Indian Ocean. *Proceedings of Indian Academy of Science* 88, 131-154.

Sarkar, A.,(1994) Occurrence and distribution of persistent chlorinated hydrocarbons in the seas around India. In *The Oceans: Physical-Chemical dynamics and human impact.* Eds. Majumdar, S.K., Miller, E.W., Forbes, G.S., Schmalz, R.F., and Panah, A.A.: 444-458.

Sarkar, S. K., Bhattacharya, B. and Debnath, S.(1994) The suitability of tropical marine bivalves as biomonitors of heavy metals in deltaic Sunderbans, north-east India. *Chemosphere*; 29(4): 759-70.

Sarkar, S. K., Bhattacharya, B., Bandopadhaya. G., Giri, S., and Debnath, S.,(1999) Tropical coastal organism as qualitative indicators of mercury and organomercury for sustainable use of living resources. *Environment, Development and Sustainability; 1*: 135-47.

Sarkar, S. K., Bhattacharya, B., Debnath, S., Bandopadhaya, G. and Giri, S., (2002) Heavy metals in biota from Sunderban wetland ecosystem, India: Implications to monitoring the environmental assessment. *Aquatic Ecosystem Health and Management;* 5(2): 207-14.

Sarkar, S. K., Bhattacharya, B., and Das, R., (2003) Seasonal variations and inherent variability of selenium in marine biota of a tropical wetland ecosystem: implications for bioindicator species. *Ecological Indicators*, vol. 2 (4). *Oxford: Elsevier Press*; p. 367-75.

Sarkar, S. K., Bilinski, S. F., Bhattacharya, A., Saha, M., and Bilinski, H., (2004) Levels of elements in the surficial estuarine sediments of the Hugli river, northeast India and their environmental implications. *Environment International* 30:1089-98.

Sarkar, S.K., Bhattacharya, A., Giri, S., Bhattacharya, B., Sarkar, D., Nayak, D.C., and Chattopadhaya (2005). Spatiotemporal distribution of benthic polychaetes (Annelida) and relationships with environmental variables in a tropical estuary. *Wetlands Ecology And Management*, Volume 13, pp. 55-67, 2005.

Sarkar, S. K., Saha, M. Takada, H., Bhattacharya, A., Mishra, P. and Bhattacharya, B. (2006) Water quality management in the lower stretch of the river Ganges, east coast of India: an approach through environmental education. *Journal of Cleaner Production.* (in press).

Selvaraj K, Ram Mohan V, and Szefer P (2004). Evaluation of metal contamination in coastal sediments of the Bay of Bengal, India: geochemical and statistical approaches. *Marine Pollution Bulletin* 49, 174-185.

Sen, M., Chakraborty, A.K., Ghosh, A.K. and Bahdopadhaya, K.K. (1994). Water quality of the Hooghly: some aspects. *Sea Explorers*, 14-22.

Shailaja, M. S., and Nair, M.,(1997) Seasonal differences in organochlorine pesticides concentrations of zooplankton and fish in the Arabian sea. *Marine Pollution Bulletin.* 44:263-274.

Shailaja, M. S., and Sen Gupta, R.,(1990) Residues of dichlorodiphenyltrichloroethane and metabolites in zooplankton from the Arabian Sea. *Current Science.* 59: 929-931.

Shailaja, M. S., and Sengupta, R.,(1989) DDT residues in fishes from the eastern Arabian Sea. *Marine Pollution Bulletin.*20: 629-630.

Shailaja, M. S., and Singbal, S. Y. S., (1994) Organochlorine pesticide compounds in organisms from the Bay of Bengal. *Estuarine Coastal. Shelf Science.* 19:219-226.

Sharma, V. K., Rhudy, K. B., Koening, R. and Vazquez, F. G. (1999). Metals in sediments of the Upper Laguna Madre. *Marine Pollution Bulletin* 38; 1221-1226.

Shaw, T. J., Gieskes, J. M. and Jahnke, R. A.. (1990). Early digenesis in differing depositionsl environments. The response of transition metals in pore water. *Geochimica Cosmochimca Acta*, 54, 1233-1246.

Sinha, S.K. and Sinha, R.K. (2001). The unholy accrual. *Down to Earth.* 49.

Snelson, J.T., (1977) The importance of chlorinated hydrocarbons in the world agriculture. *Ecotoxicology Environmental Safety*, 1: 17-30.

Stamoulis, S., Gibbs, R. J. and Menon, M. G., (1996). Geochemical phases of metal in Hudson River estuary sediments. *Environment International,* 22, 185-194.

Subramanianm, V., Jha, P. K. and Van Grieken, R. (1988). Heavy metals in the Ganges estuary. *Marine Pollution Bulletin,* 19, 290- 293.

Sunil Kumar, R. (1996). Distribution of organic carbon in the sediments of Cochin mangroves, south west coast of India. *Indian Journal of Marine Science*, 25:274-6.

Tanabe, S., Iwata, H., and Tatsukawa, R., (1994) Global contamination by persistent organochlorines and their ecotoxicological impact on marine mammals. *Science and. Total Environment.* 154:397-403.

Taylor, S. R. (1964). Abundance of Chemical elements in the continental crust: a new table. *Geochimica et Cosmochimica Acta*, 28, 1278-1285.

Taylor, S. R. and Mclennan, S. M. (1985). The continental crust: Its composition and Evolution 1985. Blackwell.

London. UNEP (United Nations Environment Programme). GESAMP: Cadmium, Lead and Tin in *Marine Environment. United Nations Environment Programme: Regional Seas Reports and Studies No. 56*, 1985,90 pp.

Tessier, A. and Campbell, P.,(1988). Partitioning of trace metals in sediments. In: Kramer, J.R., Allen, H.E. (Eds.), *Metal Speciation: Theory, Analysis and Application. Lewis Publishers*, Chelsea, MI, USA, pp. 183-199.

Tiller, K. G., Merry, R.H., Zarcinas, B.A. and Ward, T.J., (1989). Regional geochemistry of metal-contaminated surficial sediments and sea grasses in Upper Spencer Gulf, South Australia. *Estuarine, Coastal and Shelf Science,* 28, 473-493.

Thompson, J., Turkian, K. K. and McCaffrey, R. J. (1974). The accumulation of metals in and released from sediments of Long Island Sound. *Annual Progress Report. to Atomic Energy Commission.*

Turekian, K. K., and Wedepohl, K. H., (1961) Distribution of the elements in some major units of the earth crust. *Bulletin of Geological Society of America.* 72; 175-92.

UNEP (United Nations Environment Programme).(1985). GESAMP: Cadmium, Lead and Tin in Marine Environment. *United Nations Environment Programme: Regional Seas Reports and Studies No. 56,* 90 pp.

USEPA (US Environmental Protection Agency), 1996. *SW-846 Test Methods for Evaluating Solid Waste Physical/Chemical Methods.* US Environmental Protection Agency, Office of Solid Waste and the national Technical Information Service, Springfield, Virginia, CD-ROM.

Valette-Silver H. J. (1993). The use of sediment cores to reconstruct historical trends in contamination of estuarine and coastal sediments. *Estuaries,* 16 (3B) , 577- 15 88.

Waldichuk, M. (1985). Biological availability of metals to marine organisms. *Marine Pollution Bulletin,* 16 ,7- 11.

Walkey, A. and Black, T. A., (1934) An examination of the Dugtijaraff method for determining soil organic matter and proposed modification of the chronic and titration method. *Soil Sci*ence, 37: 23-38.

Wang, Q.C., Shen, W.G. and Ma, Z.W., (2000). Estimation of mercury emission from coal combustion in China. *Environmental Science and Technology,* 34: 2711-2713.

Winkels, H.J., Vink, J.P.M. and Beurskens, J.E.M., (1993). Distribution and geochronology of priority pollutants in a large sedimentation area, River Rhine, the Netherlands. *Applied Geochemistry,* S2, 95-101.

Woodwell, G. M., and Craig, P. P., (1971) Johnson, H.A. DDT in biosphere: Where does it go? *Science* 174:1101-1107.

Young, D. R., Young, C. S. and Hlavka, G. E. (1973). Sources of trace metals from highly-urbanized Southern California to the adjacent marine ecosystem. In *Cycling and Control of Metals,* pp. 21 – 39. US Environmental Research Center, Cincinnati, OH.

Zhang, G., and Min, Y. S., (1999) Mai, B.N.; Sheng, G.Y.; Wang, Z. S. Time trend of BHCs and DDTs in a sedimentary core in Macas estuary, Southern China. *Marine Pollution Bulletin.* 39:326-330.

Zingde. MD.,and Desai. BN., (1979). Mercury in Thane Creek, Bombay Harbor. *Marine Pollution Bulletin.* 12, 237.

Zwolsman, J. J. G., Berger, G. W., and van Eck, G. T. M., (1993) Sediment accumulation rates, historical inputs, post depositional mobility and retention of major elements and trace metals in salt marsh sediments of the Scheldt estuary, SW Netherlands. *Marine Chemistry* 44, 73-94.

In: Environmental Research Trends
Editor: Marvin A. Cato, pp. 105-132
ISBN: 978-1-60021-556-8

Chapter 3

SOIL MAGNETIC SUSCEPTIBILITY MEASUREMENTS FOR CHARACTERISING HEAVY METAL PATTERNS IN INDUSTRIAL AREAS

M. Ragosta[1], M. Macchiato[2], R. Caggiano[3], R. Coppola[3], M. D'Emilio[3] and A. Loperte[3]
[1]Dipartimento di Ingegneria e Fisica dell'Ambiente, Università della Basilicata, Potenza Italy
[2]Dipartimento di Scienze Fisiche, Università Federico II, Napoli Italy
[3]Istituto di Metodologie di Analisi Ambientale CNR, Tito Scalo (PZ) Italy

ABSTRACT

The development of innovative monitoring strategies, able to provide detailed information on temporal and spatial evolution of contaminants in superficial soil, combining fast, not invasive and low cost acquisition procedures for in situ monitoring pollution, is a crucial task for environmental research. In this context, electromagnetic parameter measurements represent suitable tools for identifying innovative experimental procedures. Particularly, magnetic susceptibility measurements of superficial soil may be used as a simple, rapid, cheap and non-destructive proxy variable to evaluate heavy metal patterns in industrial, urban and agricultural areas. We developed an experimental procedure to carry out field surveys for characterizing heavy metal patterns in superficial soil. This procedure is based on in situ soil magnetic susceptibility measurements supported by analytical determinations of metal concentrations. The well defined experimental protocol allows to reduce intrinsic environmental biases and to standardize the measurement procedure making simpler data interpretation.

In this chapter we present a review of case studies concerning the characterization of contamination patterns in different industrial areas of Basilicata region in Southern Italy. Basilicata is an inner region, rather unpolluted, in which the main land use is agriculture The investigated sites show different geomorphologic and anthropogenic features, so they represent optimal sites for studying the potentiality of this kind of approach in different environmental conditions.

1. INTRODUCTION

Soil has always been important to humans and their health, providing a resource that may be used for different aims. Soil significantly influences a variety of functions such as plant growth and the cycling of water that sustains the human population; so, scientists, having different cultural background and expertise, focus their attention in the development of new monitoring strategies devoted to soil contamination and soil quality (Arshad and Martin, 2002; Llobet et al., 2002; Navas and Machin, 2002; Karlen et al., 2003).

It is well established that many anthropogenic activities (MSWI, industrial plants, mining activities, vehicular emissions) may release metals into the environment (Morselli et al., 2002; Alvarez et al., 2003; Al-Khashman, 2004). Generally, for determining metal concentrations and their spatial distribution in surface and subsurface soil, analytical techniques are used (Barahona and Iriarte, 2001; Viard et al., 2004; Clemente and Bernal, 2006). These techniques are expensive and require long times for the sample chemical pre-treatment and analysis. So, the development of innovative methods for monitoring pollutant concentrations in surface and subsurface soil is a crucial task for environmental research. Actually the main purpose is to develop monitoring strategies able to provide detailed information about the temporal and spatial evolution of contaminants in subsoil, combining fast, not invasive and low cost acquisition procedures for in situ observations.

Recently, the application of electromagnetic techniques to environmental monitoring demonstrated the reliability of these methods in different fields: for mapping geological structure of subsoil (Suzuki et al., 2000), for monitoring waste disposal (Aristodemou and Thomas-Betts, 2000), for characterizing contaminated sites (Atekwana et al., 2000; Nadim et al., 2000; Steeples, 2001; Hanesch and Scholger, 2002). The electromagnetic methods allow us to define images of subsoil providing information about location and nature of buried targets. Moreover they may put in evidence the presence of pollutants in soil and in groundwater (Sharma, 1996 and reference therein).

In the last years, a strong interest has been turned to magnetic susceptibility measurements for environmental applications (Durza, 1999; Petrovsky et al., 2001; Hanesch and Scholger, 2002; Boyko et al., 2004). Properties of magnetic minerals such as magnetic susceptibility, isothermal remanent magnetization (IRM), saturation of isothermal remanent magnetization (SIRM), anhysteretic remanent magnetization (ARM) may provide information about the amount, the grain size and the mineralogy of the magnetic fraction within the sediments and they are generally used for paleoclimatic analysis and paleoceanographic studies (Dinares-Turell et al., 2003; Thouveny et al., 2004). In the investigation of contamination phenomena, magnetic susceptibility values of superficial soil may be reliable information. This soil parameter mostly depends on ferrimagnetic mineral concentration. These minerals have origin by desegregation of parent rocks during pedogenesis, also bacterially mediated, by lithogenic processes and by anthropogenic activities. In particular, atmospheric depositions on soil are the principal anthropogenic cause of magnetic enhancement of soil. Dusts and fly ashes produced by many anthropogenic activities (traffic, power plant, metallurgical settlements, domestic heating) are rich of ferrimagnetic particles. Moreover, magnetic susceptibility values of various types of sediments and materials seem to be influenced by the presence of some pollutants like heavy metals (Heller et al., 1998;

Kapicka et al., 2000; Muxworthy et al., 2001; Jordanova et al., 2003; Lecoanet et al., 2003; Ng et al., 2003; Desenfant et al., 2004; Jelenska et al., 2004).

In this context we have recently investigated on the use of magnetic susceptibility measurements in superficial soil as a simple, rapid, cheap and non-destructive proxy variable able to evaluate concentration levels of heavy metals. A monitoring method based on this parameter may give results directly in situ and in real time. It may be easily combined with other geophysical and analytical investigation techniques. Moreover it may be carried out both in zones with different anthropogenic impacts (industrial areas, urban areas, agricultural zones) and in remote zones, in which it is not possible to install other monitoring systems (Chianese et al., 2006; D'Emilio et al., 2006a; D'Emilio et al., 2006b).

The relationships between magnetic particles and pollutants are complex: they depend on environmental availability of pollutants, on soil genesis and on other parameters such as substrate characteristics, presence of humic layer and soil moisture. For this reason, it was necessary to develop a well defined experimental protocol, in order to reduce the indeterminations and to standardize the measurement procedure. In this way it is possible to reduce the intrinsic environmental biases and to make simpler data interpretation (D'Emilio et al., 2006b). Particularly we study investigation depth of probes and measurement reproducibility; we evaluate the influence of external parameters, as vegetation layers and water content; we test the combined use in situ of different probes. Finally, we carried out some laboratory tests and in situ analysis aimed to evaluate the correlations between metal concentrations and soil magnetic susceptibility values.

Here we present a review of case studies concerning the characterization of contamination patterns in different industrial areas of Basilicata region in Southern Italy. The investigated sites (pre-treatment oil plant area in Agri valley, S. Nicola di Melfi industrial area, Potenza industrial area and industrial sites in Basento valley) show different geomorphologic and anthropogenic features and they represent optimal sites for studying the potentiality of this kind of approach.

2. Study Area

Basilicata is an inner region of Southern Italy (figure 1). It is a small region, mostly mountainous, with two short coastlines on the Ionian Sea and on the Tyrrhenian Sea. The region, due to massive emigration in the past, is under populated, the economy is mostly based on agriculture. Industrial development is low, and industrial settlements are located in well defined small areas. In the follow we describe the four sites in which the field surveys were carried out, focusing geomorphologic and anthropogenic features. We would note that the first two study cases concern industrial areas, located in naturalistic zone, where, starting from ninety, there was a rapid increase of anthropogenic activities. On the contrary the last two cases focus sites in which the industrial plants are located in urban areas or industrial zones operating since the seventies.

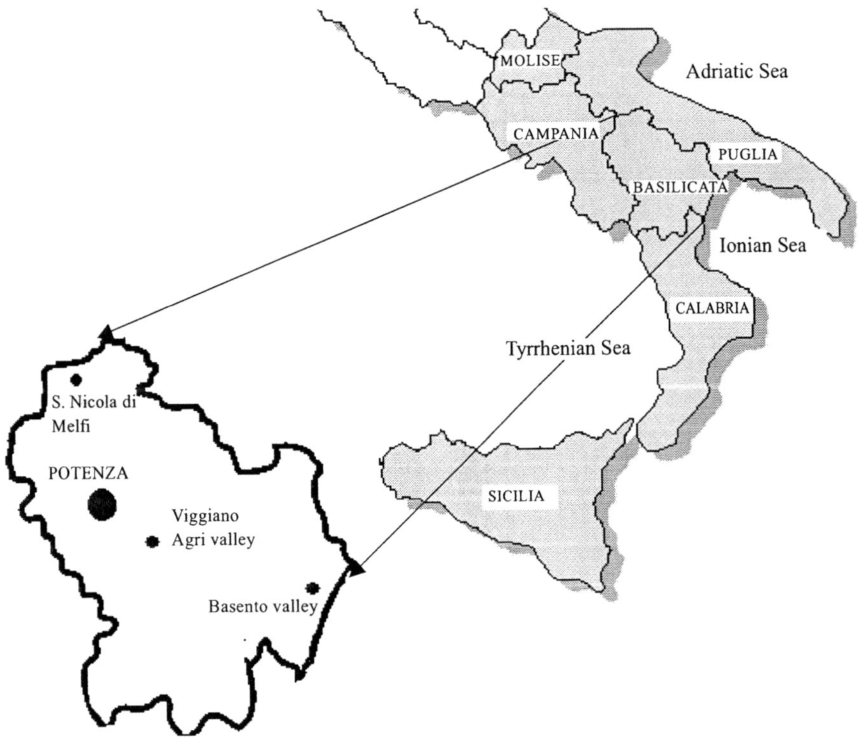

Figure 1. The four investigated sites in Basilicata region (southern Italy).

2.1. Viggiano Pre-Treatment Oil Plant Area in Agri Valley

The Agri valley extends over 1860 km^2 from the northwest to the southeast of Basilicata region, in the middle of the Lucano Apennine, and is crossed by the Agri river. This area consists of 29 municipalities with a total of 73075 inhabitants. This area is an ecologically intact region from a naturalistic point of view, features that led to the establishment of the Natural National Park of Agri valley (Italian laws 394/91 and 426/1998).

The main economic activities include agriculture (cereal and forage cultivation, breeding and horticulture, olive tree, vine and citrus cultivation), industry (food and drink, buildings) and, above all, oil-mining activities. As a matter of fact, Agri valley represents the largest Italian oil field (with estimated reserves of 420 million barrels equivalent). In 1997 about 11352 barrels per day of crude (9.5% of the national production) were produced, pre-treated at the Viggiano pilot oil plant and then transported to the refinery, located in a neighboring region, by tank truck. Actually it is estimated a daily production of 154 000 barrels and a pipeline ensures a straight connection between the pre-treatment oil plant and the refinery (Pietrapetosa et al., 2003).

From the geological point of view, the Agri valley is an intermountain basin formed during Quaternary times along the axial zone of the chain and is it filled by middle Pleistocene alluvial deposits. The pre-Quaternary bedrock is constituted by Mesozoic-Cenozoic shallow-water slope carbonates, mainly outcropping along the western side of the basin, thrust on coeval pelagic succession that crop out mainly along the eastern flank of the valley. Toward east and southeast the bedrock is formed by Tertiary siliciclastic sediments that occupy the southern part of the valley.

2.2. S. Nicola di Melfi Industrial Area

This industrial area is a part of Vulture basin, a valley along the Ofanto river in the north of Basilicata region. This is a rather unpolluted zone where the main land use is agriculture (wheat is the most spread cultivation and there are many milk cow farming). There are two urban sites Melfi (13000 inhabitants) and Lavello (16000 inhabitants) far from industrial area about 10 km (Caggiano et al., 1998). From a geological point of view, alluvial deposits mainly characterize the area of S. Nicola di Melfi. It is surrounded by low hills varying from 137 m to 508 m a.s.l.

The extension of industrial settlement of S. Nicola di Melfi is about 10 km^2. In this area there are 54 operating plants (alimentary products, stoneware products, industrial plants for printed and painted motor vehicle components). The most important is FIAT-SATA automotive plant. It has 5012 employs and it is scheduled for producing 1800 vehicles per day, operating from 1993. Moreover since from October 2000 a waste pre-treatment plant and an integrated platform for incineration (SATA FENICE project) combining waste management and electric energy production are operative. The incinerator has two ovens. A rotating oven in which 20000 t for year of toxic and harmful waste produced by SATA (and its induct) and 14400 t for year of toxic waste coming out from Basilicata region are collected; a grid oven in which 30000 t of MSW coming from the entire region are collected. All these industrial activities determine a high volume of vehicular traffic in the surrounding areas.

2.3. Potenza Industrial Area

The industrial area is located at the bottom of the urban agglomeration (Potenza with about 69000 inhabitants, is the most important urban site of the region). The extension of industrial settlement is about 2 km^2. The area is at 660 m a.s.l. and the Basento river crosses it. Besides the closeness to the urban centre, this industrial district is located in the neighborhood of areas in which the main land use is agriculture. From a geological point of view, the industrial area of Potenza is mainly characterized by alluvial deposits, whereas the low hills surrounding the area are characterized by polygenic conglomerates with sandy intercalation. The soil of the area contents a high percentage of clay. The activities in the area mainly consist of commercial sector, mechanic and steel industries. They represent a direct source of environmental pollution, producing gaseous, liquid and solid wastes as well as an indirect cause for the increased vehicular traffic in the surrounding areas. As regards the characterization of point sources, most of activities consist of small factories, except for iron and steel and mechanic sectors. Taking into account both these kinds of sources, CO_2, SO_2, NO_x and CO are the most significant emissions arising from combustion processes, while VOC and TSP emissions are mostly due to mechanic and building activities (D'Emilio et al., 2006a).

2.4. Industrial Area in Basento Valley

The industrial area of the Basento valley covers about 37.5 km^2 and includes the industrial settlements of three different towns: Salandra, Pisticci, and Ferrandina. The investigated zone is an alluvial area crossed by the Basento River and is surrounded by low hills. The highway S.S. Basentana delimits this area and so the volume of the commercial traffic causes a non-negligible contribution to the total emission rate. From 1958 to 1980, the most important anthropogenic activities characterizing this area were represented by chemical industrial plants for the production of tissues and chemical compounds. In particular, these plants turned out vinyl chloride (CVC and PVC), a highly toxic and noxious substance (ETERNIT), determining a scattered asbestos contamination and serious effects on the working staff's health. Actually, this area is characterized by the presence of about 300 different industrial activities including many chemical industrial plants for the production of polymeric yarns, synthetic tissues, nylon films, polymeric additives, and PVC bottles. Concerning the metallurgic industries, there is the production of mechanical components, rolled iron and electrical motors. There are other industrial activities related to building materials and leather tanning. The industrial settlements are surrounded by areas in which the main land use is agriculture. Finally, the presence of waste dumps in Basento valley industrial area may produce local soil pollution phenomena. In addition, all the activities in the investigated area (industrial plants, commercial traffic, and agricultural practices) may cause heavy metal soil pollution. From a geologic point of view, the investigated area is located in the Bradanic trough, a Pliocene–Pleistocene foredeep deposit basin. The middle-Pleistocene closing deposits of the Bradanic trough are primarily conglomerates and marine sands arranged onto several marine terraces. The main rivers of the region (Bradano, Basento, and Cavone rivers), with their beds having NW–SE direction, section the marine terraces and the underlying lower and middle–Pleistocene clay and silt. Particularly The industrial areas of Ferrandina and Pisticci are located into the Basento valley and precisely onto the west side stream terraces. These are formed by Holocene alluvial conglomerates and sands with composition that is predominantly siliceous (Chianese et al., 2006).

3. METHOD

3.1. Experimental Equipment

All the in situ measurements are carried out in georeferenced grids. We use a GPS mapping and GIS data collection/ maintenance system, Trimble geoexplorer, for georeferencing sampling points. This instrument is extremely portable, because it is small and lightweight. The precision of the instrument is 1-5 m, after differential correction.

The magnetic susceptibility measurements are performed using a portable meter (Bartington MS2) with three probes: MS2F and MS2D probes for in situ investigations and MS2B probe for laboratory analysis. The MS2F probe, with a diameter of 15 mm for in situ high-resolution surface measurements, is a miniature probe for the stratigraphic study of exposed geological and archaeological sections; it is also used where difficult surface conditions are present. The shape of the field generated by this probe is a dipole field. The

MS2D probe, with a diameter of 185 mm for in situ surface measurements, is designed for rapid assessment of the magnetic susceptibility of the land surface. The shape of the field generated by this probe is a toroid. Both the probes measure volume magnetic susceptibility, expressed as a dimensional value $x10^{-5}$ SI. The MS2B probe is a double frequency magnetic susceptibility meter for laboratory measurements. This sensor can operate at two different frequencies (0.465 Hz, low frequency and 4.65 Hz, high frequency) for studying the susceptibility frequency dependence. Particularly, for each examined soil sample, it allows to collect information on the concentrations of magnetic minerals and on the presence of submicrometer ferromagnetic grains (superparamagnetic grains). In the follow, we indicate with K_F, K_D, K_B(lf) and K_B(hf) the measurement of magnetic susceptibility carried out with the three different probes MS2F, MS2D and MS2B (low frequency/high frequency) respectively.

Metal concentrations measurements are performed by means of a Varian 200 Atomic Absorption Spectrophotometer with a GTA 100.

3.2. Experimental Procedure

In each investigated area, a preliminary screening survey is carried out. According to geological and anthropogenic features, a georeferenced sampling grid is defined. For each sampling point, four independent soil magnetic K_D susceptibility measurements are carried out at the vertices of a square with $L = 1$ m and the mean value was assume as the magnetic susceptibility of the sampling point. In order to minimize the intrinsic variability of measurements and to improve their reproducibility, it is necessary to remove the vegetation layer of the soil during the field survey. The water content has not negligible effects on magnetic susceptibility values and so, during field survey, it is useful to measure soil humidity and to take into account this parameter for data interpretation (D'Emilio et al., 2006b). Moreover it is possible to operate combining different probes. K_D and K_F integrated in situ measurements allow us to make hypotheses about the contamination deepness. In this case, for K_F measurements, it is possible to follow the same experimental procedure used for K_D measurements. Instead K_D and K_B integrated measurements may give information about the presence of ferromagnetic grains. In this case, for K_B laboratory measurements, we use the soil samples collected and pre-treated with the experimental procedure illustrated in the follow for metal determination.

During the field surveys, we collected four soil samples at the vertices of the same square in which we carried out magnetic measurements and we mixed them for obtaining a homogeneous sample. The soil samples were sealed in a polythene bag and frozen for following analysis. Regarding the determination of metal concentrations with AAS techniques, for the sample pre-treatment, we apply the protocol for the metal total fraction evaluation. 1.5 g of sieved soil was added with 3 ml of HF, 6 ml of HNO_3 and 0.25 ml of hydrogen peroxide 30% and digested with a microwave system (Palumbo et al., 2000; D'Emilio et al., 2006b). In the follow, all the metal concentrations are expressed in mg/kg.

3.3. Field Survey Plan

In this paragraph, we present the plan of the four study cases, illustrating the sampling grid features and the measured parameter set. A plan summary is shown in table 1.

In the field survey carried out in the Viggiano pre-treatment oil plant area (Agri valley), we define a sampling grid including three transects, with 22 sampling points (figure 2). This grid covers an area of 6.5 km^2 all around the oil pre-treatment plant. In each sampling point K_D values, Cd, Co, Cr, Cu , Fe, Mn, Ni, Pb and Zn concentrations were measured.

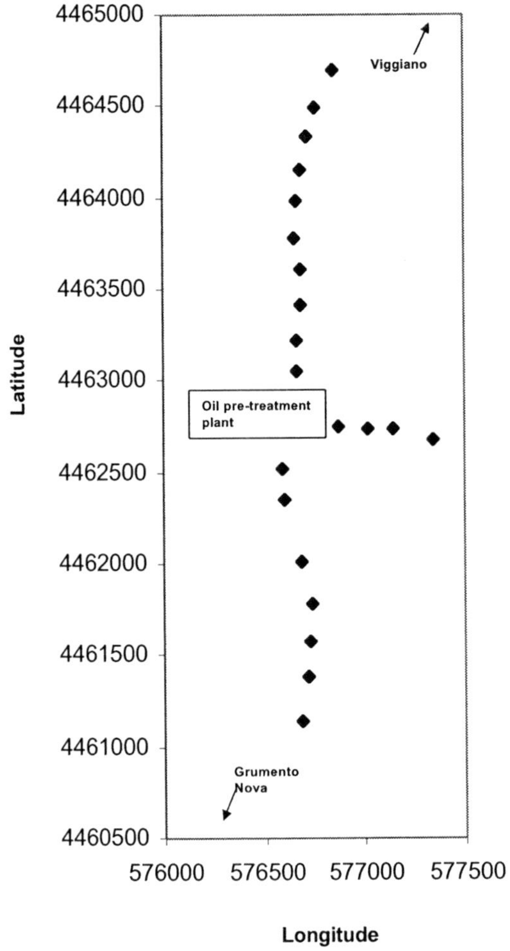

Figure 2. Sampling grid in Viggiano pre-treatment oil plant area in Agri valley.

In S. Nicola di Melfi industrial area the grid (figure 3) includes 103 georeferenced sampling points set in radial direction, along eight transects (A,B,...,H), starting from SATA plants; it covers an area of about 200 km^2. In each sampling point K_D value is measured and in each collected sample K_B(lf) and K_B(hf) are measured and Cd, Co, Cr, Cu , Fe, Mn, Ni, Pb and Zn concentrations were determined.

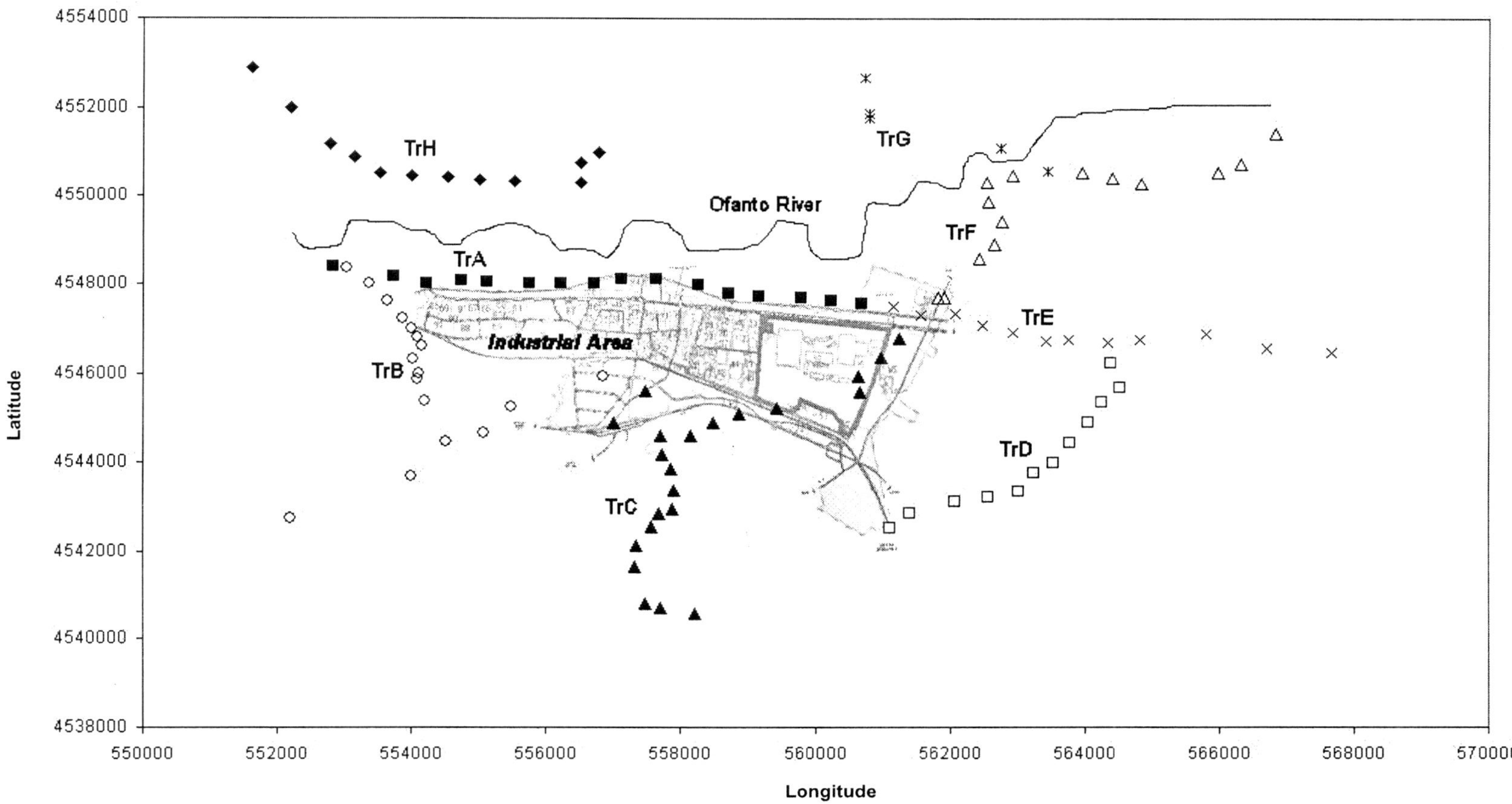

Figure 3. Sampling grid in S. Nicola di Melfi industrial area.

Table 1. A summary plan of the four study cases

	Screening survey							Detail maps	
	A *(km^2)*	n	K_D	K_F	K_B	Metal Conc.	*Cluster*	K_D	K_F
Viggiano pre-treatment oil plant area in Agri valley	6.5	22	•			Cd. Co. Cr. Cu. Fe. Mn. Ni. Pb. Zn	•		
S. Nicola di Melfi industrial area	200	103	•		•	Cd. Co. Cr. Cu. Fe. Mn. Ni. Pb. Zn	•		
Potenza industrial area	2	57	•			Pb			
Zone A								•	•
Zone B								•	•
Zone C								•	•
Industrial area in Basento valley	37.5	75	•	•	•	Co. Cu. Ni. Pb. Zn			
Ferrandina site								•	•
Pisticci site								•	•

In this first two study cases, we determine the total fraction of a large set of metals, in order to investigate their correlation structure and to point out the statistically significant relationships among metals and magnetic parameters. On the contrary, in the following cases, we operate an a priori choice, selecting a limited set of metals on the bases of the characteristics of the potential sources of emissions.

In the industrial area of Potenza city, a preliminary screening is carried out on 57 georeferenced sampling points distributed along three transects in an area of 2 km^2 (figure 4). For each sampling point, we measure K_D value and in the correspondent soil sample we determine Pb concentration. In a following step, on the base of preliminary screening results, we individuate three sub areas for detailed investigations (A, B and C zones). In these sub-areas, regular grids are defined and, for each sampling point, integrated measurements of magnetic susceptibility (K_D and K_F-values) are carried out.

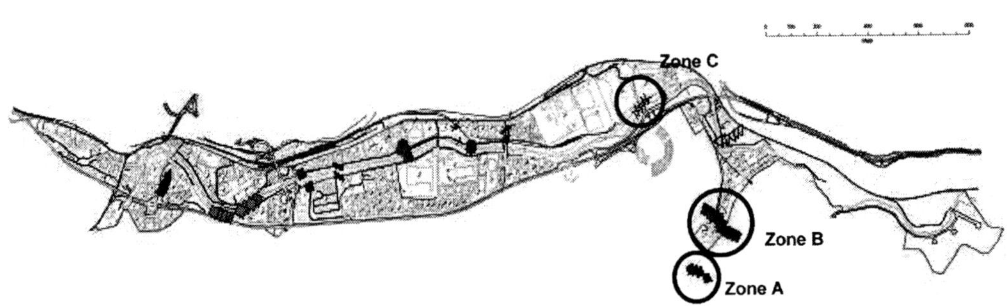

Figure 4. Sampling points overlapped on topographical map of industrial area of Potenza. Black circles indicate the zones in which detail magnetic maps are carried out.

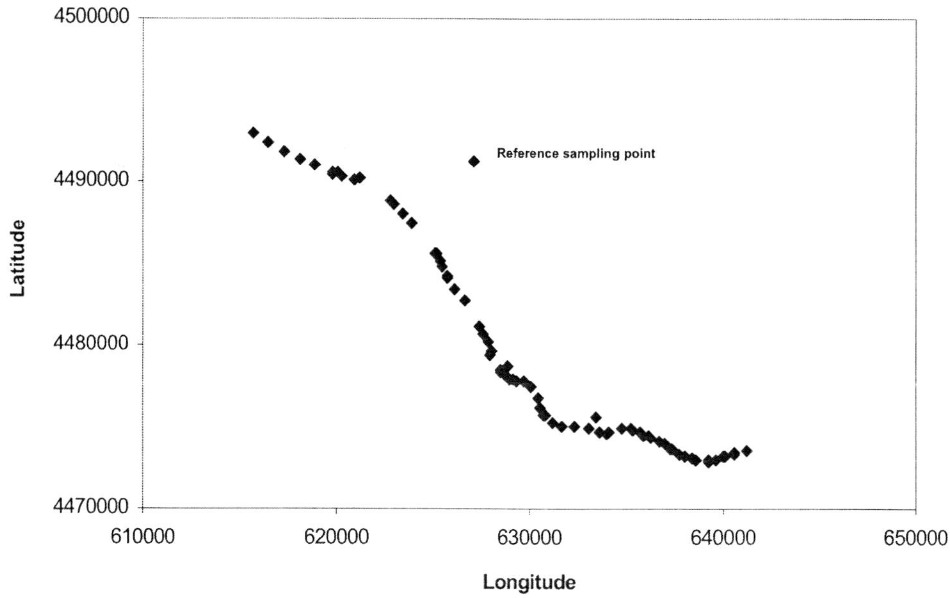

Figure 5. Sampling grid in Basento valley.

In the industrial areas of Basento valley, a preliminary screening is carried out on 75 sampling points distributed along one transect in an area of 37.5 km^2 (figure 5). For each sampling point, we measure K_D, K_F and K_B values and we determine Co, Cu, Ni, Pb and Zn concentrations. In a following step, starting from the preliminary survey analysis, we carry out two detail maps in Ferrandina and Pisticci sites respectively, defining regular grids and measuring K_D and K_F values.

3.4. Data Analysis

At first, in all the examined cases, we carry out an explorative statistical analysis of the different measured variables. The explorative parameters are m = mean value; sd = standard deviation; $CV\%$ = percentage variation coefficient; md = median; r = range (*min-max*). Moreover, we define as anomalies the values of magnetic susceptibility that exceed the 90th percentile.

Regarding the MS2B data (K_B(lf) and K_B(hf)), it is possible to define the frequency dependent magnetic susceptibility χ_B as (Whorm, 1998)

$$\chi_B = \frac{(K_B(lf) - K_B(hf))}{K_B(lf)} * 100$$

The values, expressed in percentage, may be discussed using the Dearing model (Dearing et al., 1996) in which soil samples with $\chi_B < 2\%$ are dominated by frequency independent coarse multidomain grains; samples with $2\% < \chi_B < 6\%$ correspond to a mixture of frequency-dependent and frequency-independent grains; sample with $\chi_B > 6\%$ are dominated by superparamagnetic grains.

In order to investigate the correlation structure among heavy metals and magnetic susceptibility values, we may apply a multivariate statistical tool. In particular, the analysis aims to characterize the correlation patterns among the descriptors: metal concentrations and magnetic susceptibility. In this case, in the multivariate procedure, we take into account the magnetic susceptibility as an endogenous variable for better highlighting the relationships among chemical and physical parameters. So, for each data matrix, we calculate the descriptor association matrix, using the Pearson correlation coefficient, and, to this matrix, we apply a clustering algorithm at complete linkage (Ragosta et al., 2003).

At the end, in the cases in which detail maps are carried out, data are collected on regular grids and the spatial patterns are determined by means of the kriging. It is a very flexible gridding method for producing maps. This interpolation method uses variogram to express the spatial variation. It minimizes the error of predicted values that are estimated by spatial distribution of the predicted values (Cressie, 1993). In the investigated cases, the measured data are interpolated using ordinary kriging and the maps are obtained with Surfer.8 ©.

4. Results

4.1. Viggiano Pre-Treatment Oil Plant Area in Agri Valley

The statistical explorative parameters of measured variables are shown in table 2. In figure 6 we represent the magnetic susceptibility values. We measure high magnetic susceptibility values near to the road that is main access to oil pre-treatment plant. Here K_D ranges from 97 10^{-5} SI to 240 10^{-5} SI. Moreover high magnetic susceptibility values are measured near a vineyard (from K_D = 54 10^{-5} SI to K_D = 76 10^{-5} SI). On the contrary the lowest magnetic susceptibility value, K_D = 17 10^{-5} SI is recorded in an uncultivated area. Regarding the metal concentrations, in specific sampling points, we measure Cu and Pb values higher than the limit established by Italian law for residential and private use soil (max(Cu) = 341 mg/kg and max(Pb) = 179 mg/kg in comparison with Italian normative limits Cu = 120 mg/kg and Pb = 100 mg/kg).

Table 2. Explorative statistical parameters of data measured in Viggiano pre-treatment oil plant area (Agri valley). Legend: *n* = number of sampling point; *m* = mean value; *sd* = standard deviation; *CV%* = percentage variation coefficient; *md* = median; *r* = range

	K_D *(x10^{-5} SI)*	Cd	Co	Cr	Ni	Cu *(mg/kg)*	Fe	Mn	Pb	Zn
n	22	22	22	22	22	22	22	22	22	22
m	121	0.39	21	48	38	63	21599	1451	68	62
sd	55	0.15	4	13	7	81	8115	382	27	19
CV%	45%	39%	20%	27%	19%	128%	38%	26%	40%	31%
md	119	0.36	21	49	37	37	22637	1611	63	56
r	17-240	0.18-0.84	12-34	23-78	21-54	9-341	2406-43000	504-1973	35-179	37-121

For characterizing the correlation pattern among descriptors, we apply the cluster procedure discriminating two homogeneous sub-groups (Fe, Ni, Mn, Co, Cd), (K_D, Cu, Zn, Pb) and one isolated descriptor, Cr concentration (figure 7).

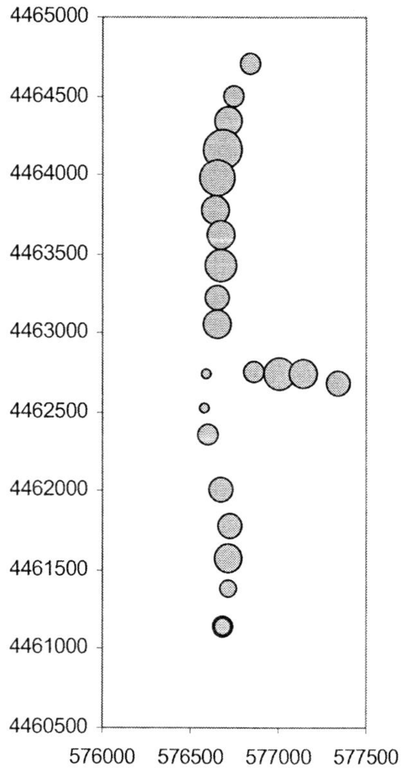

Figure 6. Magnetic susceptibility values measured in Viggiano area (bubble area is proportional to K_D value).

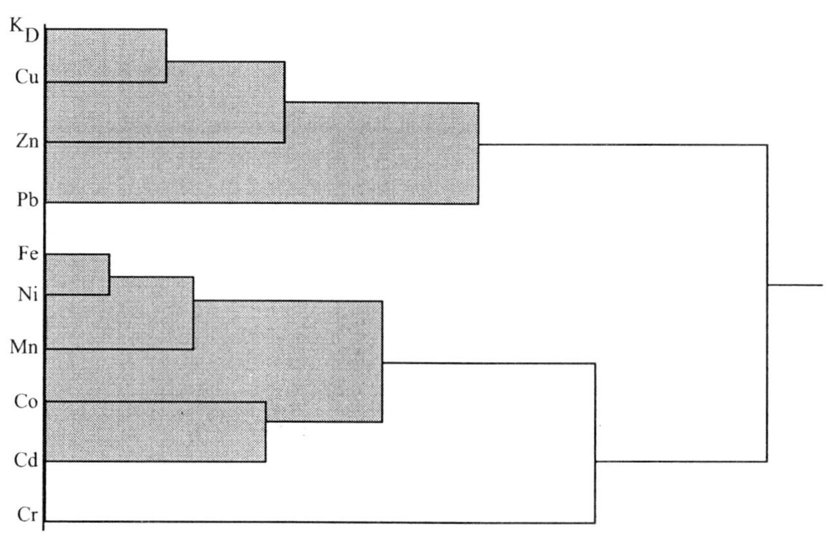

Figure 7. Dendrogram of clustering procedure applied to descriptor association matrix obtained starting from data measured in Viggiano area.

The link among K_D values, Cu, Zn and Pb concentrations agrees with results presented in literature (Heller et al., 1998; Xie et al., 2001, Hanesh et al., 2002; Desenfant et al., 2004). Furthermore it is in agreement with the metal contamination pattern observed in this area. In fact high values of K_D are related to high concentrations of some metals suggesting a relationships between presence of ferromagnetic minerals and specific emission sources. Particularly, in this case, they may be identified in the traffic and in the agricultural practices (verdigris is widely used for preventing grape disease in many vineyards).

4.2. S. Nicola di Melfi Industrial Area

In this area different annual field surveys were carried out starting from the nineties when the FIAT-SATA start-up phase was initiated. Here we present only data collected in the autumn 2002 field survey. In table 3 parameters of explorative analysis are shown. Regarding K_D values (figure 8), the range is 27-458 10^{-5} SI and mean value is 146 10^{-5} SI comparable with magnetic measures carried out in other European countries. In this case K_D has the same order of urban and/or uncultivated soils. The anomalies are localized all around the industrial plants and near the main roads. K_B(lf) and K_B(hf) values are strongly correlated with K_D measurements (Pearson's coefficient are higher than 0.75 in both cases) and they show spatial patters similar to K_D spatial distribution.

Table 3. Explorative statistical parameters of data measured in S. Nicola di Melfi industrial area

	K_D	K_B(hf)	K_B(lf)	Cd	Co	Cr	Cu	Fe	Mn	Ni	Pb	Zn
		(x10^{-5} SI)						*(mg/kg)*				
n	103	96	96	103	103	103	102	103	103	103	103	103
m	146	248	255	0.54	27	49	30	21981	981	41	70	65
sd	98	154	159	0.13	5	15	10	6296	336	8	15	12
CV%	67%	62%	62%	25%	17%	31%	35%	29%	34%	20%	22%	18%
md	127	229	235	0.51	27	48	28	22769	950	41	67	65
r	27-458	37-844	38-863	0.29-0.87	17-36	12-92	10-76	4713-32820	412-1973	21-69	46-143	33-114

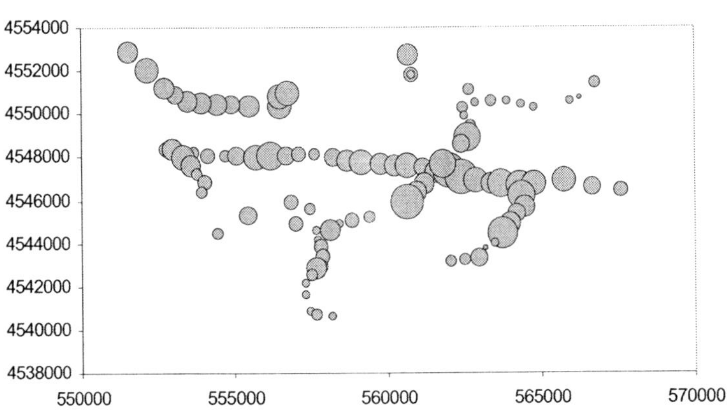

Figure 8. Magnetic susceptibility values measured in S. Nicola di Melfi industrial area

Regarding the correlation structure among metal concentrations and magnetic susceptibility, we apply the clustering procedure and we point out two groups of descriptors (Cr, Ni, Zn, Co) and (Fe, Mn, Cu, Cd, Pb, K_D) (dendrogram is shown in figure 9), which well describe the two main sources of pollution present in the investigated area. The first is representative of industrial plant emissions that, in this area, are mainly due to chromium- and zinc-plating of automotive components. The second is representative of traffic emissions mixed to natural content of heavy metals in the superficial soil. Magnetic susceptibility is included in this second cluster indicating a stronger relationship with this kind of emission profile.

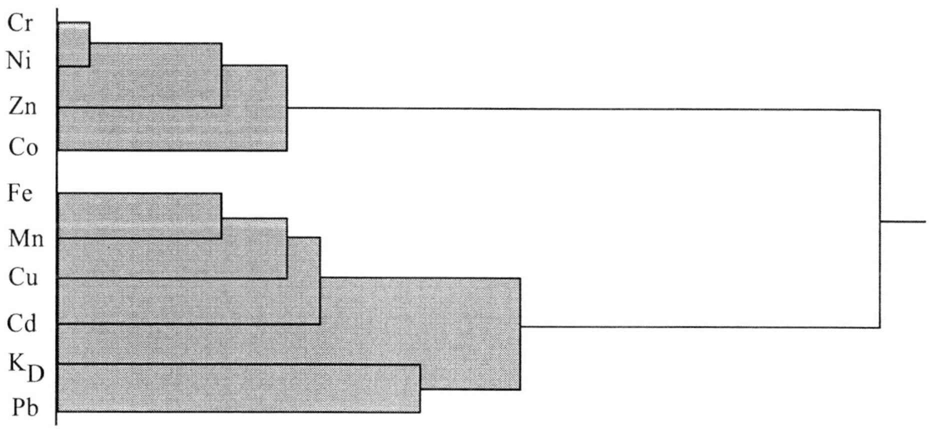

Figure 9. Dendrogram of clustering procedure applied to descriptor association matrix obtained starting from data measured in S. Nicola di Melfi area.

Due to lithological variability of the investigated area we discuss the data dividing the whole data set in eight parts, one for each transect. Data are shown in table 4.

Transects A and E cross the industrial settlement and, from a geological point of view, they are uniform, characterized by sandy-clayey soil. Here the susceptibility values are higher than the mean value calculated on the whole area (in E transect the first tow anomalies are present) and they are higher than values measured in other zones of Basilicata with similar geological features, indicating a significant anthropogenic contribute. In this sub-area, χ_B shows high values ($\chi_B > 3.0$) confirming the significant presence of magnetic grains. Regarding the heavy metal concentrations, we measure high values for the metals included in the first cluster (Fe, Mn, Cu, Cd, Pb) according to magnetic susceptibility pattern. This result may give a suggestion about the weight of the different contamination sources. In fact in this sub-area, the traffic seems to be the main anthropogenic source of emissions.

Transect B and a part of transect C are characterized by sandy soil and they show low values of magnetic susceptibility in agreement with agricultural land use. On the contrary in the part of transect C nearest the SATA plant, we may observe high values of magnetic susceptibility. Analogous behavior is shown by χ_B. It ranges from 0.7% to 4.5%, showing different behavior among the different sampling points. In this zone Cd, Cr, Ni and Zn concentrations are higher than the mean values calculated on the whole area, indicating a significant contribute to soil pollution due to industrial plant emissions. This result is confirmed also by the application of a pollutant diffusion model, based on stack emissions, which puts in evidence the presence of maximum deposition zones in this sub-area (Bonfiglio et al., 1998).

Table 4. Explorative statistical parameters of data measured in S. Nicola di Melfi industrial area calculated for each transect

Transect		K_D ($x10^{-5}$ SI)	χ_B (%)	Cd	Co	Cr	Cu	Fe (mg/kg)	Mn	Ni	Pb	Zn
A	*m*	193	3.1	0.73	23	48	37	27213	1223	38	77	64
(*n*=16)	*md*	196		0.76	23	47	32	28174	1173	36	68	56
	sd	66		0.10	4	11	16	3455	206	7	26	17
B	*m*	135	3.0	0.47	27	47	27	23616	1217	43	62	63
(*n*=11)	*md*	136		0.50	29	50	28	24210	1200	43	61	62
	sd	58		0.06	4	11	4	5540	346	5	10	8
C	*m*	121	2.0	0.49	29	58	28	20606	1054	47	70	69
(*n*=22)	*md*	88		0.49	31	59	27	22217	962	49	69	70
	sd	99		0.09	4	20	7	6652	404	9	10	10
D	*m*	185	2.4	0.43	27	50	27	20495	1048	44	62	62
(*n*=21)	*md*	173		0.36	25	45	26	20801	928	41	63	63
	sd	101		0.13	4	18	7	5637	431	11	9	8
E	*m*	293	3.6	0.66	25	40	29	21629	951	37	79	61
(*n*=12)	*md*	292		0.66	24	41	30	23319	986	36	75	60
	sd	138		0.07	6	14	7	6625	232	6	15	10
F	*m*	98	2.5	0.53	30	52	25	18329	748	44	69	66
(*n*=14)	*md*	97		0.52	29	48	27	20004	572	42	67	71
	sd	42		0.07	5	14	8	5800	295	7	11	12
G	*m*	110	2.1	0.51	27	29	15	14860	573	29	70	50
(*n*=5)	*md*	97		0.44	26	28	13	13935	526	31	72	51
	sd	40		0.11	4	10	6	6935	164	5	7	11
H	*m*	201	3.3	0.52	29	51	39	25451	928	41	71	74
(*n*=12)	*md*	205		0.51	28	52	37	25580	936	41	71	75
	sd	40		0.03	4	6	8	1695	77	3	6	7

Transect D is located far from the main industrial plants and it is characterized by two different geologies and by two different land uses. The northern part, from a geological point of view is similar to A and E transects and it is near to the main road connecting the industrial area with Lavello, the most important urban settlements of the Vulture basin. Also in this case we measure an anomaly and the values of susceptibility are higher than values characterizing sandy-clayey soil, supposing a significant increase of magnetic minerals duo to traffic emissions. On the contrary, the southern part of the transect is in a woody area and the soil is characterized by sand and gravel of the riverbed. Here K_D is about 123 10^{-5} SI. Regarding χ_B values, we may observe different values according to K_D pattern. In this transect the metal concentrations do not show a specific pattern, confirming the heterogeneity of the sub area.

F and G transects are located in the nearness of the Ofanto river, far from industrial settlement and horticulture is the main land use of this sub-area. F transect is characterized by sandy-clayey sediments and gravels of the riverbed, whereas G transect is characterized by polygenic calcareous and arenaceous conglomerates. Here magnetic susceptibility is very low in comparison with the values observed in the entire investigated area, highlighting low contamination level of soil. This result is according to metal pattern. In fact the metal concentrations are lower than the mean values calculated on the all sampling points. Particularly the metals correlated with magnetic susceptibility show very low concentration.

H transect, as the previous two transects, is located in the nearness of the Ofanto river, far from industrial settlement. It is characterized by sandy-clayey sediments and gravels; the wheat crop is the main land use and some farms are located in this zone. The magnetic susceptibility is very high (K_D = 201 10^{-5} SI and χ_B = 3.3) and we may suppose that is due to the use of pesticides and fertilizers for agricultural practices. The metal concentration values show an anomalous behavior. In fact the metals showing high concentrations (Co, Cr, Cu, Fe, Zn) belong to both the clusters, suggesting the presence of specific contamination sources.

4.3. Potenza Industrial Area

In table 5, we show the explorative statistical analysis of magnetic susceptibility and lead concentration values measured in the preliminary screening. The values range from 8 10^{-5} SI to 636 10^{-5} SI. Variations of the same magnitude order were found in another European industrial area (Hanesch et al., 2003). The soil of Potenza industrial area is a clayey one. Generally, this kind of soil shows low magnetic susceptibility values comparable to minimum observed values (K_D = 8 10^{-5} SI) (Desanfant et al., 2004; Gattacceca et al., 2004). The investigated area is characterized by a uniform geology, consequently we may suppose that the observed variations in magnetic susceptibility values are mainly ascribable to anthropogenic input of magnetic material. Particularly we found three anomalies (K_D = 636 10^{-5} SI, K_D = 240 10^{-5} SI, K_D = 98 10^{-5} SI) near the bus parking and three anomalies (K_D = 178 10^{-5} SI, K_D = 173 10^{-5} SI, K_D = 166 10^{-5} SI) near to the iron and steel industrial plant.

Table 5. Explorative statistical parameters of magnetic susceptibility values and Pb concentrations measured in the industrial area of Potenza during the preliminary screening

	K_D *($x10^{-5}$ SI)*	Pb *(mg/kg)*
n	57	57
m	72	89
sd	78	57
CV%	108%	63%
md	49	61
r	8-636	46-233

Regarding the lead concentrations, we measure higher values (*m* = 89 mg/kg, *max* = 233 mg/kg) than the levels observed in areas far from urban zones (in Agri valley *m* = 68 mg/kg *max* = 179 mg/kg; in S. Nicola di Melfi industrial area *m* = 70 mg/kg, *max* = 143 mg/kg).

Moreover the concentration spatial pattern is very heterogeneous. Particularly we observe high level of Pb in the area of the first three anomalies.

Table 6. Explorative statistical parameters of variables measured in the three different sub-areas individuated in the industrial area of Potenza. Legend: *A* = sub-area extension

	Zone A		Zone B		Zone C	
A	30 *m* x 10 *m*		25 *m* x 35 *m*		20 *m* x 60 *m*	
n	21		48		15	
	K_D	K_F *(x10^{-5} SI)*	K_D	K_F *(x10^{-5} SI)*	K_D	K_F *(x10^{-5} SI)*
m	16	14	40	31	134	343
sd	7	7	24	23	37	159
CV%	44%	50%	60%	74%	28%	46%
md	16	12	32	26	138	322
r	5-34	6-38	9-129	9-159	52-184	108-780

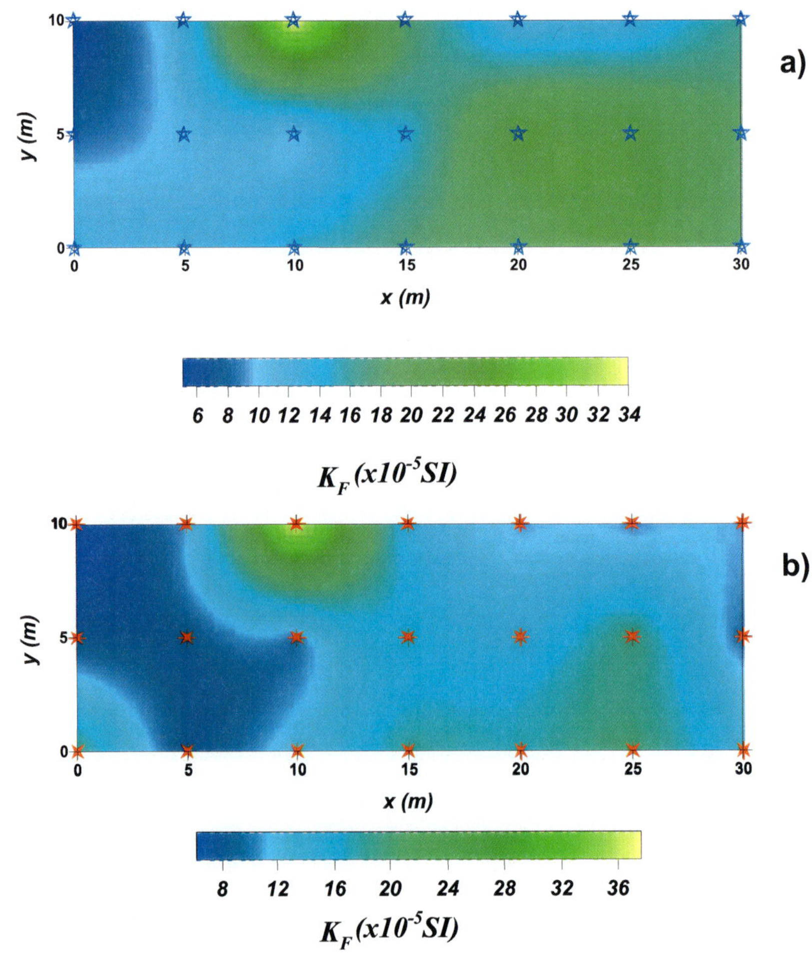

Figure 10. Detail maps carried out in the zone A of Potenza industrial area: a) K_D values; b) K_F values.

On the bases of these results, we carried out three detail maps: in the minimum value area (zone A), in the parking bus area (zone B), in which we observe high values for both measured parameters, and in the iron and steel plant area (zone C) in which we measure only high K_D values. In the first two sub-areas, we define regular grids with cells of 5 m^2 and measured magnetic susceptibility values with different probes (K_D values and K_F values). On the contrary, in the third sub-area, we define an irregular grid because of specific morphological features of the site. Also in this case we measure K_D and K_F values. Explorative analysis of data is shown in table 6. Maps obtained by means of the kriging procedure are shown in figures 10-12.

The zone A is a rural area located far both from industrial/commercial settlements and streets with a high volume of traffic. Moreover, this is an untilled area and this peculiarity prevents soil direct contaminations coming from fertilizers or pesticides. From the maps (figure 10), we note that magnetic susceptibility values measured by means of both the probes show low variability and similar distributions. In fact, K_F values range 6-38 10^{-5} SI, and K_D values range 5-34 10^{-5} SI. Moreover, correlation coefficient between the two data sets is $\rho = 0.87$ (with a statistical significance of 5%). The good agreement between the data measured at different investigation depths highlights a uniform distribution of magnetic susceptibility values of different soil layers. On the basis of these results, we may suppose that magnetic susceptibility values of soil are not due to anthropogenic influences, and, therefore, this site may be considered to be a good background site.

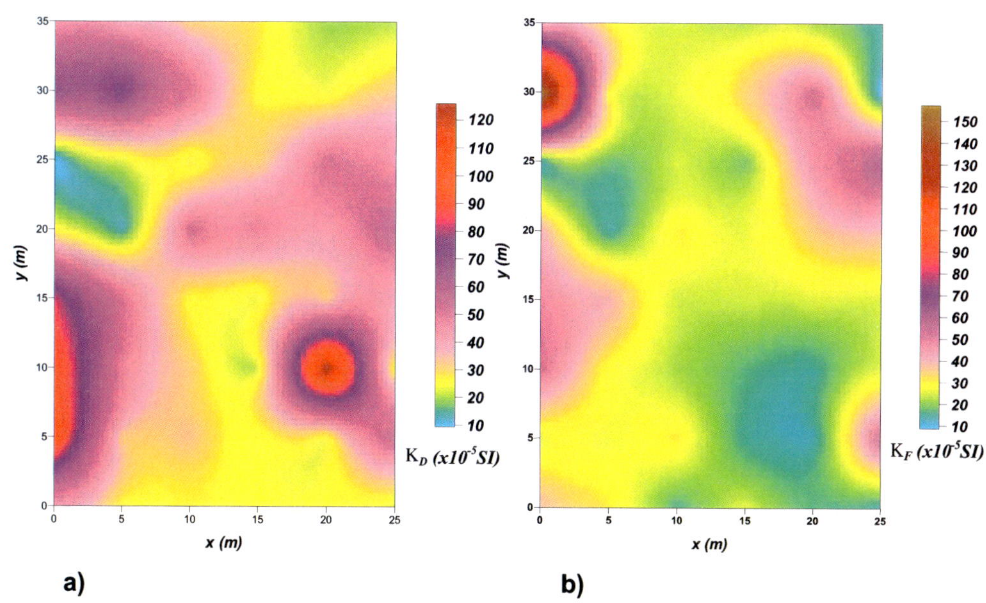

Figure 11. Detail maps carried out in the zone B of Potenza industrial area: a) K_D values; b) K_F values.

Concerning with soil parameters, we observe that, during the filed survey, the water content is quite low with a limited range of variability on the spatial grid, while the soil temperature shows a uniform spatial pattern. The comparison between magnetic susceptibility and humidity spatial distributions puts in evidence that we measure higher magnetic susceptibility values in the areas of lower humidity. It confirms that an increase in soil humidity may cause a decrease in soil magnetic susceptibility values, suggesting a not negligible effect of this parameter on soil magnetic properties (D'Emilio et al, 2006b).

In the zone B (figure 11), near the parking bus, we note that magnetic susceptibility values show a large range of variability $r(K_F)$ = 9-159 10^{-5} SI, and $r(K_D)$ = 9-129 10^{-5} SI. Moreover, magnetic susceptibility mean values ($m(K_F)$ = 31 10^{-5} SI and $m(K_D)$ = 40 10^{-5} SI) are higher than K-values recorded in the background area. From the observation of the maps, we may note that there is not a good agreement between the data measured by means of the two probes. This may be ascribable probably to different distributions of contaminants in the different soil layers. Particularly, being K_D values higher than K_F values, we may suppose that magnetic susceptibility is influenced by anthropogenic contamination sources, such as the emissions coming from busses.

In the zone C (figure 12), near the steel and iron plant, we measured very high magnetic susceptibility values ($m(K_F)$ = 343 10^{-5} SI and $m(K_D)$ = 134 10^{-5} SI). Moreover, we noted that the variation coefficient shows the lowest values (28% for K_D values and 46% for K_F values). It is clear that in this area the presence of a specific industrial source has a relevant influence on the magnetic susceptibility soil values. In this case, from maps, we may note a good agreement between the data measured by means of the two probes: higher magnetic susceptibility values are recorded in the same areas. Moreover in this case K_F values are higher than K_D values, so we may suppose that, in comparison with the zone B investigation, the contamination is mainly superficial.

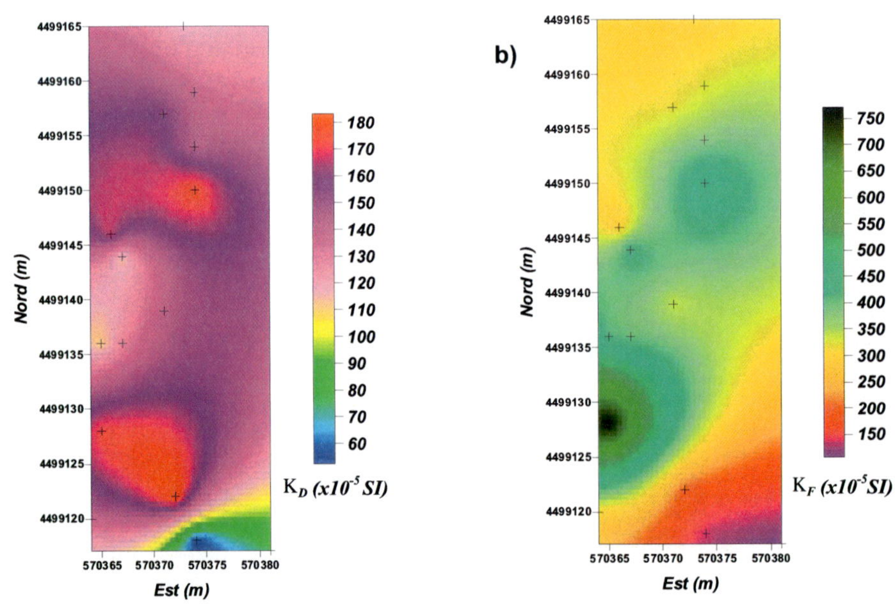

Figure 12. Detail maps carried out in zone C of Potenza industrial area: a) K_D values; b) K_F values.

4.4. Industrial Area in Basento Valley

During the first phase of the investigation, a preliminary screening of the area is carried out measuring magnetic susceptibility throughout the length of the Basentana freeway, starting from the industrial site of Salandra up to the industrial site of Pisticci. Seventy-five sampling points are individuated and measurements are carried out by means MS2D and MS2F probes. Data are shown in table 7 and in figure 13. Contemporaneously soil samples are collected for in laboratory K_B measurements and metal concentration determinations.

Table 7. Explorative statistical parameters of variables measured in the Basento valley during the preliminary screening

	K_F	K_D	K_B(lf)	K_B(hf)	Co	Cu	Ni	Pb	Zn
			(x10^{-5} SI)				*(mg/kg)*		
n	75	75	75	75	75	75	75	75	75
m	14	14	22	21	29	21	44	76	51
sd	10	11	14	13	5	6	8	33	16
CV%	70%	76%	62%	63%	18%	30%	19%	43%	31%
md	12	12	19	18	29	19	44	67	48
r	5-69	5-89	8-103	8-101	17-43	10-57	27-65	45-267	28-121

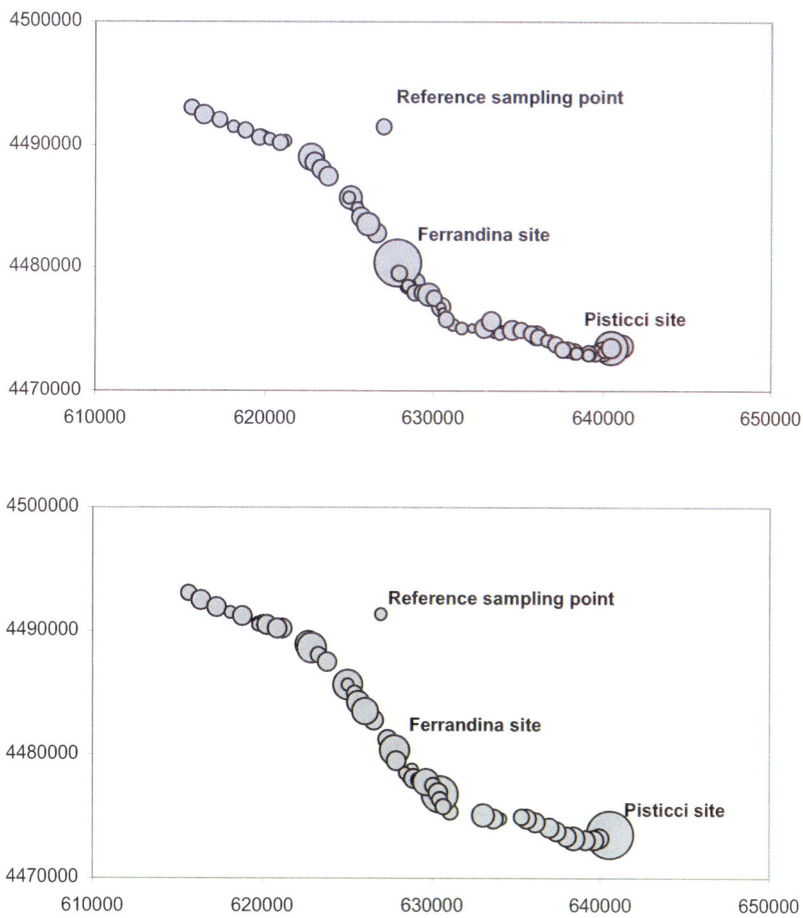

Figure 13. Magnetic susceptibility values measured in the industrial area of Basento valley: a) K_D values; b) K_F values.

As it may be observed, the magnetic susceptibility values measured with the three different sensors are strongly correlated. They vary from 5 10^{-5} SI to 103 10^{-5} SI along the entire transect with only two anomalies in correspondence to the areas surrounding the main industrial plants. In particular, the anomalies are recorded near the Ferrandina industrial plant (K_D = 89 10^{-5} SI), and near the Pisticci industrial plant (K_D = 45 10^{-5} SI). Lower values are present in correspondence to the agricultural areas far from the industrial sites with a mean value of about 11 10^{-5} SI.

The lowest value K_D = 8 10^{-5} SI is recorded, in an area very far from the Basentana freeway, to have a reference parameter of a non polluted area. Analogous measurements are carried out by using MS2F and MS2B probes, highlighting similar magnetic susceptibility value distributions with the highest value recorded near the Pisticci industrial plant and near the Ferrandina industrial plant.

Regarding concentrations of Co, Cu, Ni, Pb, and Zn, measured in soil samples collected along the transect, they show a pattern similar to magnetic measurements. Particularly Cu and Zn concentrations are correlated both with K_F and with K_D magnetic susceptibility measurements (the values of correlation coefficient are ρ(Cu-K_F) = 0.4, ρ(Cu-K_D) = 0.4, ρ(Zn-K_F) = 0.7, ρ(Zn-K_D) = 0.5 with a significance level α=5%). On the contrary lead concentrations are correlated only the magnetic measurements carried out in the superficial soil (ρ(Pb-K_F) = 0.4 with α=5%), indicating different emission patterns for the three metals.

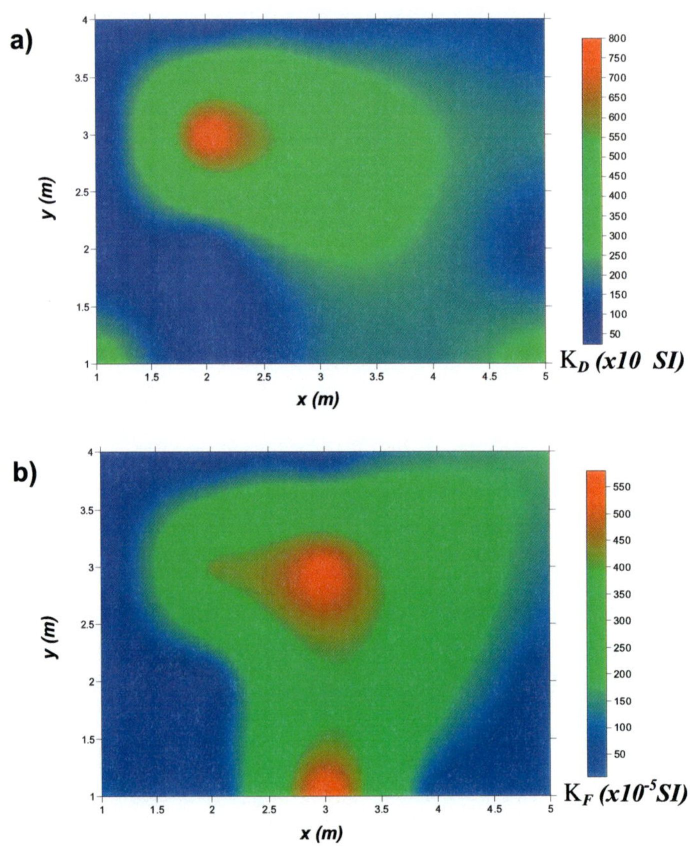

Figure 14. Detail maps carried out in Ferrandina site: a) K_D values; b) K_F values.

In the sub-areas in which the highest values of magnetic susceptibility are observed, two detail maps are carried out measuring on regular grids K_D and K_F values. Regarding the Ferrandina sub-area, we define a sampling grid covering an area of 150 m x 120 m with a sampling step of 30 m. The parameters of explorative analysis are shown in first part of the table 8 and the interpolated maps are shown in figure 14. We note that both the measured parameters show the highest value in the middle part of the investigated zone and that $max(K_F) = 814\ 10^{-5}$ SI is higher than $max(K_D) = 592\ 10^{-5}$ SI. Moreover K_F maximum zone is larger than K_D maximum zone. This is probably due to a very confined and superficial contamination of ferromagnetic pollutants.

Table 8. Explorative statistical parameters of variables measured in the Ferrandina site and in Pisticci site

	Ferrandina site		Pisticci site	
A	150 *m* x 120 *m*		90 *m* x 60 *m*	
n	20		24	
	K_D	K_F *(x10^{-5} SI)*	K_D	K_F *(x10^{-5} SI)*
m	137	188	33	16
sd	200	217	72	24
CV%	145%	115%	218%	150%
md	77	102	30	21
r	16-592	16-814	26-369	26-130

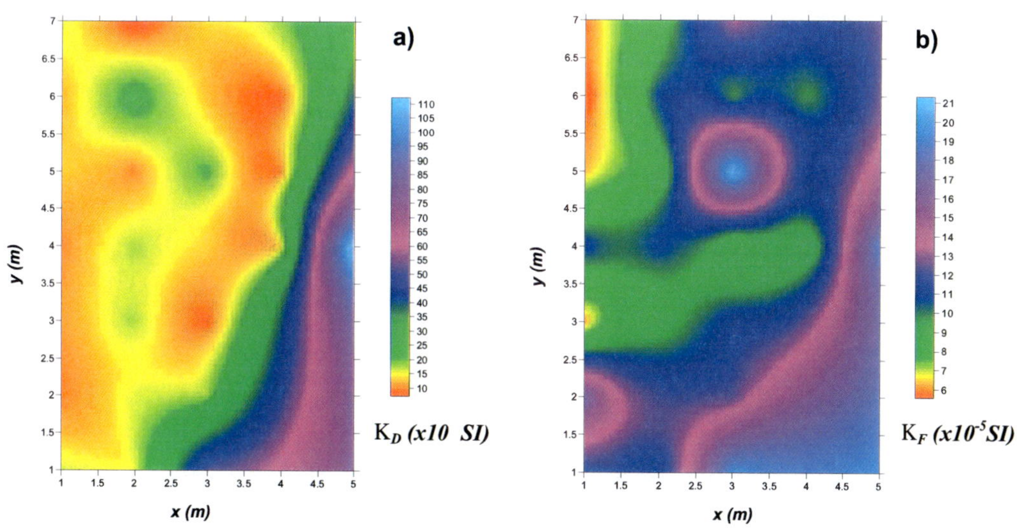

Figure 15. Detail maps carried out in Pisticci site: a) K_D values; b) K_F values.

Regarding the Pisticci sub-area, the investigations are carried out on a regular grid covering an area of 90 m x 60 m with a sampling step of 15 m. The parameters of explorative analysis are summarized in the second part of the table 8 and the maps are shown in figure 15. Both the maps present a long maximum zone in the middle part; here we record $max(K_D) = 369\ 10^{-5}$ SI that, in this case, is higher than $max(K_F) = 130\ 10^{-5}$ SI. On the contrary, only K_D values show a circular maximum zone in which it is about 113 10^{-5} SI. On the base of these observations, we may suppose that, in this case, the contamination is located to a depth higher than 8 cm.

5. CONCLUSION

We have presented a review of four study cases concerning the characterization of heavy metal patterns in different industrial areas by means of magnetic screening. Measures of magnetic susceptibility, carried out in situ and in laboratory and analytical determination of metal concentrations are used to investigate the contaminant patterns, taking into account both the different geomorphologic features of the investigated sites and the different potential emission sources. The investigated cases confirm the reliability of the monitoring technique based on the measurement of magnetic parameter in superficial soil. They highlight that magnetic susceptibility measurement represents a useful proxy variable for investigating the metal patterns.

We would underline that the use of a well established experimental procedure for carrying out the field survey, allows to reduce the fluctuations due to many external factors and to point out the influence of the specific geomorphologic and anthropogenic features of each examined area.

The results point out the strong link among magnetic susceptibility values and metals in superficial soil coming from specific pollution sources.

At the first the traffic emissions and the consequent soil metal depositions seems to be positively correlated to magnetic susceptibility. In fact, in the cases in which the correlation structure is analysed by means multivariate techniques, lead concentration and magnetic susceptibility are belonging the same homogeneous sub-group of descriptors. In other cases in which a metal selection is made up a priori, the correlation between the two descriptors is statistically significant. Moreover, this result is confirmed by analysis of magnetic measurements carried out with different probes operating at different depths. In many cases we point out a marked relationship between lead concentration and magnetic susceptibility measured in the more deep soil layers.

Regarding the emissions of industrial origin, the study puts in evidence the necessity to integrate both measurements of different parameters and observations carried out by means of different techniques, taking into account also the specific features of the examined case. In fact in many cases the only preliminary screening is not able to point out the impact of a specific plant and it is necessary to individuate specific sub-areas and to carry out the survey with an appropriate experimental procedure.

Regarding other pollution sources, we note that agricultural practices may have a significant influence on metal levels and on magnetic mineral presence in the soil. In the examined cases, we note a relationship among high levels of some metals, occurrence of magnetic anomalies and specific agricultural practices in which fertilizer, pesticides or other specific pigments are used.

In conclusion, we would underline the necessity to collect preliminarily detailed information on the site to have been investigated and to plan carefully the monitoring survey. Moreover in order to better characterize the studied pollution phenomenon and to interpret the collected data, we suggest the application of integrated experimental procedure in which different chemical, physical and/or biological parameters are measured and in which different techniques are applied. The integrated approach may give additional, useful and valid support for the development of monitoring innovative strategies.

ACKNOWLEDGEMENTS

This research was partially funded by MIUR (PON 2002–2005) in the framework of the project *Sviluppo dei Distretti Industriali per le Osservazioni della Terra* and partially funded by the Basilicata Region Authority in the framework of the projects *Monitoraggio delle aree urbane e sub-urbane di Potenza and Matera* and *Caratterizzazione speditivi dell'inquinamento del suolo nelle aree industriali di Ferrandina e Pisticci (Val Basento) attraverso tecniche elettromagnetiche non invasive.* We would thank Mrs. S. Sabia for her technical assistance.

REFERENCES

Al-Khashman, O. A. (2004). Heavy metal distribution in dust street dust and soils from the work place in Karak Industrial Estate, Jordan. *Atmospheric Environment, 38*, 6803–6812.

Alvarez, E., Fernandez Marcos, M.L., Vaamonde, C. and Fernandez-Sanjurjo, M.J. (2003). Heavy metals in the dump of an abandoned mine in Galicia (NW Spain) and in the spontaneously occurring vegetation. *The Science of the Total Environment, 313*, 185–197.

Aristodemou, E., and Thomas-Betts, A. (2000). DC resistivity and induced polarisation investigations at a waste disposal site and its environments. *Journal of Applied Geophysics, 44*, 275–302.

Arshad,. M.A., and Martin, S. (2002). Identifying critical limits for soil quality indicators in agro-ecosystems *Agriculture, Ecosystems and Environment, 88*, 153–160.

Atekwana, E.A., Sauck, W.A., and Werkema, Jr., D.D. (2000). Investigations of geoelectrical signatures at a hydrocarbon contaminated site. *Journal of Applied Geophysics, 44*, 167–180.

Barahona, E., and Iriarte, A. (2001). An overview of the present state of standardization of soil sampling in Spain. *The Science of the Total Environment, 264*, 169–174.

Bonfiglio, A, Macchiato, M., Minervini, L., Ragosta, M., Santangelo, R. (1998). Pollutant diffusion model and GIS integration procedure for evaluating atmospheric emissions in industrial areas. In Brebbia, C.A., Ratto, C.F. and Power, H. (Eds), *Air pollution VI.* (WIT Press CMP pp .849-858) Southampton, UK.

Boyko, T., Scholger, R. and Stanjek, H. (2004). Topsoil magnetic susceptibility mapping as a tool for pollution monitoring: repeatability of in situ measurements. *Journal of Applied Geophysics, 55*, 249-259.

Caggiano, R., Macchiato, M.F. and Ragosta, M. (1998). Background level characterization of heavy metals soil concentrations in an industrial area of Southern Italy, *Il Nuovo Cimento C, 21*, 49-63.

Chianese, D., D'Emilio, M., Bavosi, M., Lapenna V. and Macchiato M. (2006). Magnetic and round probing radar measurements for soil pollution mapping in the industrial area of Val Basento (Basilicata region, Southern Italy): a case study. *Environmental Geology, 49*, 389-404.

Clemente, R. and Bernal, M.P. (2006). Fractionation of heavy metals and distribution of organic carbon in two contaminated soils amended with humic acids. *Chemosphere*, 64: 264-273.

Cressie, N. (1990) *Statistics for Spatial Data.* Iowa State University, Wiley-Interscience Publication, Wiley, New York.

Dearing, J.A., Dann, R.J.L., Hay, K., Lees, J.A., Loveland, P.J., Maher, B.A. and O'Grady, K. (1996). Frequency-dependent susceptibility measurements of environmental materials. *Geophysical Journal International, 124*, 228–240.

D'Emilio, M., Coppola, R., Loperte, A., Macchiato, M., and Ragosta, M. (2006a). Pollution levels in the industrial area of Potenza (Southern Italy). *Fresenius Environmental Bulletin, 15*, 36-42.

D'Emilio, M, Chianese, D., Coppola, R., Macchiato, M. and Ragosta, M. (2006b). Magnetic susceptibility measurements as proxy method to monitor soil pollution: development of experimental protocols for field surveys. *Environmental Monitoring and Assessment,* in press.

Desenfant, F., Petrovsky, E. and Rochette, P. (2004). Magnetic signature of industrial pollution of stream sediments and correlation with heavy metals: case study from south France. *Water Air and Soil Pollution, 152*, 297-312.

Dinares-Turell, J., Hoogakker, B.A.A., Roberts, A.P., Rohling, E.J., and Sagnotti, L. (2003). Quaternary climatic control of biogenic magnetite production and eolian dust input in cores from the Mediterranean Sea. *Palaeogeography, Palaeoclimatology, Palaeoecology, 190*, 195–209.

Durza, O. (1999). Heavy metals contamination and magnetic susceptibility in soils around metallurgical plant. *Physics and Chemistry of the Earth, 24*, 541–546.

Gattacceca, J., Eisenlohr, P. and Rochette, P. (2004). Calibration of in situ magnetic susceptibility measurements. *Geophysical Journal International, 158*, 42-49.

Hanesch, M. and Scholger, R. (2002). Mapping of heavy metal loadings in soils by means of magnetic susceptibility measurements. *Environmental Geology, 42*, 857-870.

Hanesch, M., Scholger, R. and Rey, D. (2003). Mapping dust distribution around an industrial site by measuring magnetic parameters of tree leaves. *Atmospheric Environment, 37*, 5125–5133.

Heller, F., Strzyszcz, Z., and Magiera, T. (1998). Magnetic record of industrial pollution in forest soils of Upper Silesia, Poland. *Journal of Geophysical Research, 103*, 17767–17774.

Jordanova, D., Veneva, L., and Hoffmann, V. (2003). Magnetic susceptibility screening of anthropogenic impact on the Danube river sediments in Northwestern Bulgaria – Preliminary results. *Studia Geophysica et Geodaetica, 47*, 403–418.

Jelenska, M., Hasso-Agopsowicz, A., Kopcewicz, B., Sukhorada, A., Tyamina, K., Kadzialko-Hofmokl, M. and Matviishina, Z. (2004). Magnetic properties of the profiles of polluted and non-polluted soil. A case study from Ukraine. *Geophysical Journal International, 159*, 104-116.

Kapicka, A., Jordanova, N., Petrovsky, E., and Usjak, S. (2000). Magnetic stability of power plant fly ash in different soil solutions. *Physics and Chemistry of the Earth (A)*, *25*, 431–436.

Karlen, D.L., Ditzler, C.A. and Andrews, S.S. (2003). Soil quality: why and how? *Geoderma, 114*, 145–156.

Lecoanet, H., Leveque, F., and Ambrosi, J.P. (2003). Combination of magnetic parameters: an efficient way to discriminate soil contamination sources (south France). *Environmental Pollution, 122*, 229–234.

Llobet, J.M., Schuhmacher, M., and Domingo J.L. (2002). Spatial distribution and temporal variation of metals in the vicinity of a municipal solid waste incinerator after a modernization of the flue gas cleaning systems of the facility. *The Science of the Total Environment, 284*, 205-214.

Morselli, L., Passarini, F. and Bartoli, M. (2002). The environmental fate of heavy metals arising from a MSW incineration plant. *Waste Management, 22*, 875–881.

Muxworthy, A.R., Matzka, J., and Petersen, N. (2001). Comparison of magnetic parameters of urban atmospheric particulate matter with pollution and meteorological data. *Atmospheric Environment*, *35*, 4379–4386.

Nadim, F., Hoag, G.E., Liu, S., Carley, R.J., and Zack, P. (2000). Detection and remediation of soil and aquifer systems contaminated with petroleum products: an overview. *Petroleum Science and Engineering*, *26*, 169–178.

Navas A. and Machin, J. (2002). Spatial distribution of heavy metals and arsenic in soils of Aragon (northeast Spain): controlling factors and environmental implications. *Applied Geochemistry, 17* 961–973.

Ng, S.L., Chan, L.S., Lam, K.C., and Chan, W.K. (2003). Heavy metal contents and magnetic properties of playground dust in Hong Kong. *Environmental Monitoring and Assessment*, *89*, 221–232.

Palumbo, B., Angelone, M., Bellanca, A., Dazzi, C., Hauser, S., Neri, R., and Wilson, J. (2000). Influence of inheritance and pedogenesis on heavy metal distribution in soils of Sicily, Italy. *Geoderma*, *95*, 247–266.

Petrovsky, E., Kapicka, A., Jordanova, N., and Boruoka, L. (2001). Magnetic properties of alluvial soil contaminated with lead, zinc and cadmium. *Journal of Applied Geophysics*, *48*, 127–136.

Pietrapertosa, F., Cosmi, C., Macchiato, M., Marmo, G. and Salvia, M. (2003). Comprehensive modeling for approaching the Kyoto targets on a local scale. *Renewable and Sustainable Energy Reviews, 7*, 249–270.

Ragosta, M, Caggiano, R., D' Emilio, M and Macchiato M. (2003). Multivariate analysis for investigating profile sources of atmospheric heavy metals emissions. *Journal of Applied Statistical Sciences, 12*, 251-263.

Sharma, P.S. (1996). *Engineering and Environmental Geophysics*. Cambridge: Cambridge University Press.

Steeples, D.W. (2001). Engineering and environmental geophysics at the millennium. *Geophysics*, *66*, 31–35.

Suzuki, K., Toda, S., Kusunoki, K., Fujimitsu, Y., Mogi, T., and Jomori, A. (2000). Case studies of electrical and electromagnetic methods applied to mapping active faults beneath the thick quaternary. *Engineering Geology*, *56*, 29–45.

Thouveny, N., Carcaillet, J., Moreno, E., Leduc, G., and Nerini, D. (2004). Geomagnetic moment variation and paleomagnetic excursions since 400 kyr BP: a stacked record from sedimentary sequences. *Earth and Planetary Science Letters*, *219*, 377–396.

Viard, B., Pihan, F., Promeyrat, S., and Pihan, J. (2004). Integrated assessment of heavy metal (Pb, Zn, Cd) highway pollution: bioaccumulation in soil, Graminaceae and land snails. *Chemosphere*, *55*, 1349–1359.

Whorm, H-U. (1998). On the superparamagnetic stable single domain transition from magnetite and frequency dependence of susceptibility. *Geophysical Journal International, 133*, 201-206.

Xie, S., Dearing, J. A., Boyle, J. F., Bloemendal, J. and Morse A. P. (2001). Association between magnetic properties and element concentrations of Liverpool street dust and its implications. *Journal of Applied Geophysics*, *48*, 83-92.

In: Environmental Research Trends
Editor: Marvin A. Cato, pp. 133-152
ISBN: 978-1-60021-556-8

Chapter 4

BIOSORPTION OF HEAVY METAL IONS FROM AQUEOUS SOLUTIONS

Jana Kadukova* and Miroslav Štofko
Technical University of Kosice, Faculty of Metallurgy,
Department of Non-Ferrous Metals and Waste Treatment,
Letna 9, 04200 Kosice, Slovakia

ABSTRACT

Water contamination by toxic metals has become a serious problem threatening the entire ecosystem including man. Metals released into the environment by technological activities tend to persist indefinitely, circulating and eventually accumulating throughout the food chain. Heavy metal ions in trace quantities are difficult to remove from aqueous streams using conventional methods such as chemical precipitation, membrane separation, ion exchange etc. The process of biosorption appears to be one of the few alternatives available for such situations. It combines the advantages of biotechnology (such as low operating cost, minimization of the volume of chemical and/or biological sludge to be disposed of, high efficiency in detoxifying very dilute effluents, and no nutrient requirements) with the advantages of conventional chemical methods e.g. resistance to the physicochemical properties of heavy metal solutions (very high or low pH, high metal concentrations) because of the use of dead biomass.

The mechanism of biosorption onto the cell surface does not depend on the cell metabolism, but is based on physical and chemical interactions between metal and function groups of the cell wall. While this is a metabolism independent process, metal binding is very quick and mostly reversible. Used biomass has all the characteristics of other sorbents (e.g. active carbon), which presents a great advantage for industrial application of this process.

Experimental results presented in this work deal with kinetics study and partially with the study of mechanisms of biosorption by materials of different origin.

The basic parameters chosen to examine and to compare biosorptive abilities of selected biomaterials were capacity and rate of the process. The process was very fast – metals were removed from the solution within the first 10 minutes by most of the

* Tel. 00421-55-6022426, Fax. 00421-55-6022428, E-mail: jana.kadukova@tuke.sk, kadukova@post.sk

biosorbents. Only some of the biosorbents significantly increased pH value of the solution resulting in metal precipitation. But precipitates can cause difficulties in metal – sorbent separation.

Probable mechanisms responsible for the biosorption is ion-exchange which offers the possibility to remove metal bound on the biosorbent particles by desorption. The advantage of this process is its selectivity. Two times the capacity was required to remove silver by alga *Chlorella kessleri* than copper and cadmium. The advantage of this process was the difference between pH value necessary for Ag biosorption (2.9) and for copper and cadmium, 5 and 5.6, respectively, helping in the process of selectivity because copper and cadmium are not removed from the solution at low pH value.

SYMBOLS

$1/n$	constant dependent on temperature, it is in a range 0.2 – 0.9
b –	coefficient related to the affinity between sorbent and sorbate
B –	constant related to the surface interaction energy
c_0	initial concentration of metal in solution
c_e	equilibrium concentration of metal in solution
c_s	saturation concentration of solution
K	constant depending on character of the adsorbent and adsorbate
K_{AB}	equilibrium constant of ion exchange reaction
n	amount of substance
x_0	initial mole fraction of metal in solution
x_e	equilibrium mole fraction of metal in solution
q	specific adsorption
Q	total concentration of binding sites on biomass
q_m	maximum adsorption
x_A	variable in ion exchange isotherm
y_A	variable in ion exchange isotherm

1. INTRODUCTION

Contamination of aquifers and surface waters by toxic metals has become a serious problem limiting plant productivity and threatening human health. Metals released into the environment by technological activities tend to persist indefinitely, circulating and eventually accumulating throughout the food chain. Presently, a great effort is made to prevent metals from entering the environment or to reclaim them from contaminated parts of the environment [1].

Conventional approaches to heavy metals removal use technologies listed in Table 1, but the majority of them, with the emissions standards tightening, become progressively more inadequate and often cause new problems with the secondary waste produced which is difficult to dispose.

On the other hand, more effective technologies are more costly and often just not feasible. This is the reason for the development of new technologies which are more acceptable from the environmental or economic point of view. Some efforts are focused on the improvment of existing technologies [2, 3], but there is a big effort to utilize natural capability of organisms to cope with negative impacts of polluting matters so using new possibilities offered by biotechnologies is preferred [4, 5].

Table 1. Conventional metal removal technologies [6, 7]

Method	Description	Comments
Chemical precipitation and filtration	Preparation of insoluble compounds followed by their filtration.	Difficult separation.
Adsorption	Species removed mainly by hydrophobic type of adsorption.	Using of conventional sorbents (activated carbon).
Ion exchange	Ions removed by exchange with harmless ions from the resin.	High capital and operating costs. Effective.
Solvent extraction	Extraction of metal ions from aqueous solution by chelating agents in non-polar solvents.	High capital and operating costs. Well established technology.
Chemical oxidation and reduction	Preparation of insoluble oxides or metal.	Require chemicals. Climate sensitive.
Micro/nano/ultra filtration	Elimination of metal precipitates and other pollutants by permeation through membranes.	High capital and operating costs. Extensive membrane preparation: solids, microorganisms, oil and grease removal.
Reverse osmosis	Elimination of metal ions by permeation through membranes.	High capital and operating costs. The necessity of pretreatment.
Electrochemical treatment	Electrolytic deposition of metals.	Very expensive. For high concentration.
Flotation	Gas bubbles attach to solid particles or aggregates containing the pollutants to be removed.	Low cost technology. Suitable for diluted and concentrate metal bearing effluents.

The capability of organisms to cope with negative impacts of polluting matters is provided by a whole series of mechanisms, many of which are still not fully understood, as they mostly involve complex biochemical processes. Nevertheless, many of them are becoming a basis for various biological methods used for water and soil cleaning from pollutants, including metals. Biological methods suitable for metals removal from water solutions or soils can be divided into four large groups [8]:

1. Bioaccumulation (intracellular accumulation) – is metal sequestration inside the cell which may be accomplished through several ways, e.g. metal binding onto intracellular compounds, metal precipitation in the cell, methylation of the metals, intracellular accumulation of metal in higher organisms etc.. This process depends on metabolism and usually is slow with low capacity.
2. Biosorption – mechanism of sorption onto the cell surface does not depend on the cell metabolism, but is based on physical and chemical interactions between metal and function groups of the cell wall. Used biomass has all characteristics of other

sorbents (e.g. active carbon), which presents great advantage for industrial application of this process [9].

3. Bioprecipitation – is based on metal precipitation in the form of poorly soluble compound (e.g. organometallic complexes, sulphides or oxides of metals) [10]. Extracellular precipitation may, but needs not to be metabolism dependent [11], in most cases it is slow, irreversible and temperature dependent [12]. Organisms frequently protect themselves from toxic metals by producing various compounds facilitating the precipitation of metals on the cell surface or in the solution (metabolism dependent mechanism).
4. Production of volatile compounds (biovolatilization) – volatile compounds are produced by reduction or methylation and then released out of the cell. This process is the part of biogeochemical cycles of many metals (Bi, Hg, Sn) and metalloids (As, Sb, Se, Te) [13].

Probably combination of several mechanisms takes place in removal of metals from waters or soils.

The process of biosorption appears to be one of the few alternatives available for waste water treatment. It combines the advantages of biotechnology (such as low operating cost, minimization of the volume of chemical and/or biological sludge to be disposed of, high efficiency in detoxifying very dilute effluents, and no nutrient requirements) with the advantages of conventional chemical methods e.g. resistance to the physicochemical properties of heavy metal solutions (very high or low pH, high metal concentrations) because of the use of dead biomass.

1.1. General Characteristics of the Biosorption

Biosorption is a metal uptake by the inactivated or dead cell or cell debris. The use of inactive biomass is one of the major advantages of biosorption. It decouples the growth and propagation of the biomass from its function as metal sorbing material. It allows the use of waste biomass without necessity to cultivate it. It offers possibility of easy recovering of the metal(s) and the reusing of biosorbent material. The potential of the biosorbents for being regenerated after each metal uptake cycle is extremely important. Metal elution studies show that it is possible to achieve significant concentration of the metallic species sorbed when it is released from the biosorbent in a much smaller volume of eluant solution. Recovery of many metals of value from these concentrates can then be easily accomplished using conventional metal winning technologies [14].

Mechanisms of cell surface sorption are independent of cell metabolism; they are based upon physicochemical interactions between metal and functional groups of the cell wall. Generally, three sorption mechanisms are responsible for the biosorption:

- physical adsorption
- chemisorption
- ion adsorption (mainly ion exchange)

In minor part other mechanisms can take part in biosorption such as complexation or microprecipitation [11]. The real mechanism of biosorption is still not known [15]. Since this process is independent of the metabolism, binding of metals is very quick, up to 1 min [16, 17] and mostly reversible [11].

1.2. Parameters Influencing Biosorption Process Equilibrium

Heavy metal biosorption is affected by several environmental factors such as pH, ionic strength, simultaneous presence of different heavy metals, nature of inorganic and organic ligands present in solution etc. [18]. Mainly temperature, pH, biomass concentration and the presence of other ions influence the process.

1. *Temperature* – there are many factors important in biosorption which are affected by temperature [19], such as metal ions stability in solution, autoprotolysis constant of water, stability of the biosorbent – metal complexes, cell wall configuration and cell wall ionization.

Generally, if the main mechanism is physical adsorption the effectiveness of the process should increase with decreasing temperature as physical adsorption is exothermic process. If the main mechanism is chemisorption, the effectiveness increases with increasing temperature. For example, the sorption of copper and nickel by *Zooglea ramigera* increased with decreasing temperature but the sorption of lead and iron by the same bacterium increased with the increasing temperature [20].

Veglio and Beolchini [11] on the basis of collected data published in literature concluded that temperature in range 20 – 35°C does not significantly influence biosorption.

2. *pH value* – is important parameter influencing biosorption performance, because it effects dissolving of metals and their behaviour in solution as well as activity of functional groups of cell wall [21].

Functional groups of cell wall are very different according to the nature of the biosorbent: carboxylic, phosphate, sulphate, ammino, amidic and hydroxylic groups are the most commonly found [22]. They are protonated or deprotonated in dependence on the pH value of solution. Increasing of pH increases negative charge of cell wall till all relevant sites in cell wall are deprotonated what help cation adsorption from solution. Usually critical pH value exists for each metal. Over it significantly grows metal sorption mainly till the pH when metal is precipitated [23].

3. *Biomass concentration* – it was found that at low biomass concentration specific adsorption increases and with the increase of biomass concentration specific adsorption decreases [24]. But Esposito et al. [22] found for cadmium biosorption by bacterium *Sphaerotilus natans* that increasing of biomass concentration (within some limits) does not affect the maximum capacity of the biosorption.

4. *Other ions presence* – the presence of a large number of metals in industrial metal-bearing solutions makes necessary to investigate their effect on the final metal uptake by individual biosorbent. For example, iron or aluminium are not considered a major environmental problem but as they are often present in waste waters it is important to study their effect on biosorption of selected toxic metal. It was found that presence of Fe^{2+} influences uranium uptake by *Rhizopus arrhizus* [11], lithium influenced cadmium uptake by *Sargassum polycystum* [25] or aluminium decreased the biosorption of Pb, Cu, Cd, Zn and Cr by *Laminaria japonica* [26].

1.3. Biosorption Equilibrium

Basic parts of biosorption process are solid phase (sorbent), liquid phase (mostly water) and ions which should be sorbed (sorbate). Because of higher affinity of sorbent to sorbate, these ions are attracted by the solid phase and bound by many different mechanisms. Whole process continues till equilibrium. The equation expressing adsorption capacity (the amount of sorbate bound on the sorbent) is:

$$q = \frac{n(x_0 - x_e)}{m} \quad [\text{mmol/kg}] \qquad (1)$$

$$q = \frac{V(c_0 - c_e)}{m} \quad [\text{mg/kg}] \qquad (2)$$

As there is very big variability in the factors influencing biosorption one of the key process for industrial design of the process is mathematical modeling of the equilibrium behaviour of the system metal - biosorbent. System one metal – biosorbent is simplified system. As the right mechanism of biosorption is still not known, there are not equations describing it exactly. The adsorption isotherms are the most often used for the description of biosorption. Volesky [15] has summarized the most often used models for such systems (Table 2).

Table 2. Mathematical models applicable in one metal system [15]

Isotherm	Equation	Advantages	Disadvantages
Langmuir	$q = \frac{bq_m C_e}{1 + bC_e}$	well interpretable parameters	unstructured monolayer sorption
Freundlich	$q = KC_e^{1/n}$	easy formulation	no equivalent to maximum adsorption
Combination (Langmuir-Freundlich)	$q = \frac{bq_m C_e^{1/n}}{1 + bC_e^{1/n}}$	Combination of above mentioned	very complicated
Radke a Prausnitz	$\frac{1}{q} = \frac{1}{aC_e} + \frac{1}{bC_e^{\beta}}$	easy formulation	empirical, it uses 3 parameters
Reddlich-Peterson	$q = \frac{aC_e}{1 + bC_e^n}$	it approaches Freundlich at high concentration	no special advantages
Brunauer (BET)	$q = \frac{BCQ^0}{(C_s - C)[1 + (B-1)C/C_s]}$	Multilayer adsorption, infection point	no equivalent to maximum capacity
Dubinin-Radushkevich	$\frac{W}{W_0} = \exp\left[-k\left(\frac{\varepsilon}{\beta}\right)^2\right]$ (adsorbed volume)	independent of temperature	unlimited behaviour in the regime of Henry law

The most often Langmuir and Freundlich isotherms are used. For more complicated systems with many ions multilayer adsorption isotherms are used, for example:

$$q = \frac{q_m B c_e}{(c_s - c_e)[1 + (B-1)(c_e/c_s)]} \tag{3}$$

Although these models are often used there are not always suitable for mathematical modeling of biosorption. As ion exchange is consider to be the main mechanism of biosorption is more suitable to use models based on ion – exchange principles [27]. Ion – exchange models represent more realistic depiction of the system, they are more complex but still do not accurately elucidate the metal – sorbent interactions [28]. The equations of ion-exchange isotherm can be expressed by formulas published by Figuiera et al. [29]:

$$q_{M1} = \frac{Q}{\left(1 + \frac{c_{eM2}}{K_{M1M2}.c_{eM1}}\right)} \tag{4}$$

$$q_{M2} = \frac{Q}{\left(1 + \frac{c_{eM1}}{K_{M1M2}.c_{eM2}}\right)} \tag{5}$$

Symbols M1 and M2 represent heavy metals or other ions, such as Ca^{2+}, K^+ and H^+, in system. Equilibrium constant K_{M1M2} can be calculated by equation:

$$K_{M1M2} = \frac{q_{M1}^{n}.c_{eM2}^{m}}{q_{M2}^{m}.c_{em1}^{n}} \tag{6}$$

Symbols m and n represents valency of ions M1 and M2. Other symbols are defined at the end of the article.

Big advantage of this model is that proton concentration (pH) is involved into the equations as independent variable, but it is not suitable model for multiionic systems.

Very effective tool in solving problems connected with biosorption offers computer simulation. Computer simulation as well as mathematical modelling facilitates a deeper understanding of the physical and chemical mechanisms that operate in metal biosorption necessary for optimization of operating conditions and reactor configuration. Of course, there is necessary to verify experimentally the predictions calculated by computer. No model can absolutely replace experiments but the development in computer hardware and software brings these efforts closer to real conditions and makes it necessary to further biosorption process study [30, 31].

1.4. Process Performance

The principle of the process of metal removal from aqueous solutions or their recovery by means of biosorption is the contact of solid and liquid phases. It includes the biosorption and desorption cycle. The most suitable set-up to study basic thermodynamic and kinetic parameters of the process is discontinuous experimental set-up. They enable obtaining data necessary for sorption isotherms necessary to describe biosorption process and to compare sorption abilities of different biomaterials [32, 33].

More data necessary for the process characteristics is possible to obtain by studying the process behaviour in biosorption columns. The most effective devices for the continuous removal of heavy metals are considered fixed packed-bed reactors/columns. Packed-bed reactor consists of a tube, usually vertical, packed with biosorbent particles. Medium can be fed either at the top or bottom of the column and forms a continuous liquid phase between the particles. Damage due to particle attrition is minimal in packed beds compared with other types of reactors. The particles should be relatively incompressible and able to withstand their own weight in the column without deforming and occluding liquid flow [34]. For this reason biosorbents are very often immobilized in carriers. The reuse of the active biosorption particles improves the economy of whole process.

Biosorption columns operate on cycles consisting of loading, regeneration and rinsing. Operation begins with packing of the column with sorption material, after that a metal-bearing liquid is passed through the column and metals are taken up from the liquid by the biosorbent. Upper part of the column comes into contact with the most concentrated solution and becomes first saturated. Relatively fresh lower part sequesters the remains of the metal from less concentrated solution. After some time, whole column is saturated with metal and must be taken out of operation. Metal is removed from biosorbent particles most often by washing with acid or alkaline solution. The cycle ends with backwashing of the column content with water to remove of remains of the regenerants and suspended solids captured in the column. The whole process can be distributed between two or more columns so while one of the columns is undergoing regeneration and rinsing, the other is being loaded with heavy metals [27].

The overall performance of flow-through columns is strongly related to the length and shape of the ion-exchange zone that develops during sorption and regeneration. This zone develops between the section of the column that is saturated with heavy metals and the section that still contains fresh biosorbent. The shorter the dynamic ion-exchange zone in the column, the larger the fully utilized sorbent portion inside the column. From the present studies [15, 27] of the biosorption process is obvious that biosorbent, very similar to synthetic ion-exchange columns is possible to prepare in different ionic forms, for example protonated (H-form) or saturated by Ca, Mg, and Na ions by washing the biosorbent with acid, inorganic salts or hydroxides.

2. Materials and Methods

2.1. Biomass Preparation and Chemicals

The biomass of *Chlorella kessleri* was obtained from Institute of Botany of the Slovak Academy of Science. The biomass was cultivated in Z-medium (ZEHNDER in STAUB 1961) which contains in 1000 ml of water: 467 mg $NaNO_3$, 59 mg $Ca(NO_3)_2.4H_2O$, 31 mg K_2HPO_4, 25 mg $MgSO_4.7H_2O$, 21 mg Na_2CO_3, 10 ml Fe-EDTA. During cultivation, biomass was aerated and lighted by 4x40 W fluorescent tubes. The biomass was dried at 55°C during 24 hours and subsequently powdered.

For biosorption capacity study 9 different sorbents were used. Plant biosorbents (birch leaves, aspen tree leaves, *Polytrichum commune*, *Sphagnum sp.* and pine wood) and animal biosorbent *Leptinotarsa decemlineata* were collected from nature, washed and air dried. Then they were homogenized and the fraction with particular size was used for the experiments. Fungus *Agaricus campestris* – field agaric was bought, air dried and homogenized. Active coal and ion-exchange resin (Katex A-IR-120) were bought and without any modification used in experiments.

Analytical grade $CuSO_4.5H_2O$, $CdSO_4$ was used for sorption experiments. Analytical grade Ag was dissolved in HNO_3.

2.2. Batch Equilibrium Biosorption Experiments

Powdered biosorbents were added into mixed solution of particular metal. pH value of solution was adjusted by adding 20 and 10% NaOH and 10% H_2SO_4. Samples for analyses were withdrawn in chosen intervals; supernatant liquid was filtered through paper filters. Only after algal biosorption the supernatant was separated by membrane filtration. The concentration of unadsorbed metal ions in filtrate was determined using an atomic absorption spectrophotometer (Perkin Elmer 3100) with an air-acetylene flame. Biosorbent-free blanks were used as controls.

2.3. Adsorption Isotherms

The general methodology for adsorption isotherms experiment is written elsewhere. For experiments in this study the methodology described by Kaduková, Virčíková [8] was used. Copper concentration in experimental solution is in Table 3.

Table 3. Initial copper concentration in solution

	1	2	3	4	5	6	7	8	9
Initial copper concentration [mg/l]	0,45	4,74	9,23	27,68	46,9	124,25	478,25	944,5	4719

Volume of experimental media was 30 ml. Biomass concentration was 2 g/l. pH value of solutions was adjusted by adding 20 and 10% NaOH and 10% H_2SO_4 and kept during whole experiment. All experiments reported here were investigated at an initial pH 4.5.

When living cells were used (bioaccumulation process) the biomass concentration was 3 g/l. Membrane filters were used to separate cells from solution. Volume of the media was 50 ml. Copper concentration in experimental solutions is in Table 4.

Table 4. Initial copper concentration in bioaccumulation experiments

	1	2	3	4	5	6
Initial copper concentration [mg/l]	5,03	9,41	24,34	493,42	4447,1	8661,9

2.4. Preparation of Samples for Electron Microscopy

Samples of cells were taken after 24 hours bioaccumulation or biosorption. The separation of the biomass from solution was achieved by membrane filtration. Cells were washed with distilled water and a few drops of cell suspension were put on the base from thick aluminium or steel foil. Sample was flame fixed and coated with thin layer of gold. Samples were examined by scanning electron microscopy with a JEOL JSM-35CE.

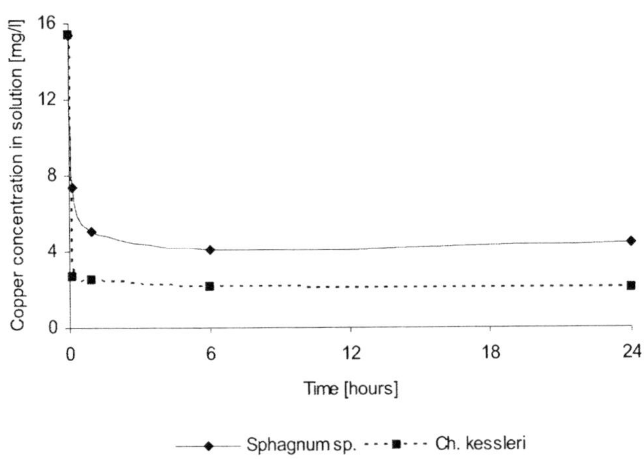

Figure 1. Decrease of the copper concentration after addition of alga *Chlorella kessleri* and peat moss *Sphagnum sp.* into experimental solution. Initial copper concentration was 15.4 mg/l. Initial biomass concentration was 1.2 g/l of algae and 1 g/l of moss.

3. Experimental Result of Biosorption

3.1. Biosorption Kinetics

The decrease of metal cation concentration during biosorption was very quick. As the example biosorption of copper and cadmium by green alga *Chlorella kessleri* and peat moss *Sphagnum sp.* is shown (Figures 1 and 2).

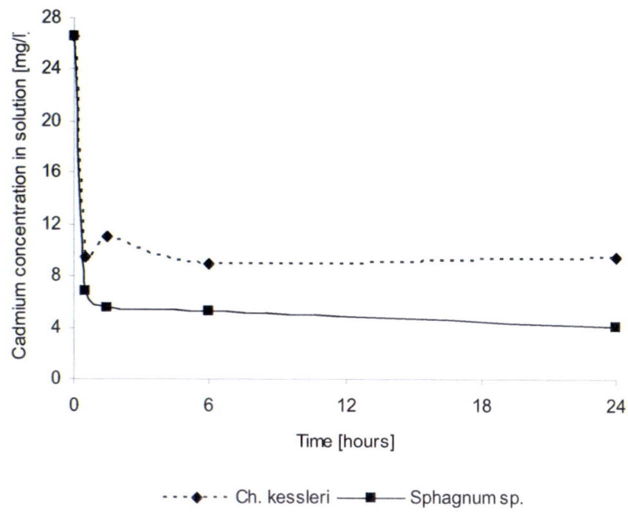

Figure 2. Decrease of the cadmium concentration after addition of alga *Chlorella kessleri* and peat moss *Sphagnum sp.* into experimental solution. Initial cadmium concentration was 26.5 mg/l. Initial biomass concentration was 2 g/l.

Most of the metals were sorbed onto biosorbent till 10 minutes after the beginning of the process. The decrease of copper concentration during first hour after addition of green alga *Chlorella kessleri* is in Figure 3. This behaviour is typical for biosorption. Some authors [16, 17] found that metal concentration after biomass adding decreases during first minute of the process. Quick decrease of metal concentration confirms that the mechanisms of metal uptake is independent of metabolisms, thus metal does not enter the cell body but binds on the surface.

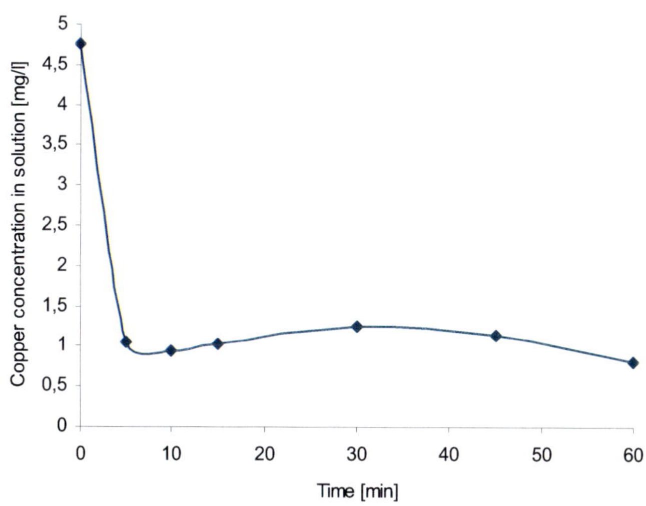

Figure 3. Decrease of copper concentration during first hour after addition of *Ch. kessleri* into experimental solution. Initial copper concentration was 4.75 mg/l. Initial biomass concentration was 2 g/l.

3.2. Equilibrium Study of the Biosorption

Equilibrium data can be analyzed using adsorption isotherms to provide a basis for the design of adsorption system. Although ion – exchange models are considered to give more realistic picture of the system, they still not accurately elucidate the metal – sorbent behaviour [28]. Hence, Langmuir and Freundlich models were used in this study. Both fitted well to the experimental data obtained in the range from very low concentration to approximately 400 mg/l of equilibrium copper concentration (Figure 4).

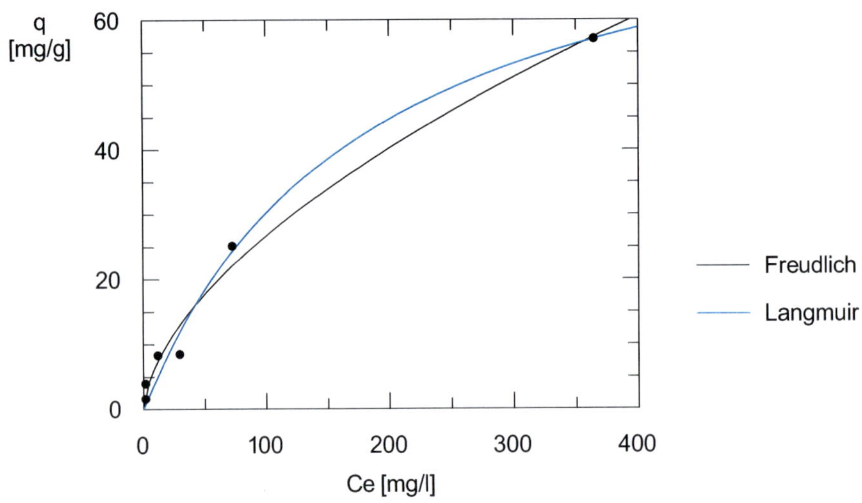

Figure 4. Experimental equilibrium data fitted to the Langmuir and Freundlich isotherms.

Above 400 mg/l of equilibrium copper concentration in solution specific adsorption – q increased quickly (Figure 5). Probably copper ions were accumulated on the surface of biomass in more layers.

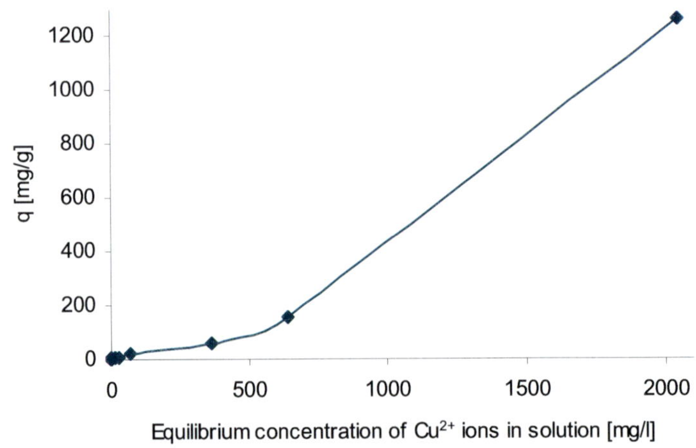

Figure 5. Adsorption isotherm for copper biosorption by *Ch. kessleri*.

3.3. Capacity of the Biosorption

It is interesting to compare equilibrium adsorption capabilities of materials of different origin for specific metal. In our study copper was chosen as a model metal (Figure 6). Equilibrium time was 24 hours.

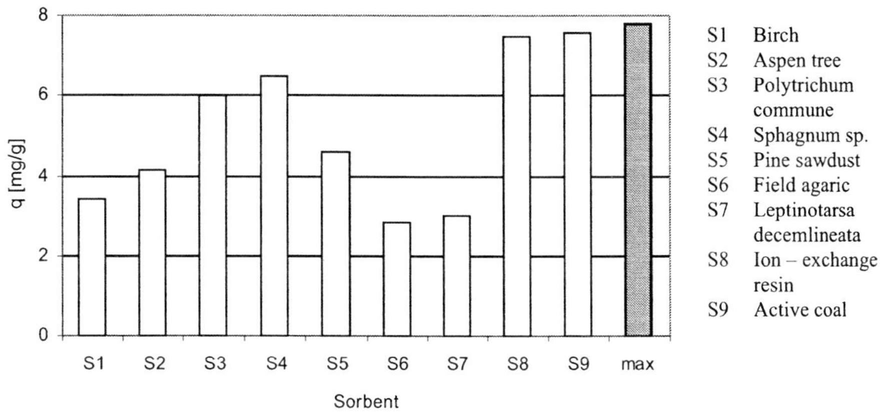

Figure 6. Comparison of equilibrium specific adsorption for studied sorbents. Dark column – max, represents theoretically calculated maximum specific adsorption for studied conditions. Initial copper concentration was 16 mg/l. Initial biomass concentration was 2 g/l.

The best sorbents for copper were commercially produced sorbents – active coal (S9) and ion – exchange resin (S8). Specific adsorption - q of active coal represents 97% of theoretically calculated specific adsorption at chosen conditions, for ion – exchange resin it was 96%. From naturally occurred sorbents only mosses exceeded the limit of 75%. Moss *Sphagnum* (S4) sequestered 83% of total copper in solution and moss *Polytrichum* (S3) 76%.

If specific adsorption calculated after 10 minutes (Figure 7) is compared the whole picture changes.

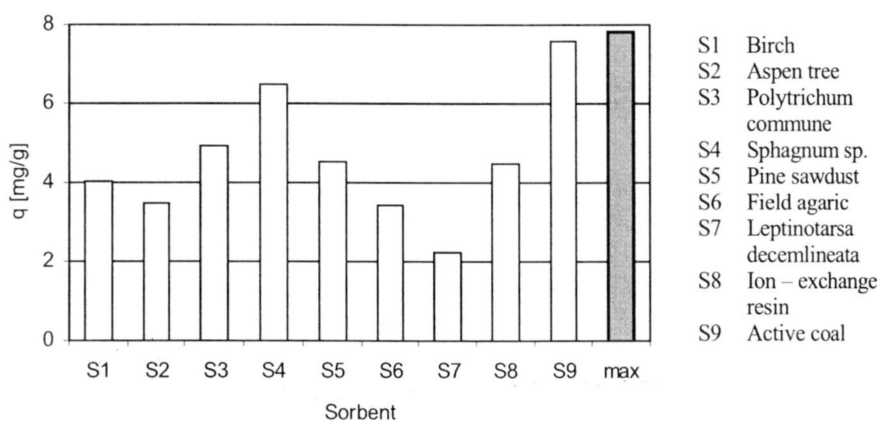

Figure 7. Comparison of specific adsorption for studied sorbents after 10 minutes from the beginning of sorption. Dark column – max, represents theoretically calculated maximum specific adsorption for studied conditions. Initial copper concentration was 16 mg/l. Initial biomass concentration was 2 g/l.

The highest capacity had the active coal and the biosorbent prepared from peat moss. Their sorption capacity was even higher than sorption capacity of ion-exchange resin used. If active coal was utilized, pH of the solution increased till 9 suggesting that copper was precipitated. Precipitation is not favourable process, because it causes problems in separation of metal from sorbent. The rate of the copper sorption onto the surface of moss offers advantage for column operation. Solution with dissolved metal does not need stay long time in the column - the continuous biosorption process can be fast, high volume of water can be cleaned up.

3.4. Selectivity of the Process

There were observed differences in the sorption of metal cations on particular plant biomass. With regard to extent of metal removal the biosorption capacity for cations Cd^{2+}, Cu^{2+} and Ag^{+} rises in order Cd < Cu < Ag. However, referring to the number of adsorbed metal-cations, more ions of Cu were sequestered than Ag^{+} ions. It is due to the higher molecular weight of Ag in relation to Cu. Consequently, for comparison of different metals, the uptake expressed on the molar or electrical equivalents basis is more appropriate. Table 5, summarizes values of all characteristic sorption parameters. Interesting for technology is amount of adsorbed metal (mg/g of dry biomass), and thus, what amount of metal was removed from solution. From Figure 8 it is obvious that sorption abilities of algal biomass are the best in case of silver (in comparison with cadmium and copper).

Table 5. Comparisons of sorption abilities dry algal biomass for cations Cd^{2+}, Cu^{2+} and Ag^{+}

	Cd	Cu		Ag
Initial metal concentration [mg/l]	26.5	15.38	27.68	48.5
Equilibrium concentration [mg/l]	9.5	2.07	11.49	14.3
pH	5.62	5	5	2.89
Equilibrium sorption [mg/g]	8.68	10.87	8.32	16.93
Equilibrium sorption [mmol/g]	0.077	0.171	0.131	0.158
Biomass concentration [g/l]	2	1.225	2	2
Duration [min]	30	10	30	90

Table 6. Comparison of sorption abilities of dry moss biomass for cations Cd^{2+}, Cu^{2+} a Ag^{+}

	Cd	Cu	Ag
Initial metal concentration [mg/l]	26.5	15.38	48.5
Equilibrium concentration [mg/l]	4.1	4.48	29.45
pH	5.62	5	2.89
Equilibrium sorption [mg/g]	10.8	10.77	8.43
Biomass concentration [g/l]	2	1.01	2
Duration [min]	30	60	30

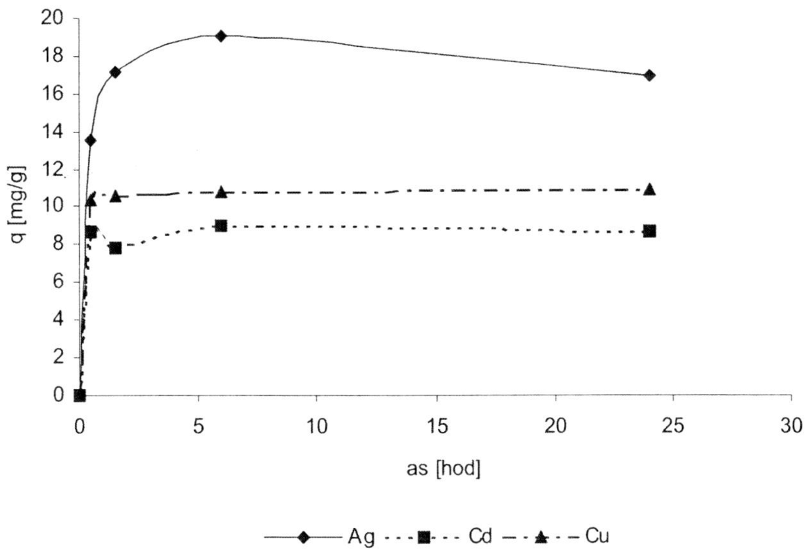

Figure 8. Comparison of sorption abilities of dry algal biomass for cations Cd^{2+}, Cu^{2+} and Ag^{+}.

The sorption values for this metal are twice as high as sorption values for Cu and Cd. The sorption capacity for copper and cadmium by this biomass is comparable. The advantage for selective removal of this metal is that required pH for silver sorption is 2.89 and for copper and cadmium sorption is 5 and 5.6, respectively. Cd and Cu will not be removed from solution with such a low pH so silver can be removed selectively.

The sorption abilities of moss biomass for cations Cu^{2+}, Cd^{2+} a Ag^{+} are presented in Table 6 and Figure 9. Compared with algal sorption, the sorption ability of moss biomass for tested metals were very similar. So, we cannot say about better biosorbent for some of the metals.

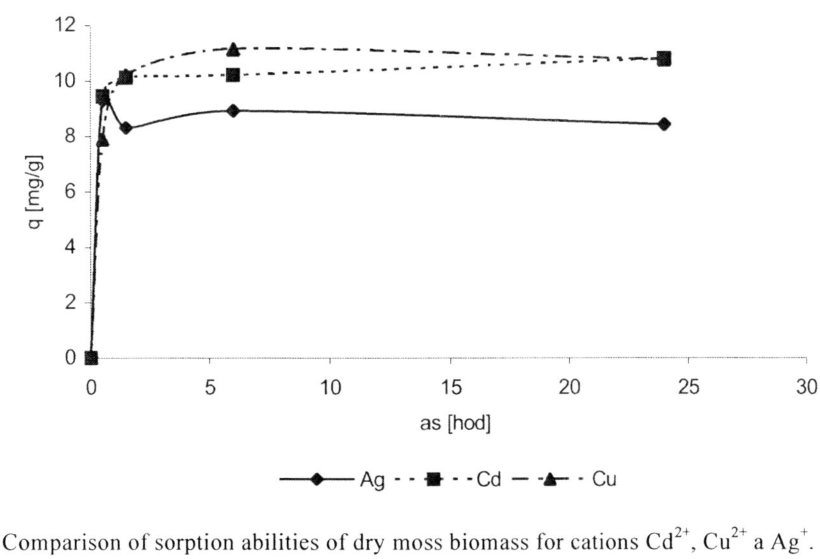

Figure 9. Comparison of sorption abilities of dry moss biomass for cations Cd^{2+}, Cu^{2+} a Ag^{+}.

3.4. Comparison of Biosorption by Living and Dead Biomass

There is always risk of heavy metal accumulation inside the cell when living biomass is used creating new waste necessary to treat. Except of it, it is more difficult to predict the behaviour of organisms in the presence of metals. It is very useful to compare sorption by living and dead biomass. It is very interesting that affinity of organisms to metals changes if it is sorbed by living biomass or dead biomass. For example, it was found following dependence for living cell of *Chlorella regularis* [35]:

$U >> Cu >> Zn \geq Ba \geq Mn \geq Co \approx Cd \geq Ni \approx Sr$

But for dead cells the affinity has changed:

$U >> Cu >> Mn \geq Ba > Zn \geq Co \geq Cd \geq Ni > Sr$

The differences were found not only for the alga but also for actinomyces *Streptomyces pilosus* or other organisms [36].

Anytime when living alga *Ch. kessleri* was used to take up copper (the process when living organisms are used in metal sequestration is called bioaccumulation) the increase of copper concentration in solution ions within first hours was observed (Figure 10).

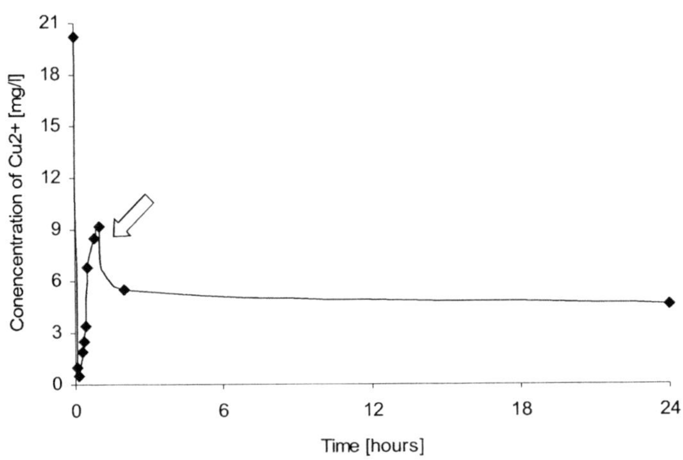

Figure 10. Decrease of the copper concentration after addition of dead cells of *Ch. kessleri*. Partial release of copper ions back into solution is marked by arrow. Initial copper concentration was 20.2 mg/l. Initial biomass concentration was 2 g/l.

Probably copper toxicity caused paralysis of cell walls and consequently partial release of Cu^{2+} ions into solution. Similar phenomenon was observed by Danihelka et al. [37]. The temporary increase of copper concentration was not observed when dead cells of *Ch. kessleri* were used (compare in the Figure 1).

The influence of copper ions on the cell surfaces was evaluated by means of electron microscope images. Figure 11a shows damage on the surfaces of living cells visible as crateriform cavities by scanning EM. Cells on the pictures have been in copper solution for 24 hours. Probably because of the damage, have cells partially lost their binding abilities resulting in higher metal concentration in solution after 24 hours than after 10 minutes of bioaccumulation. If dry cells are used in the process there is very rarely cell surface damage present (Figure 11b) in solutions with high copper concentration, as suggested by electron scanning microscope photos [38].

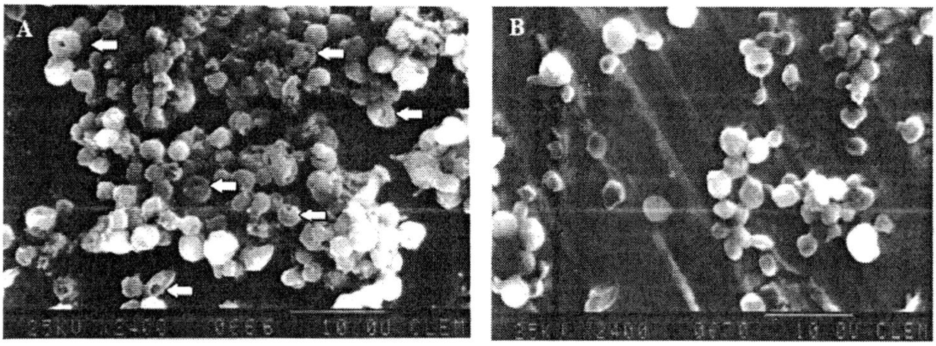

Figure 11. Scanning electron microscope image of *Chlorella kessleri* cells after bioaccumulation of copper by living cell (A) and after biosorption of copper by dead cells (B). Arrows show some of the damages of the cells caused by copper. Magnification 2400x. [38].

Another difference between bioaccumulation and biosorption of copper is illustrated on Figure 12. From the comparison of adsorption isotherms obtained for *Chlorella kessleri* it is obvious that specific adsorption (q) is about six times higher when dead cell are used. High copper concentrations cause serious damage to the cell walls of living cells so they loose their binding abilities resulting in remarkably lower uptake capacity [38].

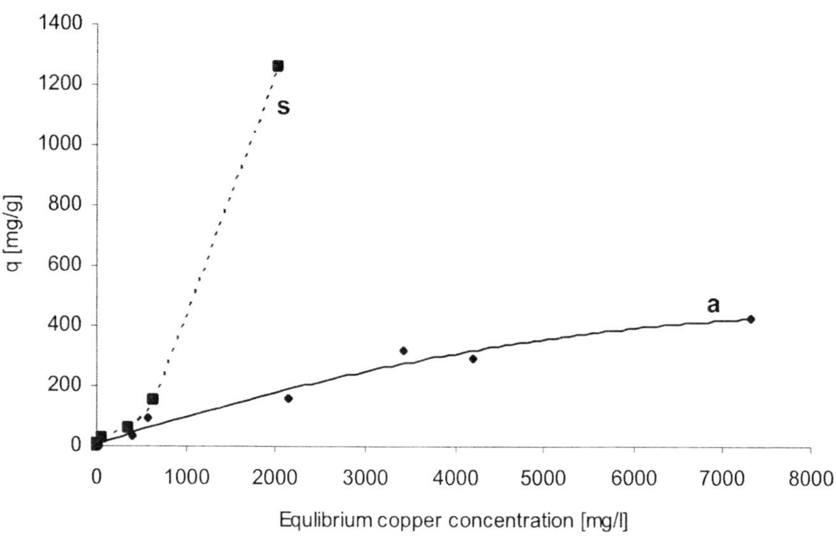

Figure 12. Comparison of adsorption isotherms for bioaccumulation (a) and biosorption (s) of copper at high equilibrium concentrations [38].

4. CONCLUSION

Contaminated wastewater from different industries, acid mine drainage or water from landfills brings about a lot of problems connected with environmental pollution. On the basis of suitable physico-chemical characteristics biological processes are considered as alternative processes to conventional clean-up methods.

Biosorption, one of the perspective methods, is based on the ability of organisms to sequester and accumulate large amounts of metals. Although practical spreading of biosorption is still hindered by a lack of information about its mechanisms, biosorption based processes are already working in several plants worldwide.

There are many advantages offered by biosorption application, such as low operation costs, comparable capacity of biomass to capture metals to commercial products, possibility to remove bound metals from biomass, possibility of the re-use of biomass, possibility to apply the method to a wide range of metal concentration which is advantageous for final treatment of waste waters or drinking water, no interference with organic materials and the rate of the process which runs under normal temperature and pressure. As with all methods, biosorption has some disadvantages mainly high expenses for research, problems with biomass separation from solution if untreated biomass is used and problems related to different characteristics of the same biomass.

Still there is a necessity for the research which would facilitate better combination of biomass, metals and outer conditions aiming in the highest effectiveness of the process.

ACKNOWLEDGEMENT

This work was supported by VEGA 1/3220/06 grant from Slovak Grant Agency. Authors thank Ing. G. Janák for making an electron microscope photographs and Mgr. Emília Lacková and Ms. Alena Otávková for the help with experimental work.

REFERENCES

[1] Volesky, B., Weber, J., Vieira, R. (1999). Biosorption of Cd and Cu by different types of *Sargassum* biomass. In: Amils, R., Balester, A. (Eds.), Biohydrometallurgy and the Environment Toward the Mining of the 21st Century, *Proceedings of the International Biohydrometallurgy Symposium IBS´99*, Elsevier, 473-481.

[2] Havlik, T., Laubertová, M., Miskufova, A., Kondas, J., Vranka, F. (2005). Extraction of copper, zinc and cobalt in acid oxidative leaching of chalcopyrite at the presence of deep-sea manganese nodules as oxidant. *Hydrometallurgy,* 77, 51-59.

[3] Miškufová, A., Halžová, I., Havlík, T., Ukašík, M. (2005). Hydrometalurgické spracovanie hliníkových sterov, *Zborník zo 4. medzinárodnej konferencie o hliníkových zliatinách Aluminium 2005*, Česká republika, 93-102.

[4] Kaduková, J., Virčíková E. (2001). Biosorpcia - perspektívna technológia čistenia kontaminovaných vôd od ťažkých kovov. *Zborník Environmentálne problémy miest*, 49-52.

[5] Luptáková, A., Kušnierová, M. (2004). Bioremediation of heavy metals contaminated industrial waste waters. *Acta Metallurgica Slovaca*, 10, 2, 183-188.

[6] Rubio, J., Schneider, I.A.H., Aliaga, W. (1996). New processes for heavy metals separation from waste water streams in the mining industry. In: Sanchéz, M.A., Vergara, F. and Castro, S.H. (Eds.), *Clean Technology for the Mining Industry*, University of Concepción, Concepción-Chile.

[7] Volesky, B. (2001) Detoxification of metal-bearing effluents: biosorption for the next century. *Hydrometallurgy*, 59, 203-216.

[8] Kaduková, J., Virčíková, E. (2003). Minerálne biotechnológie III., *Biosorpcia kovov, prvé vydanie,* VŠB TU Ostrava.

[9] Kaduková, J., . Štofko, M. (2006). Environmentálne biotechnológie pre hutníkov, *prvé vydanie, Equilibria*, Košice.

[10] Raraz, A. G. (1995). Biological and Biotechnological Waste Managment in Materials Processing. *JOM*, February 1995, 56-62.

[11] Veglio, F., Beolchini, F. (1997). Removal of metals by biosorption: a review. *Hydrometalurgy*, 44, 301-316.

[12] Kotrba, P., Ruml, T., Macek, T. (1994). Vazba kadmia buňkami mikroorganismů a rostlin. *Chemické listy*, 88, 642-649.

[13] Michalke, K., Wickenheiser, E.B., Mehring, M., Hirner, A.V., Hensel, R. (2000). Production of Volatile Derivatives of Metal(loid)s by Microflora Involved in Anaerobic Digestion of Sewage Sludge. *Applied and Environmental Microbiology*, 66, 7, 2791-2796.

[14] Volesky,B. (1990). *Biosorption of heavy metals*. CRC Press, Boca Raton, Florida.

[15] Volesky, B. (2004). *Sorption and Biosorption*, BV Sorbex, Inc., Montreal.

[16] Bustard, M., Donnellan, N., Rollan, A., McHale, A.P. (1997). Studies on the biosorption of uranium by a thermotolerant, ethanol-producing strain of *Kluyveromyces marxianus*. *Bioprocess Engineering*, 17, 45-50.

[17] Delgado, A., Anselmo, A.M., Novais, J.M. (1998). Heavy metal biosorption by dried powdered mycelium of Fusarium flocciferum, *Water Environment Research*, 70, 3, 1998, 370-376.

[18] Pagnanelli, F., Esposito, L., Toro, F., Veglio, F. (2003). Metal speciation and pH effect on Pb, Cu, Zn and Cd biosorption onto *Sphaerotilus natans*: Langmuir-type empirical model. *Water Research*, 37, 627-633.

[19] Bedell, G.W., Darnall, D.W. (1990). Imobilization of Nonviable, Biosorbent, Algal Biomass for Recovery of Metal Ions. In: Volesky, B., *Biosorpton of Heavy Metals*, CRC Press, Boca Raton, Florida, 313-326.

[20] Sağ, Y., Kutsal, T. (1995). Biosortpion of heavy metals by *Zooglea ramigera*: use of adsorption isotherms and comparison of biosorption characteristics. *The Chemical engennering Journal,* 60, 181-188.

[21] Veglio, F. (2006). Adsorption and Biotechnology: Recent Application in Biosorption. *EBA*, Brasil.

[22] Esposito, A., Pagnanelli, F., Lodi, A., Solisio, C., Veglio, F. (2001). Biosorption of heavy metals by *Sphaerotilus natans*: an equilibrium study at different pH and biomass concentrations. *Hydrometallurgy*, 60, 2, 129-141.

[23] López, A., Lázaro, N., Priego, J.M., Marqués, A.M. (2000). Effect of pH on the biosorption of nickel and other heavy metals by *Pseudomonas fluorescens* 4F39. *Journal of Industrial Microbiology and Biotechnology*, 24, 146-151.

[24] Fourest, E., Roux, J. C. (1992). Heavy metals biosorption by fungal mycelial by-products: mechanisms and influence of pH. *Applied Microbioliology and Biotechnolology,* 37, 399-403.

[25] Yun, Y.S., Volesky, B. (2003). Modelling of Lithium Interferencein Cadmium Biosortpion. *Environmetal Science and Technology*, 37, 3601-3608.

[26] Lee, H.S., Suh, J.H., Kim, I.B., Yoon, T. (2004). Effect of aluminium in two-metal biosortpion by an algal biosorbent. *Minerals Engineering*, 17, 487-493.

[27] Kratochvil, D., Volesky, B. (1998) Advances in the biosorption of heavy metals. *TIBTECH,* 16, 291-300.

[28] Kim, S.H., Song, H., Nisola, G.M., Ahn, J., Galera, M.M., Lee, C.H., Chung, W.J. (2006). Adsorption of lead(II) ionas using surface-modified chitins. *Journal of Industrial and Engineering Chemistry*, 12, 3, 469-475.

[29] Figuiera, M.M., Volesky, B., Ciminelli, V.S.T., Roddick, F.A. (2000). Biosorption of metals in brown seaweed biomass, *Water Research*, 34, 1, 196-204.

[30] Csontó, J., Polák, M., Kaduková, J. (2001). Artificial Life Simulation of Living Alga Cells and its Sorption Mechanisms, *Journal of Medical Systems*, 25, 3, 221-231.

[31] Csontó, J., Palko, M. (2002). *Umelý život*, ELFA, Košice.

[32] Khudhair, A.M. (2001). Batch biosorption of zinc by using bones of spent sheep. New Trends in Mineral Processing, *Proccedings from international conference*, Ostrava, 167-177.

[33] Chovancová, I. (2001). Uptake of trivalent chromium by biosorbent of algae of *Chlorella vulgaris.* New Trends in Mineral Processing, *Proccedings from international conference*, Ostrava, 493-499.

[34] Doran, P. M. (1995). *Bioprocess Engineering Principles*, Academic Press, London, 340-341.

[35] Borovec, Z. (1989). *Bioakumulace kovů a dočisťování vod baktériemi, houbami a řasami,* Rudy, 12, 37, 375-382.

[36] Unz, R.F., Shuttleworth, K.L. (1996). Microbial mobilization and immobilization of heavy metals. *Current Opinion in Biotechnology*, 7, 3, 307-310.

[37] Danihelka, P., Chovancová, I., Špinková, K. (1997). Biotechnology for heavy metals elimination. Mine Planning and Equipment Selection. In: Strakoš, Kebo, Farana, Smutný (Eds.), *Proccedings of VI. International Symposium, Balkema.* Rotterdam, 839-842.

[38] Kaduková, J., Virčíková, E. (2005). Comparison of differences between copper bioaccumulation and biosorption, *Environment International*, 31, 2, 227-232.

In: Environmental Research Trends
Editor: Marvin A. Cato, pp. 153-204
ISBN: 978-1-60021-556-8

Chapter 5

CONTAMINATION TRAPS AS TRANS-FRONTIER MANAGEMENT CHALLENGES: NEW RESEARCH ON THE IMPACT OF REFUSE DUMPS ON THE CONSERVATION OF MIGRATORY AVIAN SCAVENGERS

Guillermo Blanco*[*,1], *Jesús A. Lemus*[2], *Oscar Frías*[3], *Javier Grande*[1], *Bernardo Arroyo*[1], *Félix Martínez*[1] and *Noemí Baniandrés*[1]

[1]Instituto de Investigación en Recursos Cinegéticos, IREC (CSIC-UCLM-JCCM), c/Ronda de Toledo s/n 13005 Ciudad Real, Spain

[2]Departamento de Ecología Evolutiva, Museo de Ciencias Naturales (CSIC) J. Gutiérrez Abascal 2, 28006, Madrid, Spain

[3]Oficina de Especies Migratorias, Ministerio de Medio Ambiente, Gran Vía de San Francisco 4, 28005 Madrid, Spain

ABSTRACT

Efforts devoted to conservation of endangered species may often be hindered by threats occurring at spatially distant areas used by individuals along their seasonal migration routes. Therefore, specific conservation programs may temporarily remain out of the control of regional conservation efforts when individuals move from areas subjected to conservation measures to those areas that are unprotected. The evaluation of the repercussions of different threats at different areas and their role in the overall conservation status and trends represents a current challenge in conservation management of many species. Conservation of migratory birds represent a clear example of these trans-frontier challenges, as conservation effort in a given area may be unsuccessful without the knowledge of threats occurring in other key areas throughout the annual cycle. We evaluated the responsibility of Spain in the contamination of migratory red

* Author for correspondence: Guillermo Blanco, Instituto de Investigación en Recursos Cinegéticos, IREC (CSIC-UCLM-JCCM), c/ Ronda de Toledo s/n 13005 Ciudad Real, Spain. e-mail: gublanco2@yahoo.es; gblanco@irec.uclm.es. Fax: + 926 295451.

(*Milvus milvus*) and black kites (*Milvus migrans*) with heavy metals and bacteria resistant to antibiotics acquired in urban and refuse dumps in which these species feed upon livestock remains. Urban and refuse dumps are sources of abundant and predictable food for kites in Spain such that these places may be crucial for the conservation of wintering red kites and floating and migrating black kites from northern and central Europe. These foraging locales may also represent contamination traps when associated with sources of industrial and urban residues and/or when the food on which kites feed upon contain high levels of detrimental contaminants, pathogens, veterinary drugs, and bacteria resistant to these drugs. We assessed the origin of non-breeding kites using refuse dumps in Spain, and evaluated the role of dumps in attracting large numbers of wintering red kites as well as floating and migrating black kites in different Spanish regions. Results indicate the presence of high levels of resistance to antibiotics in faecal bacteria isolated from wintering red kites associated with carcass dumps of livestock treated intensively with antibiotics, especially fattening pigs. On the contrary, low levels of antibiotic resistance were found in areas where wintering red kites mostly feed upon natural prey and free-range livestock carcasses. Nestlings of both species showed high levels of bacterial resistance associated with a diet primarily composed by carrion from stabled livestock treated intensively with antibiotics. Increasing numbers of floating and migrating black kites from northern Europe foraging at a refuse dump associated with a solid-waste incinerator may be increasingly at risk from exposure to airborne contaminants. Detrimental effects of using these contamination traps in Spain may be contributing to the declining trends of local and wintering, floating and migrating kites breeding in distant areas. Global conservation plans for threatened species should consider simultaneously the impact of variable threats at different areas, especially in highly mobile species.

INTRODUCTION

The increasing impact of human intrusions on wildlife habitat has contrasting dimensions in different parts of the world. These differences depend on multiple factors regarding human population density and the degree of development, and the impact of human activities such as agriculture, livestock farming, industry and urbanization (Diamond 1989, Meffe and Carroll 1997, Redford and Richter 1999, McKinney 2002, Jha and Bawa 2006). Wildlife inhabiting different local areas and regions of the world may thus be differently influenced by the most predominant and influential human activities in each area. Taking into account that wildlife populations are generally spatially structured, the impact of human activities may promote spatially distinct influences leading to contrasting features in local populations. These contrasting human impacts on spatially separated subpopulations may have variable consequences on the dynamics, demography and persistence of local populations as a part of metapopulations (Pulliam et al. 1994, Watkinson and Sutherland 1995) depending on the effects, magnitude and type of impact, as well as on features of the metapopulations, including number of subpopulations, population density, degree of fragmentation, connectivity, etc. (Pulliam 1988, Hanski and Gilpin 1997).

Metapopulation theory has been increasingly applied in attempts to understand population structure and dynamics in species of conservation concern, primarily breeding segments of holarctic bird metapopulations negatively affected by habitat fragmentation due to human activities (Opdam 1990). Much less effort has been made to identify demographic constraints affecting migratory birds under the view of metapopulation theory (Esler 1999).

Efforts devoted to conservation of endangered species may often be obscured by threats occurring at spatially distant areas used by individuals during their seasonal movements. Therefore, specific conservation programs may temporarily remain 'out-of-control' when individuals move from good-quality patches yielding a demographic surplus of individuals to poor-quality patches yielding a demographic deficit. These patches may or may not correspond to areas subjected to conservation measures or unprotected areas, respectively. The evaluation of the repercussions of different threats at different areas and their role in the overall conservation status and trends of particular species represent a current challenge in conservation management of many species. Conservation of migratory birds represents a clear example of these trans-frontier challenges, as conservation effort in a given area may be unsuccessful without the knowledge of threats occurring in the remaining key areas (reviewed in Rappole et al. 1993).

Classical metapopulation and source-sink models are generally organised as a system in which habitat patches of high quality are sources of surplus immigrants towards sink patches of poorer quality where productivity may not compensate for mortality (Pulliam et al. 1994, Hanski and Gilpin 1997). These classical models generally do not incorporate potential constraints affecting individuals of different subpopulations when joining together distinct spatial areas throughout the annual cycle as in migratory species. This may be due to difficulties in the determination of demographic independence or panmixia in migratory species when individuals belonging to demographically and spatially independent subpopulations may occur together in a geographically distant area during some period of the annual cycle (Esler 2000). Consideration of spatial, temporal and behavioural factors influencing demographic independence of subpopulations may be required to apply the conceptual framework of metapopulation theory to conservation of migratory species. In practice, information on the interactions between spatial, temporal and behavioural mechanisms acting in the determination of the degree of subpopulation demographic independence is very difficult to obtain, especially in abundant and widely distributed species. This has frequently led to the consideration that particular areas in which species of conservation concern are abundant should be units of conservation priority (Caughley and Gunn 1995, Wright 1996). However, when this consideration is applied without information about population structure at larger spatial scales it may not greatly contribute to species conservation, but rather may introduce uncontrolled demographic influences on population dynamics that can promote undesirable and detrimental consequences for the long-term viability and conservation of the species (Janzen 1986, Caughley and Gunn 1995, Pressey et al. 1996, Sutherland 1998, Baillie et al. 2000, Gaston et al. 2002). For instance, areas of conservation priority may be demographic sinks, primarily attractive sinks (Delibes et al. 2001), when individual animals maladaptatively perceive as good habitats those that are objectively rich in food but ultimately of bad quality due to the impact of high intentional or unintentional human-induced mortality (Battin 2000, Schlaepfer et al. 2002).

Human-induced mortality through persecution and harvesting may influence distribution patterns of populations by promoting a despotic distribution of individuals (sensu Fretwell 1972), where habitat quality may be determined by survival prospects against persecution rather than by habitat quality *per se* (McCullough 1996, Whitfield et al. 2004). Thus, mortality due to human harvesting and persecution may influence the organisation of a metapopulation system with sources of dispersing individuals and sinks of mortality (Fox and Madsen 1997, Hanski and Gilpin 1997). In some species, individuals may be mostly persecuted in habitat optimal for foraging and reproduction and thus, areas in which they prefer to reside, which may lead to the creation of attractive sinks (Kenney et al. 1994, Delibes et al. 2001). In addition, habitats preferentially occupied may be not optimal for breeding due to landscape and climate features negatively affecting productivity and hence they may function as classical sinks where productivity is less than mortality and immigration. This may arise primarily due to the ability of individuals of some species to avoid direct human-induced mortality such as persecution (Ferrer et al. 1990, Whitfield et al. 2004) despite it leading to preferential occupation of habitats or patches of poor quality in terms of food abundance and breeding output. Alternatively, individual animals cannot learn persecution avoidance but remain in poor-quality areas where abrupt orography or isolation has historically precluded human presence and disturbance (Donázar et al. 2002, Blanco and Montoya 2004) while conspecifics are extirpated in optimal habitats (Carter 2001). Other population limiting factors acting as sources of mortality such as contaminants and pathogens may not easily be perceived by individuals as a risk in the short-term because their detrimental effects may surface at a much later time after infection and the uptake of contaminants.

Emerging infectious disease of wildlife has gained a growing interest due to the increasing impact of human intrusions in wildlife habitat on human disease emergence and resurgence (Hubálek 1994, Daszak et al. 2000, Friend et al. 2001, Bengis et al. 2004). A growing worldwide interface between humans, domestic animals and free-ranging wildlife represents a global dimension of these intrusions. Specifically, zoonotic enteric pathogens may be transmitted between different species, including humans through the ingestion of foods contaminated with animal faeces and wildlife through scavenging on meat remains found in livestock carcass and domestic refuse dumps (Friend and Franson 1999, Thorns 2000). Antibiotics used to combat infection with pathogenic bacteria, but also to improve growth and feed efficiency and for use as prophylactic treatments, have been overused and misused throughout the world resulting in antibiotic resistance in many commensal and pathogenic bacteria (O'Brien 2002, Mateu and Martin 2001). Antibiotic resistance is a major health threat impacting both human and animal health, especially when multi-drug resistance occurs in food-borne pathogens such as *Salmonella* and *Campylobacter* (Angulo et al. 2004; Mølbak 2004). Intensification of livestock production and management systems has been accompanied by an increase in the use of antimicrobial agents and other drugs to prevent higher rates of disease in conditions of intensive livestock production which include high stocking densities and rapid growth (McEwen and Fedorka-Cray 2002). These practices are associated with increased levels of bacterial resistance (O'Brien 2002; Mateu and Martin 2001), a factor the World Health Organization considers to be one of the major health problems in human and veterinary medicine (Angulo et al. 2004; Mølbak 2004). The emergence of bacterial resistance in wildlife and the spread of these strains within ecosystems may primarily occur through avian scavengers such as vultures, gulls, kites and corvids, which are directly associated with dumps on which livestock carcasses are disposed, and expoited by these scavenger species (Friend and Franson 1999). Environmental contamination with resistant pathogenic bacteria may therefore primarily occur in areas where livestock production is intensive and where the sanitary conditions for elimination of meat residues is unsatisfactory (Blanco et al. 2006). Areas of intensive livestock production are usually exploited by avian scavengers as a source of superabundant and predictable food. Livestock carcass dumps and other refuse dumps play a key role in the conservation of threatened avian scavengers in Asia, the Mediterranean and probably Africa due to the long history of exploitation of dumps as feeding opportunities that the historical transformation of landscapes by agricultural development has allowed (Donázar et al. 1997). Today, feeding opportunities provided by human activity may be crucial for the conservation of many threatened avian scavengers (Blanco 1996, Donázar *et al.* 1997; Serrano 1999, Blanco et al. 2006). This role is especially apparent in Spain where a high proportion of scavengers primarily depend upon food resources found at livestock carcass dumps (Donázar *et al.* 1997). However, information is lacking about the health effects of scavenging on meat remains found in dumps, including the potential role of infection by pathogenic bacteria, the ingestion of residues of veterinary drugs and the effects of acquiring bacterial resistant to these drugs (Houston and Cooper 1975; Winsor *et al.* 1981, Friend and Franson 1999, Rodrigues *et al.* 2003). This knowledge may be important in order to reconcile the elimination of livestock carcasses with the conservation of scavengers, which represents a current major challenge for conservation biology in Spain, the stronghold for the declining avian scavengers in Europe (Donázar et al. 1997).

Urban and refuse dumps are sources of superabundant and predictable food for avian scavengers in Spain such that these places may be crucial for the conservation of local breeding populations of scavengers species, as well as for wintering, floating and migrating individuals from northern and central Europe (Blanco 1996). These foraging places may also represent contamination sites when associated with sources of industrial and urban residues and/or when the food upon which scavengers feed contain high levels of detrimental contaminants, pathogens, veterinary drugs, and bacteria resistant to these drugs. Human-induced environmental contamination is a relatively novel stressor, in ecological time, now negatively affecting individuals and entire ecosystems around the world due to the global dispersion of contaminants by air, water and human movements of commercial products (Iwata et al. 1993, Oehme 1991). Ubiquitous environmental contaminants are increasingly being considered as factors that influence almost any aspect of the ecology and behaviour of free-living animals through their effects on physiological pathways, especially on health state and survival, population dynamics, and the conservation status of wildlife on a global scale (Fry 1995, Newton 1998).

For the purposes of this study, we considered dumps of domestic refuse and livestock carcasses as 'contamination traps', defined as places identified as good foraging habitat by individual animals despite ultimately being dangerous foraging places in the short and long term. Specifically, refuse dumps may be a source of superabundant spatially and temporally predictable food (Donázar et al. 1997), but they may also be a source of contaminants and pathogens that animals cannot perceive as a risk. The inability or difficulty to identify these places as dangerous but, on the contrary, as good feeding habitats, could be exposing individuals at an increasing risk as a consequence of rapid conversion of refuse remains and livestock carcass from suitable to risky food for scavengers in the modern world. This may arise primarily as a consequence of rapid anthropogenic changes in the production, treatment and elimination of livestock meat and other human food remains due to rapid human population growth and concentration in urban areas. These changes have probably occurred in the last century and therefore are out of the scope of adaptive mechanisms developed in the evolutionary history of scavengers to exploit carrion (Singh and Sherikar 1999, Blount et al. 2003, DeVault et al. 2003). The concept of contamination traps is similar to that of 'evolutionary traps' defined for introduced animals and plants since species potentially 'captured' in the trap have not had time, in evolutionary terms, to adapt or have necessarily no form to develop adaptive 'escape' mechanisms in their recent evolutionary history (Schlaepfer et al. 2002). This consideration has important ramifications in conservation biology because it implied that animal species negatively affected by these traps created by human activities would not be able to escape extinction without human help depending on the degree of demographic independence of subpopulations (Sutherland 1998, Esler 1999, Reed 1999, Battin 2000).

By definition, contamination traps are attractive to individual animals that may concentrate in foraging, breeding and roosting associations in or close to these locations. In addition, in contamination traps the risk of pathogen transmission and the bioacumulation of contaminants further increase as a consequence of their preferential prolonged use, especially under an absence or scarcity of alternative good-quality habitat and food sources. Contamination traps, as defined here, may be considered a specific extension of the concept of attractive sinks under the perspective of metapopulation theory. However, the concept of contamination traps does not make *a priori* assumptions about demographic relationships between subpopulations connected into metapopulations, primarily because it applies to pathogens and contaminants that may have a variable effect on individuals depending on their health condition at different temporal and spatial scales. Thus, contamination traps may not necessarily be a demographic sink under the classic metapopulation view but rather a dangerous place which may have detrimental health effects with demographic consequences. For instance, although behavioural mechanisms such as natal philopatry may promote demographically independent subpopulations during the breeding season, co-occurrence during other part of the annual cycle can influence demography of all subpopulations similarly. Contamination traps influencing health state and mortality rates through the infection with pathogens and contamination with pollutants may contribute to indirectly determine demographic panmixia. In this case, subpopulations, in the metapopulation context, do not exist because of a lack of independence of extinction probabilities (Esler 2000).

We evaluated the responsibility of Spain in the contamination of migratory red (*Milvus milvus*) and black kites (*Milvus migrans*) with heavy metals and bacteria resistant to antibiotics acquired in urban and refuse dumps in which they feed upon livestock remains. Urban and refuse dumps attract large numbers of wintering, floating and migrating kites. These foraging places may represent contamination traps when associated with sources of industrial, urban and farming residues, implying that the food upon which kites feed or the places in which they obtain food may become detrimental for health and survival in the short and long term. We assessed the origin of wintering red kites and migrating black kites recovered in Spain, as well as, the winter destination of black kites ringed as nestling in Spain. We evaluated whether different degrees of intensification in livestock management influenced the prevalence of bacterial pathogens and the resistance of bacterial species from the digestive microflora to twelve selected antibiotics commonly used in livestock farming and human medicine. Increasing numbers of wintering, floating and migrating red and black kites from northern Europe, as well as local breeding kites foraging at refuse dumps in Spain, may be increasingly at risk from exposure to food-borne resistant pathogens and air-borne contaminants. Detrimental health effects derived from using these contamination traps in Spain may be contributing to the declining trends of local and wintering, floating and migrating kites breeding in distant areas.

Taking into account the precarious European conservation status of both black and red kites, global conservation plans for these and other endangered migratory species should consider simultaneously the impact of variable threats at different areas. Here, we assess potential influences of long-distance movements of kites on contamination with heavy metals and infection with, and transmission of, resistant pathogens between spatially distant regions corresponding to breeding, migrating and wintering areas used throughout the annual cycle. Our main focus was to highlight whether spatial and temporal co-occurrence of individuals from potentially demographically independent subpopulations may have a major impact on conservation of the whole metapopulation or species (in the case of demographic panmixia). This approach aimed to contribute to a more thorough evaluation of the impact of long-distance movements of kites on the exposure to spatially and temporally different health risks derived from human activities. We are especially interested in exploring the processes determining local abundance as influenced by abundance and movements of individuals at wider spatial and temporal scales. This intends to contribute to the increasing efforts which emphasize that conservation practice based upon establishing protected isolated areas within a matrix of rejected habitat for conservation purposes can prevent the achivement of the most successful results (Meffe and Carroll 1997, Sutherland 1998, Gaston et al. 2002). Instead, global measures involving conservation of the wider countryside through the application of universal measures of environmental care (e.g. Kyoto protocol) could be more fruitful, albeit more difficult to implement without suitable knowledge. The increasing detrimental effects of pathogens and contaminants derived from human activities are of special concern in this approach because of their potential global dispersion and generally uncontrolled and unknown effects in the emergence of new health problems worldwide, both in wildlife and humans (Daszak et al. 2000, Friend et al. 2001).

Material and Methods

Study Species

Red Kite

This species has a range mainly restricted to Europe. They are usually associated with open farmed areas with small woods or isolated trees for nesting in a wide range of lowlands and mid-mountain habitats, avoiding large forest patches, high mountain and arid areas (Carter 2001, Blanco and Montoya 2004). The species has suffered a sharp decline throughout their breeding and wintering range during the last decade. This decline has been attributed to rodenticide poisoning, human persecution, especially through illegal poisoning (Carter 2001, Blanco and Montoya 2004), and the progressive loss of many carrion dumping sites (Donázar *et al.* 1997, Blanco et al. 2006), especially after the bovine spongiform encephalopathy crisis (Tella 2001). Populations from northern and central Europe are partially migratory with wintering quarters mainly in Spain (this chapter), where wintering individuals shared habitat and range with the Spanish sedentary local population. During the winter they exploit livestock carcases and other refuse dumps as a source of superabundant and predictable food concentrating large numbers of individuals (Viñuela *et al.* 1999, Serrano 1999). In other areas in which there are no refuse dumps, they may form smaller wintering populations feeding on wild prey and carrion of un-stabled livestock grazing freely (Blanco et al. 1990; Ortega and Casado 1991; Viñuela *et al.* 1999, Blanco et al. 2006).

Black Kite

This is a widespread species throughout the temperate and tropical parts of the Old World and Australasia. European birds, *M. m. migaris,* are migratory, moving to the tropics in the winter, while the traditionally considered African subspecies *M. m. parasiticus* is resident. Recent DNA studies suggest that the yellow-billed African races of Black Kite, *M. m. parasiticus* and *M. m. aegyptius,* differ significantly from Black Kites in the Eurasian-Australasian clade and could be considered a separate species, the Yellow-billed Kite, *Milvus aegyptius* (Johnson et al. 2005). European wintering individuals shared habitat and range with the sedentary local African population. Global population trends have not been quantified, although there is evidence of a population decline (Tucker and Heath 1994). In Europe, it is locally abundant, although well monitored populations showed marked numeric fluctuations due to unknown causes as well as large to moderate decline (Blanco and Viñuela 2004, Sergio et al., 2005). The decline suffered in some areas of the Mediterranean region since the end of the nineteenth century has been mainly attributed to illegal persecution through poisoning, habitat alteration, contamination and reduction of food supplies, although global trends and major threats are generally unknown (Blanco and Viñuela 2004). European breeding populations are concentrated in Spain, France and Switzerland (Tucker and Heath 1994), while migrating individuals cross Spain to reach their wintering African quarters through the Strait of Gibraltar (Bernis 1980). European black kites are present in their breeding areas from March to early September. They nest in isolated pairs or in small loose colonies in a variety of open, ecotone and heterogeneous habitats, primarily wetlands in Europe, and are often observed in and around outback towns in Spain and Africa. This is a highly gregarious species, forming small groups of breeding individuals and large local concentrations of floaters during the breeding season. Large and even huge flocks of many thousands of birds are usually attracted to refuse dumps during migration (Blanco 1994, this chapter), also communally roosting in their surroundings, both in Spain and Africa, and especially during grasshopper plagues in the African Sahel. Black kites are opportunistic predators of small-sized prey, primarily young, injured and sick individuals, as well as facultative scavengers of the same wild prey and livestock carcasses and refuse from municipal dumps (Blanco 1997). Therefore, they are especially sensitive to contamination from toxic residues from agriculture, industry, urban settlements and livestock farming operations.

Study Areas and Study Populations

Red Kite

Faecal microflora and bacterial antibiotic resistance of wintering red kites was obtained from communally roosting sites in three areas in Spain: Segovia and Central Madrid provinces in Central Spain and Extremadura (Cáceres Province) in western Spain.

a) Segovia. Red kites from the selected communal roosts (Torredondo and Escobar de Polendos) in this province form part of a large wintering population exploiting livestock carcasses (especially swine) dumped at sites near rearing farms (Sunyer and Viñuela 1994; Blanco et al. 2006). This area has traditionally been used by large numbers of wintering red kites gathering in more than 20 known communal roosts and, therefore, it is considered to be one of the main strongholds for wintering red kites from central Europe (Viñuela et al. 1999). This population suffered a dramatic decline from 1994 (5500-5800 individuals, Viñuela et al. 1999) to 2000 (1530 individuals, Contreras 2001) that continues today (authors'unpublished data) as part of the decline of the breeding European populations wintering in the Iberian Peninsula (Thiollay 2001). Segovia Province also had one of the highest breeding densities of red kites recorded in Spain during the national census of 1994 (Viñuela et al. 1999) but since then, partial censuses have confirmed a major population decline (Blanco and Montoya 2004, this chapter). Most nestlings sampled belong to the population in this area. Data on adult wintering individuals and nestlings was mainly obtained from the 'Piedemonte' region (oak forests and grasslands in the bottom of Sierra de Guadarrama mountains), the 'Llano' region (extensive cultivations at low altitude and flat terrain), and 'Tierra de Pinares' region (pine forests at low altitudes and flat terrain) (see Blanco and Montoya 2004 for details). The sampled communal roosts were located in poplars (*Populus X cultivar*) and included between 60 and 150 individuals.

b) Central Madrid Province. Red Kites studied in this province were mainly located in the municipal districts of Alalpardo-Valdeolmos and its surroundings. This constitute the only wintering group in this area, formed by about 110-170 individuals depending on year (G. Blanco unpublished data). They feed mostly on European wild rabbits *Oryctolagus cuniculus*, which are present in large numbers (Blanco et al. 2006). This area comprises extensive open farmland devoted to dry cereal crops and scrubland promoted for game rearing and hunting. This area is surrounded by an increasingly urbanized area spreading from the city of Madrid, the largest human concentration in Spain. The sampled communal roost was located at a poplar (*Populus X cultivar*) plantation. No breeding pair nested in this area.

c) Extremadura. This is an extensive area where wintering red kites have been recently estimated at about 6400 individuals gathering in about 30 communal roosts, which represent about 40% less than the population censused ten years previously (Junta de Extremadura 2005). The wintering population shared the area with a breeding population of about 200 pairs in 2005, which represents a decline of about 80% in ten years (Junta de Extremadura 2005). Wintering red kites in this area feed upon a combination of prey depending on the location of the four sampled communal roosts: 'Torrecilla de la Tiesa' roost site (wild prey, primarily rodents and birds, and sheep carrion), 'Escurial' roost site (prey associated with irrigated crops and grasslands, mostly waterbirds, rats and some sheep carrion), 'Valdesalor' roost site (mostly sheep carrion and voles) and 'Navalmoral de la Mata' roost site (mostly urban refuse from a large municipal dump and sheep carrion). Only three nestling were sampled from this area in 2005. The habitat in which these roosts were located was mainly composed of oak (*Quercus*) dehesas, grassland, pastures and irrigated crops. The sampled communal roosts were located in poplar (*Populus X cultivar*) and *Eucalyptus* plantations, as well as in oaks (*Quercus rotundifolia*), and gathered between < 50 and > 500 individuals.

Black Kite

Faecal microflora and bacterial antibiotic resistance was obtained from communal roosts used by floating non-breeding individuals during the breeding season in two areas (Extremadura and southeast Madrid). In these two areas with the addition of the Segovia Province we also sampled nestlings. In addition we sampled faecal bacteria and associated antibiotic resistance from wintering individuals and local yellow-billed kites in the Sahel (Mauritania and Senegal).

a) Segovia. Black kites in this province are locally abundant, although their numbers showed great annual fluctuations due to unknown causes with a slight declining trend. The habitat used overlaps almost completely with that used by red kites (see above). Population size remains unknown but it can be subjectively estimated at several hundred pairs. There exist large (> 100 individuals) communal roosts used by the floating non-breeding population during the breeding season, associated with large municipal dumps of domestic refuse as well as with livestock carcass dumps. Migrating kites use the same communal roosts during the summer. However, no data was collected on floaters due to time limitations. Data from nestlings was mostly obtained from nests located in the Natural Park de las Hoces del río Duratón, which is located in the 'Tierra de Pinares' region (see above). The diet of nestlings was mostly composed of piglet and poultry carrion, and a lesser proportion of wild prey, primarily birds.
b) Southeast Madrid. The study population in this region is located in a protected area within the Regional Park of southeastern Madrid, central Spain (mainly Rivas Vaciamadrid and San Martín de la Vega districts). This area is mostly devoted to agriculture and gravel extraction. In the study area, breeding black kites nest in riparian forests along the Manzanares and Jarama rivers and feed mostly on European rabbits, *Oryctolagus cuniculus* (Blanco 1997). The study area is close to a large rubbish dump where tons of domestic refuse are deposited each day and where black kites forage regularly (Blanco 1994), especially floaters and migrants (Blanco 1997). The rubbish dump is close to a municipal solid-waste incinerator that began operating in 1995. The black kite population was composed of about 50-90 breeding pairs depending on the year, as well as of an increasing number of floaters and migrating individuals exploiting refuse in the large municipal dumps of the city of Madrid and its surroundings (Blanco 1994, 1997, this chapter). We sampled faecal microflora of floaters at the communal roost (Rivas-Vaciamadrid), which was located in the riparian forest mostly composed of elms (*Ulmus minor*) along the Manzanares river. Nestlings were sampled from nests located in the same riparian forest along the Manzanares and Jarama rivers. There are no wintering red kites in this area, and only one breeding pair of red kites has recently settled in the area.
c) Extremadura. Data from this area was obtained from a single communal roost used by floaters (Cáceres Province) located in a *Eucalyptus* plantation. Black kites in this region have been estimated in about 2500 breeding pairs and about 4700 floaters gathered in 18 communal roosts during the breeding season of 2005 (Junta de Extremadura 2005). No previous objective estimation of the population exist and hence it is not possible to assess population trends accurately (Blanco and Viñuela 2004). Breeding pairs in this area primarily feed upon wild prey, especially pigeons

and corvids among other birds, fish, reptiles, American crabs (*Procambarus clarkii*) and sheep carrion. Floaters were associated with large accumulations of refuse disposed at municipal dumps, farm surroundings and slaughterhouses and hence their diet is based on remains of livestock carrion rather than wild prey (Blanco 1997). Habitat overlapped with that used by red kites (see above), mainly dehesas of *Quercus* species.

d) Sahel. Wintering black kites from Europe (*Milvus migrans*) and local breeding kites (*Milvus aegptius*) usually shared habitat and foraging grounds in the Sahel region during the winter. Therefore, faecal samples collected cannot be assigned to individual species and for that reason were pooled as Sahel kites. Samples were collected in Mauritania (Kiffa, Aioum el Atrous and Gabou) and Senegal (Dakar) during October 2005. Samples collected in Kiffa correspond to kites foraging in piles of livestock remains from a slaughterhouse. Habitat surrounding this locality was composed of open grassland with sparse shrubs. In this locality the proportions of each kite species were 29, 38 and 33% (n = 104), respectively, for *M. migrans*, *M. aegyptius* and juveniles presumably of both species that were unassigned due to difficulties in field identification (F. Hiraldo, pers. com.). Samples from Aioum el Atrous were also collected in dumps from a slaughterhouse outside the village, which was surrounded by sandy desert and dunes. In this locality the proportions of each kite species were 49, 20 and 31% (n = 67), respectively, for *M. migrans*, *M. aegyptius* and juveniles. Samples from Gabou were collected from kites foraging in stony desert. No data about proportions of each kite species were recorded. Samples from Dakar were collected in the centre of the city, where most kites observed were *M. aegyptius* (429 of 522 individual kites, 82%). The remainder of observed kites correspond to *M. migrans* (<1%) and unidentified juveniles and subadults (17%).

Potential Contamination Traps in the Study Areas

a) Segovia. In this province two factors arise promoting the possible existence of contamination traps for avian scavengers: the large populations of these species (Martínez et al. 2004, Blanco et al. 2006) and a long livestock farming heritage providing abundant livestock carrion for scavengers. Avian scavengers have long exploited the high food availability that livestock rearing has provided, especially carrion from unstabled livestock grazing freely, mostly sheep and cows. However, recent changes in farming systems and health regulations have made carrion of unstabled livestock scarce or inaccessible since the bovine spongiform encephalopathy crisis (Donázar et al. 1997, Tella 2001), thereby reducing the potential availability of this historic food source for scavengers. This recent scarcity of carrion from free-range livestock may be promoting the increased use by kites and other scavengers of dumps containing carrion from stabled livestock reared under intensive conditions, including those from confined animal feeding operations in which antimicrobials are intensively used, especially fattening swine and poultry (Blanco et al. 2006). In this sense, it is noteworthy that Segovia has the highest concentration and number of fattening pigs in Spain. Although livestock carcass dumps provide a superabundant and predictable food supply for avian scavengers, these places may be a source of pathogens associated with rotten meat and other agents with detrimental effects reducing health and hence, survival prospects (Blanco et al. 2006). Specifically, the use of antimicrobial agents and other drugs in livestock may have direct negative effects on the health and survival of scavengers if ingested as toxic residues (Friend and Franson 1999) or may have indirect negative effects by facilitating the development of microbes with increased capacity to cause disease in these scavengers. Withdrawal times for antimicrobials are intended to prevent harmful drug residuals in food that humans consume (McEwen and Fedorka-Cray 2002). However, these waiting periods do not apply to carcass and other waste disposal in dumps. Indirect effects on vulture health and survival derived from consuming carrion of intensively medicated livestock may include the acquisition of antimicrobial-resistant bacteria (McEwen and Fedorka-Cray 2002) and the alteration of the normal protective flora by acquisition of a transient flora including pathogenic bacteria (Winsor et al. 1981, Rodrigues et al. 2003, Schneitz 2005, Blanco et al. 2006). These circumstances may create a contamination trap because the high availability of carrion from stabled livestock may be attracting large number of kites that may then suffer the consequences derived from these detrimental pathogens and resistant bacteria. Because the role of pathogen acquisition from livestock carrion in Segovia has been demonstrated previously (Blanco et al. 2006), we focused on the role of refuse dumps containing carrion from stabled livestock in the acquisition of antibiotic resistant bacteria to further complement our assessment of possible contamination traps.

b) Southeast Madrid. The existence of a large rubbish dump in which tons of domestic refuse is deposited each day in this area have attracted increasing numbers of migrating and floating black kites (this chapter). The rubbish dump is close to a municipal solid-waste incinerator that began operating in 1995. This incinerator has lead to a long-standing conflict between conservationists and management agencies from the regional government with great social repercussion given its implications in human health and environmental contamination. This conflict intensified especially after a functional failure of the incinerator in December 2000, when the incinerator expelled into the atmosphere several hundred kilograms of heavy metals, also exceeding the legal level of dioxins and furans emissions. In fact, the area is highly polluted with organochlorines, PCBs, dioxins and heavy metals from agricultural, industrial, urban and incineration activities (Jiménez et al. 1996, 2000). A previous study assessed the relationships between the distance from black kite nests to the incinerator and metal concentrations in nestling blood and failed eggs. Results showed that lead levels increased as the distance to the incinerator decreased, suggesting a generalised exposure to this contaminant in the form of dust from fly ash emitted to the air and uptake via inhalation (Blanco et al. 2003) with detrimental effects on health (Blanco et al. 2004). Additional contamination sources are the dump itself and the highly contaminated rivers and surrounding irrigated lands. These highly contaminated rivers may also contribute to create the contamination trap due to the close association of nesting and foraging black kites with rivers and other wetlands. The increasing number of floating and migrating birds using the dumps in the area may be contaminated with heavy metals and other pollutants creating contamination traps for floating individuals from other populations in Spain and central Europe, which may eventually be recruited as breeders into the local population. This potential trap may also negatively affect the local population, through detrimental effects on health, especially of nestlings (Blanco et al. 2004, authors' unpublished data).

Origin, Destination and Phenology of Wintering, Migrating and Floating Kites

Recovery data and observations in Spain of foreign red (n = 358) and black kites (n = 110) marked as nestlings in central Europe from 1905 to 2005 were obtained from the Ringing Office of the Spanish Ministry of the Environment. We used data from recoveries of metal rings, primarily from dead birds, and observations of individuals banded with wing tags and colour rings read using spotting scopes, primarily from live birds. In addition, we analyzed the recoveries in Africa (n = 52) of black kites marked as nestlings in Spain from 1905 to 2004 (n = 8652), distinguishing between those recovered in the north of the Sahara (pre-Saharan, including those recovered in the Sahara) and in the south of the Sahara (trans-Saharan). We only considered those recoveries with precise location and date (< one week of error), such that sample size varies somewhat between analyses because these data were lacking for several recoveries.

Phenology of foreign red kite recovered in Spain was established in three periods: post-nuptial migration-wintering, floating, and pre-nuptial migration-wintering. This distinction was made considering that most foreign red kites recovered in Spain form part of the wintering population compared to very few migrating individuals crossing the strait of Gibraltar to pass the winter in north Africa (Bernis 1980). There is no trans-Saharan recovery of red kites ringed as nestlings in Spain; identification in a previously published old recovery in Ghana (Díaz et al. 1996) was probably confused with black kites at the time of ringing. These periods were thus established considering a wintering peak in December (corresponding to the end of arrivals from Europe) and January (corresponding to the start of departures to Europe). Therefore, the post-nuptial migration-wintering included wintering months (November-December) and those corresponding to arrivals (August-October), while the pre-nuptial migration-wintering included January and months of departure to the breeding range in central Europe (February-April). The floating period was then restricted to May-July, when breeding pairs were fully involved in reproduction both in Spain and central Europe. In the case of black kites, we considered as foreign floaters those individuals ringed as nestlings in Europe (excluding Spain) and recovered in Spain during the breeding season, restricted to May and June to avoid including late migrants during their pre-nuptial arrival from wintering quarters (Bernis 1980). The remaining recoveries of foreign nestlings from Europe in Spain were considered as belonging to migrating individuals both during the spring (pre-nuptial migration: February- April) and the summer-autumn (post-nuptial migration: July-November).

Previously, a restriction in the area used in Spain during winter by red kites was reported, primarily in the southern half of the Iberian peninsula (Southern Plateau and Guadalquivir valley), based on recoveries of nestlings ringed in central and northern Europe (De Juana et al. 1988). This restriction was suggested being due to the irruption of Common Voles (*Microtus arvalis*) in the Northern Plateau since the latest seventies (Sunyer and Viñuela

1994). We assessed whether this historic change in the winter distribution continues nowadays by analyzing the distribution of recoveries in three time periods: 1905-1970 (n = 68), 1971-1990 (n = 89) and 1991-2005 (n = 104). We evaluated the role of the distribution of food for wintering kites by considering the relative abundance of two main food kinds: natural prey (primarily common voles) and livestock remains disposed in carcass dumps (Blanco et al. 2006). Common voles are distributed in the Northern Plateau but absent in the Southern Plateau and Ebro Valley (González-Esteban and Villate 2002) while livestock remains from carcass dumps are especially abundant and increasing in Northern Plateau and Ebro Valley due to growing livestock operations, primarily stabled livestock (Contreras 2001, Soto-Largo 2002, authors' unpublished data). We thus specifically tested the hypothesis that recent changes and current distribution of wintering red kites in Spain were determined by their increasing attraction to livestock refuse dumps (Blanco et al. 2006). For this analyses we considered three main wintering areas (see Figure 1): Northern Plateau (primarily Castilla-León), Ebro Valley (Navarra, Aragón and Cataluña) and Southern Plateau (Castilla-La Mancha, Madrid, Extremadura and Andalucia).

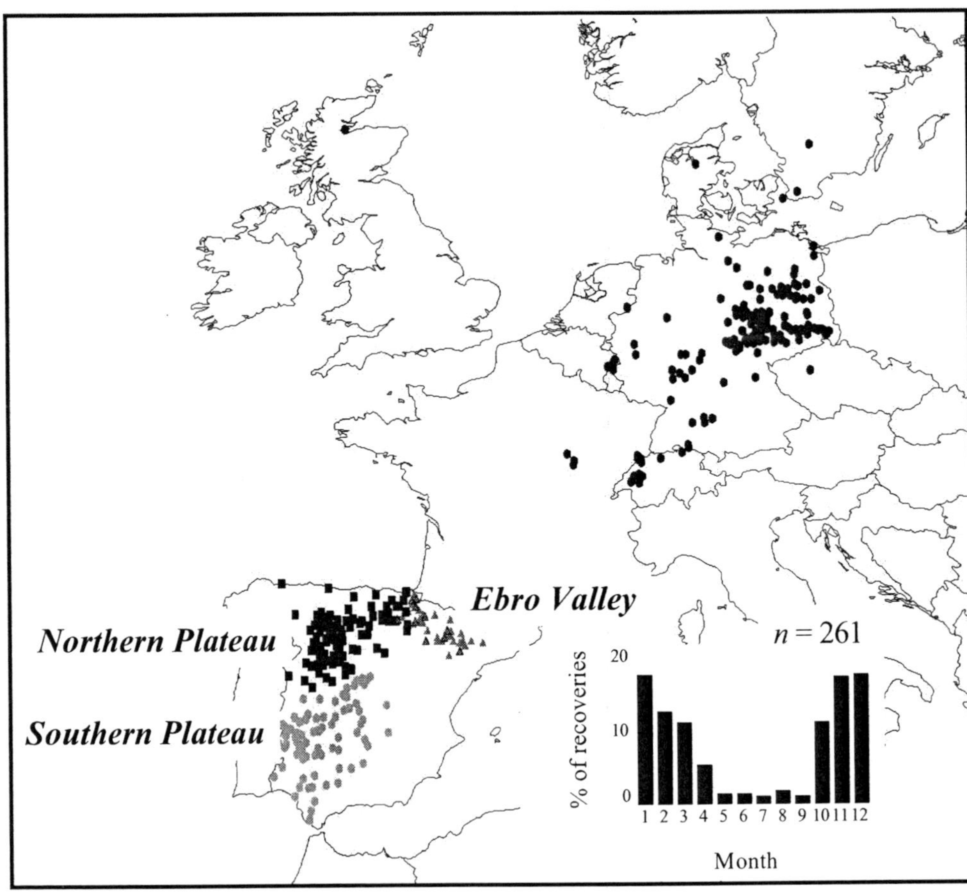

Figure 1. Location of recoveries in Spain (black squares: Northern Plateau, grey points: Southern Plateau, grey triangles: Ebro Valley) of wintering red kites ringed as nestlings in central and northern Europe (black points). Bars represent the percentage of recoveries in each month.

Population Monitoring and General Fieldwork

Breeding Pairs and Nestlings

In Segovia study area, monitoring was conducted for the breeding populations of red and black kites inhabiting the Natural Park of Las Hoces del Río Duratón from 1991 to 2006. In southeastern Madrid, complete population monitoring of black kites was conducted in 1984-1988, 1995-1997 and 2001-2006. In addition, during 1989, 1990 and 1998-2000 the number of breeding pairs was only monitored in the northern part of the study area, corresponding to the riparian forest of the Jarama River before its confluence with the Manzanares River (North Jarama River, hereafter). This monitoring was mainly conducted as part of a long-term monitoring program aimed to assess population trends and threats to these species.

All breeding pairs in the study areas were monitored using telescopes to avoid disturbance, and their nests were located. The exact position of nests (latitude and longitude) was recorded by using Global Positioning System and later mapped on geo-referenced aerial photographs by using ArcView software.

Nests were reached by climbing before and during laying to determine laying dates in order to subsequently access the nests again to sample nestlings at an adequate age (about 30 days after hatching). Laying date was defined as the date the first egg in the clutch was laid, which was determined from nests found during laying by backdating using a laying interval of two days. For nests accesed during incubation and after clutch completion, laying dates were calculated by combining both observational data of the beginning of incubation and by estimating the age of the chicks from equations relating the size of the eighth primary feather with the age of chicks of known age. Nestlings were banded with metal bands and with a unique combination of PVC bands. They were weighed to the nearest 1g and measured for several morphological variables about 30 days after hatching (just prior to nest departure). Nests were monitored until nestlings fledged to determine breeding success and number of nestling fledged.

About 3 ml of blood was extracted with heparinized syringes from the brachial vein of each chick, and blood was then transferred to Eppendorff vials. Blood samples were transported in a cooler to our laboratories on the same day of collection, and were frozen at -20C or -70C until analysis. Blood was used both metal analyses (see below).

Communally Roosting Kites

Wintering red kites were counted in the communal roosts at sunset during the study years to assess possible trends in the number of individuals gathered. Counts were conducted using telescopes from vantage points to avoid disturbance of the birds. Transects across the roost surroundings and observations of the direction of arriving and departing birds from the roosts were conducted in an attempt to identify the places and sources of food used by kites during the day. Also, the diet of kites was assessed by determining the content of regurgitated pellets collected beneath the trees used as communal roost sites.

Floater and migrating black kites were counted in the population located in Southeast Madrid (see Blanco 1994). About 2 hr before sunset we began to observe and count the kites perched on the roost trees. Later we counted the kites entering the communal roosts from the dumps. Observations were conducted from several vantage points located between the roost and the dumps, and lasted until all the kites were settled on the trees. The location of the roost was the same across years, although an associated and closely located roost was occupied in several years at the confluence of Jarama and Manzanares rivers, especially by migrating birds. The kites foraged mainly in the dumps of Valdemingomez and Pinto. Birds from the Valdemingomez dump arrived by crossing the Manzanres river from the north, while kites from the Pinto dump arrived following the course of the same river from the west. The number of floaters was determined during the breeding season (May-15 July) from 1990 to 2006. At this time, the communal roost contained mainly non-breeding birds, both subadults and birds with definitive plumage (Blanco 1994). Because the number of floaters may vary depending of census data during the breeding season, probably due to both annual variations and the highly mobile nature of floaters, it is generally difficult to determine the size of the floating population with accuracy. Therefore, several counts were conducted in the roosts during the breeding season, and the mean roost size was used in the analyses.

Several floaters from the population of Southeast Madrid were captured during the breeding season of 2006 with mist-nets mounted at the communal roost to determine the concentration of circulating lead (see below). A sample of blood was collected and treated as described above for nestlings. Migrating kites were counted using the same methodology during the first or second week of August, when the maximum abundance peak was reached (Blanco 1994).

Microbiological Sampling

Faecal microflora of communally roosting individuals was assessed by sampling fresh faeces found beneath the trees used by communally roosting kites. Faeces were sampled with sterile microbiological swabs that were subsequently inserted into tubes containing Amies transport medium. Sampling was conducted early in the morning, shortly after the kites left the roost. Samples were transported to the laboratory in a chilled container on the same day as collection and were processed within one to two hours of arrival.

Samples of communally roosting red kites were obtained from wintering individuals (January-March 2004-2006); a total of 80 samples were collected in two communal roosts located in Segovia, 33 samples in a single roost located in Madrid (see Blanco et al. 2006) and 72 samples in four different communal roosts located in Extremadura region, in Cáceres Province (Torrecillas de la Tiesa n = 12, Escurial n = 21, Navalmoral de la Mata n = 18, and Valdesalor n = 21).

Samples of adult and subadult black kites corresponded to communally roosting individuals forming part of the non-breeding, non-territorial floating segment during June-July; a total of 36 faecal samples were collected in a communal roost located in Cáceres, Extremadura in June 2005 and another 57 samples (19 and 38 in June 2005 and 2006 respectively) in a roost located in south east Madrid (Rivas-Vaciamadrid). Samples from the Sahel were mostly collected at livestock carcass dumps in the locations of Kiffa (n = 5). Aioum et Atrous (n = 11) and Gabou (n = 4) in Mauritania and in the city of Dakar (n = 3) in Senegal.

Microflora composition of red and black kites nestlings was assessed by sampling cloacal bacteria. Nests were accessed by climbing during May-June (red kites) and June-July (black kites) when nestlings were about 25-35 days old. Cloacal microflora was sampled with sterile microbiological swabs that were subsequently inserted into tubes containing Amies transport medium. Samples were transported to the laboratory within four to five days after collection and processed within one to two hours following their arrival at the laboratory. Overall, we sampled red kite nestlings in Segovia during 2004 (n = 64) and 2005 (n = 17), and in Extremadura (Cáceres Province) during 2005 (n = 3). Black kite nestlings were sampled from nests located in southeast Madrid in 2004 (n = 53) and 2005 (n = 20), Segovia in 2004 (n = 23) and 2005 (n = 8) and in Extremadura during 2005 (n = 37).

Microflora Culture and Identification

For the Enterobacteriae, samples were cultured on 5% sheep blood agar, chocolate agar, and McConkey agar, to avoid *Proteus sp* overgrowth. Plates were incubated at 37°C, in normal atmospheric, microaerophilic (10% CO_2) and anaerobic conditions respectively, for 24 hr All suspect colonies were then subcultured on appropriate medium and identified as species by using multi-substrate identification (API 20 E; Bio Merieux). Thermolabile species such as *Campylobacter sp.* could not be isolated by sampling faeces from communally roosting birds. Isolates of *Campylobacter* from cloacal samples of nestlings were identified at species level by flagellin gene A PCR/restriction fragment length polymorphism and pulsed-field gel electrophoresis.

For *Salmonella*, samples were cultured using two selective media: SM ID agar (BioMerieux) and xylose-lysine-deoxycholate (XLD) agar (Difco). A selective enrichment procedure was performed on cloacal samples. Pre-enrichment was carried out in peptone broth (Difco), and then 1 ml was cultured onto selenite brilliant green broth (Difco) and subcultured on solid media SM ID agar and XLD agar. Plates were incubated at 37°C, either in normal atmospheric and microaerophilic (10% CO_2) conditions respectively, for 24 hr except for the peptone broth, which was incubated at 37°C for 16 hr. Confirmation of the identification of *Salmonella* species were performed using standard methods. *Salmonella* serological typing was performed at the Laboratorio Central Veterinario (Algete, Madrid, Spain).

Antibiotic Susceptibility Testing

Antibiotic susceptibility and resistance testing for all the isolates were performed by disk diffusion assay. For cloacal samples of nestlings we selected *Salmonella* sp., *Campylobacrer jejuni*, *Clostridium* sp., *Serratia marcescens*, *Enterobacter cloacae*, *Enterobacter* sp., *Escherichia coli*, *Enterococcus faecalis*, *Enterococcus faecium*, *Staphylococcus epidermidis* and *Corynebacterium* sp. because of their potential pathogenic effects and high prevalence (especially *E. colli*, *E. faecalis*, *E. faecium*) allowing us to assess antibiotic susceptibility in a greater number of individuals. For faecal samples of full-grown individuals of both kite species, we selected *Salmonella* sp., *Klebsiella* sp., *Enterobacter cloacae*, *Escherichia coli*, *Pseudomonas aeruginosa*, *Enterococcus faecalis*, *Enterococcus faecium*, *Staphylococcus epidermidis*, *Bacillus* sp. and *Corynebacterium* sp. Differences between nestlings and full-grown individuals in bacterial species selected were due to the nature of samples (cloacal and faecal samples respectively) allowing or not the isolation of particular bacterial species from their intestinal microflora. Inoculates of the microorganisms were prepared from 24hr cultures. Bacterial suspensions in normal saline were adjusted to the Mc Farland n° 1 turbidity standard. In all cases, blood agar was used, except with Salmonella isolates which were cultured in SM ID agar.

We looked for susceptibility to twelve antibiotics chosen because of their use in humans and farm animals, including ciprofloxacin (CP), gentamicin (GN), enrofloxacin (EN), marbofloxacin (MA), amoxicillin (AX), amoxicillin-clavulanic (AC), ampicillin (AP), cephalexin (CE), lincomycin (LI), neomycin (NE), oxitetracycline (OT), and penicillin G procain (PG). Each bacterial isolate was classified as either susceptible, intermediate or resistant, depending on the growth inhibition diameter although ultimately intermediate isolates were considered as susceptible to be conservative and in order to simplify the analyses as less than 1% of isolates were intermediate. Disk contents and growth inhibition zones were in accordance with criteria set by the National Committee for Clinical Laboratory Standards for bacteria isolated from animals. All antibiotic resistance tests were performed for all bacterial isolates and pooled to compare the profile of antibiotic resistance between areas.

Analysis of Lead

Previously we found that concentrations of essential metals in nestling blood showed a normal distribution and mean values were similar to those recorded for birds in uncontaminated areas (Blanco et al. 2003). However, without a metal speciation analysis, it was not possible to determine what proportion of these elements was due to contamination. Levels of non-essential lead were normally distributed, values ranging from background levels in the environment to those causing sub-lethal effects in raptors (Blanco et al. 2003). This suggests a generalised exposure to this contaminant and potential toxicity considering that lead at very low concentrations can negatively affect metabolism, especially during growth (Burger and Gochfeld 1997, Spahn and Sherry 1999). Lead intoxication can be occurring at sub-clinical levels, affecting different aspects of physiological function long before it is apparent from altered behaviour in affected animals (Pain 1996). The potential for problems due to lead poisoning may even increase after the birds fledge due to their intensive use of the landfill before migrating (Blanco 1997).

Because previous analyses showed that lead was the major contaminant in black kites, we focused on this element to assess the potential impact of the dumps of the city of Madrid creating a contamination trap. Analyses of lead were carried out by Atomic Absorption Spectrometry using a Perkin Elmer longitudinal AC Zeeman (AAnalyst 600) AAS equipped with a transversely heated graphite atomizer (Perkin Elmer Hispania, S.A., Madrid, Spain). The detection limit of the methodology was 0.17 ng/ml. All specimens were analysed in batches with method blanks, known standards, and reference material, DORM-2 (Dogfish liver) from NRCC, and Seronorm®, Trace Element, Whole Blood 2 from Sero AS, Asker, Norway. In 2006, determination of lead concentration from blood samples was conducted directly in the field using a portable leadcare (ESA. Inc.) just after bleeding the nestlings in order to improve logistics and reduce costs. The detection limit of the methodology was 0.1 ppb.

RESULTS

Origin and Destination of Wintering, Migrating and Floating Kites

Red Kite

Red kites ringed as nestlings in central and northern Europe and recovered in Spain during the winter were mainly from Germany (Table 1, Figure 1). Phenology of recovery data showed a peak during the winter, primarily in December and January (Figure 1), although there were recoveries in all months (Figure 1), which suggests that a proportion of red kites from central Europe may spend the breeding season as floaters in Spain. According to age, red kites in their second year (first wintering) were recovered in Spain mostly during late winter and early spring, when they may be returning to their natal areas, although a proportion of these remained as floaters in Spain during the breeding season (Figure 2). The proportion of recoveries during the late winter and pre-nuptial migration decreased with age, while those during post-nuptial migration and early winter increased (Figure 2).

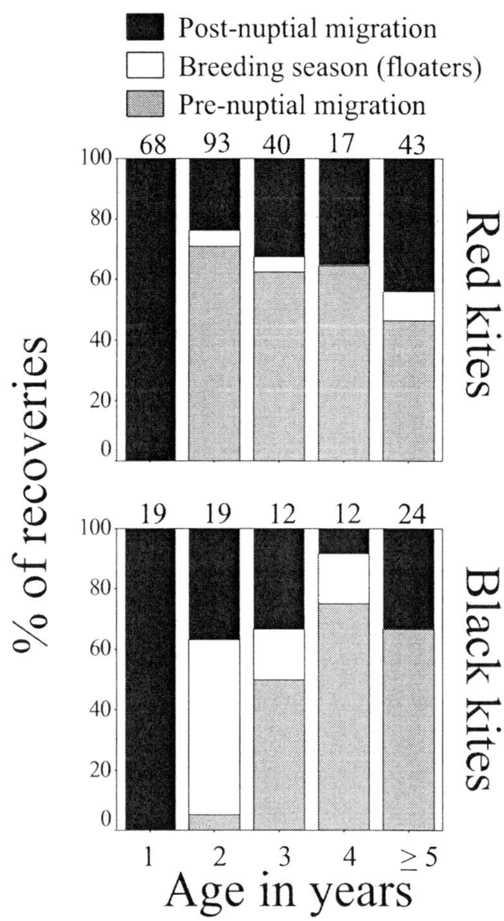

Figure 2. Age distribution according to each seasonal period of recoveries in Spain of foreign red and black kites ringed as nestlings in central and northern Europe. Sample size is shown above bars.

This suggests a progressive advance in the departure from wintering areas to breeding grounds with age or a progressive advance in the dates of arrival to the wintering areas with age, but most likely both patterns simultaneously (Figure 2).

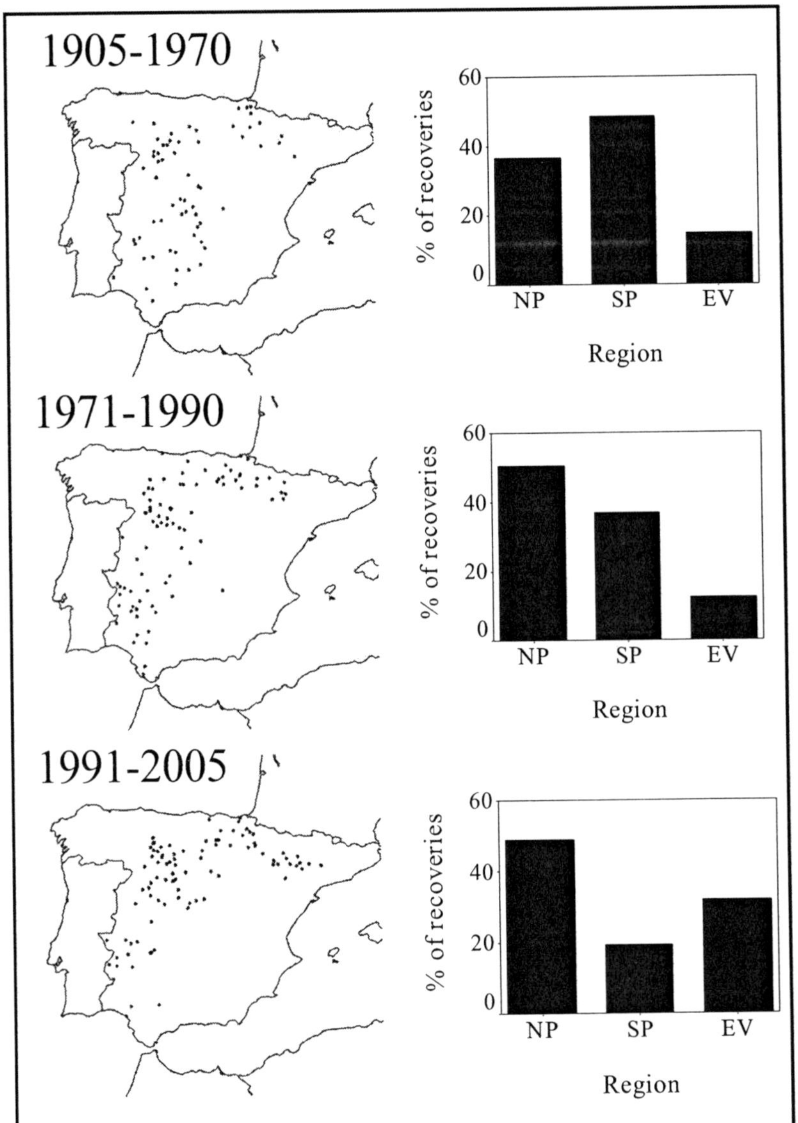

Figure 3. Location of recoveries in Spain of wintering red kites ringed as nestlings in central and northern Europe according to three periods: 1905-1970 (n = 68), 1971-1990 (n = 89) and 1991-2005 (n = 104). Bars represent the percentage of recoveries according to wintering regions (NP: Northern Plateau, SP: Southern Plateau, EV: Ebro Valley) in each period.

We confirmed that the previously reported historic restriction of the wintering area of foreign red kites nowadays continues (Figure 3), especially in the Southern Plateau, including Extremadura and Andalucia, as indicated the decreasing proportion of recoveries with time, especially during the last decade (Figure 3). On the contrary, the proportion of recoveries increased until 1970 in the Northern Plateau and from 1990 onwards in the Ebro Valley

(Figure 3). These differences between periods in the proportion of recoveries were statistically significant ($\chi^2_4 = 23.550$, $P < 0.0001$).

Black Kite

Black kites recovered in Spain during their pre-nuptial and post-nuptial migration were mostly originally ringed in Switzerland and Germany (Table 1, Figure 4). Floaters recovered in May and June were originally ringed in these countries as well (Table 1, Figure 5). The phenology of recoveries clearly showed a pre-nuptial migratory peak in March-April and a post-nuptial migratory peak in August-September (Figure 4). In addition, a considerable proportion of recoveries corresponds to foreign black kites outside the migratory period (May-June, Figures 4 and 5), which demonstrates that a fraction of floaters present in Spain during the breeding season were originally from central Europe.

Table 1. Numbers (% in brackets) of foreign recoveries in Spain of wintering and floater red kites and migrating and floater black kites ringed as nestlings by origin country

	Red kites		Black kites	
Country of origin	floaters	wintering	floaters	migrants
Sweden	-	5 (1.4)	-	-
Denmark	-	1 (0.3)	-	-
Great Britain	1 (9.1)	-	-	-
Poland	-	1 (0.3)	1 (6.3)	1 (1.1)
Czech Republic	-	1 (0.3)	-	-
Germany	9 (81.8)	301 (86.7)	4 (25.0)	26 (27.7)
Belgium	-	8 (2.3)	-	-
Switzerland	1 (9.1)	25 (7.2)	11 (68.7)	59 (62.8)
France	-	5 (1.4)	-	7 (7.4)
Portugal	-	-	-	1 (1.1)
Total	11	347	16	94

Monthly distribution by age classes of foreign recoveries in Spain of black kites ringed as nestlings in central Europe showed that juveniles (first year) were recovered during the months just after fledging, during their first post-nuptial migration to Africa (Figure 2). Birds in their second year of life were frequently recovered in May and June, in the middle of the breeding season of local breeding pairs. These individuals may form part of the floating population and they probably do not return to the natal areas in the same season. Once two-year old individuals have realized their second post-nuptial migration to Africa, the proportion of their recoveries during this period decreased and almost all subsequent recoveries occurred during the pre-nuptial migration (Figure 2). From three years old onwards, most recoveries also occurred during the pre-nuptial migration. Overall, the date of recovery during the pre-nuptial migration decreased with age (Figure 6), such that individuals of two and three years of age were recovered on average on dates (May 31 and April 22, respectively) that may be considered outside the migratory period but within the breeding season of local breeding pairs.

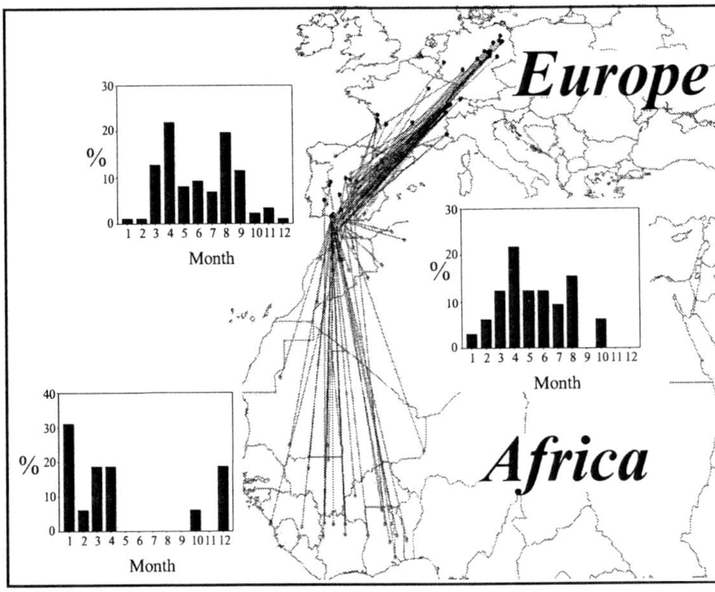

Figure 4. Location of recoveries in Spain (grey points in Spain) of black kites ringed as nestlings in central and northern Europe (black points), and recoveries in Africa (grey points in Africa) of black kites ringed as nestlings in Spain (black points in Spain). All recoveries were represented except those for foreign floaters occurred during the breeding season (May-June). Bars represent the percentage of recoveries in each month in Spain (n = 86), as well as pre-Saharan recoveries in Africa (n = 32) and trans-Saharan recoveries (n = 16), including recoveries of floaters during the breeding season.

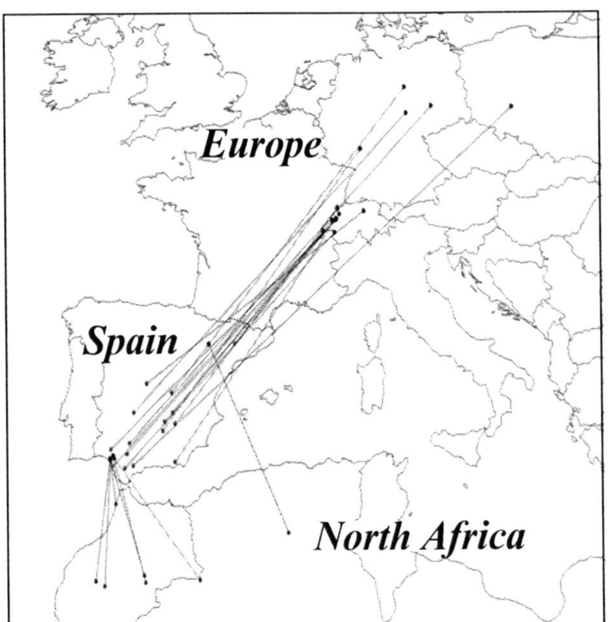

Figure 5. Locations of black kites ringed as nestlings in central and northern Europe (black points) and recovered as floaters during the breeding season (May-June) in Spain (grey points in Spain), as well as recoveries as floaters in Africa (grey points in Africa) of black kites ringed as nestlings in Spain (black points in Spain).

Birds of four or more years of age were recovered on average on dates that fit well with the pre-nuptial migration to central Europe (April 8 and March 26 respectively, Figure 6). This indicates that foreign black kites forming part of the floating population during the breeding season in Spain were primarily composed of individuals of two years of age, with a lesser proportion of individuals three and four years old.

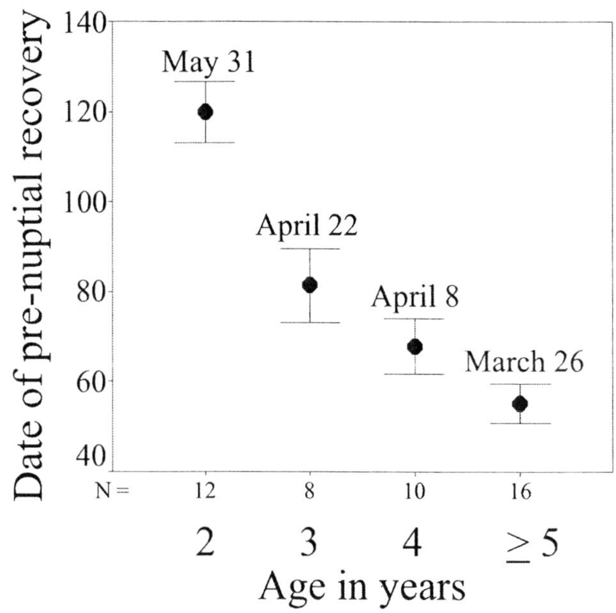

Figure 6. Mean ± SE date of pre-nuptial recovery in Spain of foreign black kites ringed as nestling in nothern and central Europe according to age. Recovery date (1 = 1st of February).

The wintering area of the black kites ringed as nestlings in Spain appears localized in western Africa primarily south of 15° latitude and in a longitudinal fringe from Guinea to Benin (Table 1, Figure 4). Trans-Saharan recoveries occurred during wintering with no recovery between May and September, although several recoveries occurred in August in the Sahara and the Sahel in Mauritania (Figure 4). African areas in the north of the Sahara, especially Morocco, were used during migration as indicated by the two main peaks of recoveries corresponding to pre-nuptial (April) and post-nuptial migration (August) (Figure 4). In addition, a considerable proportion of black kites ringed as nestlings in Spain was recovered in May-June in Morocco and Algeria (Figure 5). This indicates that, as in the case of foreign floaters recovered in Spain, a proportion of individual black kites from Spain spent the breeding season as floaters in areas distant to their natal areas. However, most individuals apparently attempt to cross the Sahara to reach their wintering quarters in western Africa in the south of the Sahel (Figure 4), as indicated by the single pre-Saharan recovery during the winter (a bird in its first-year in January in Morocco). In fact, trans-Saharan recoveries during the winter included birds of all ages (n = 4, 7, 4 and 1 recoveries for individuals in their first, second, third and fourth or more years of life, respectively). This contrasts with the age distribution of individuals ringed as nestlings in Spain and recovered as floaters in Morocco and Algeria (n = 8), probably after wintering in the south of the Sahara, which primarily included birds in their second year (n = 6 of 8, 75%). The remaining recoveries correspond to floaters in their third year of age (12.5%, n = 8) or older (12.5%, n = 8).

Trends of Local Breeding Populations Associated with Contamination Traps

Segovia

Breeding populations of red and black kites showed contrasting trends in the Natural Park of Las Hoces del Río Duratón, Segovia Province (Figure 7). Red kites showed a clear and dramatic population crash in the study area over the last decade, going from twelve pairs in 1992 to a single pair in 2006 (Figure 7). The black kite population showed a fluctuating number of breeding pairs during the study period, with no clear population trend (Figure 7).

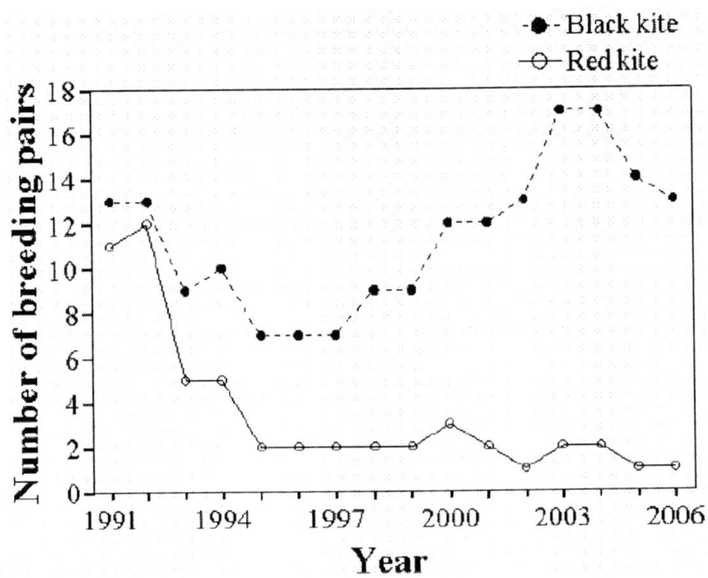

Figure 7. Annual trends in the number of black and red kite breeding pairs in the Natural Park de las Hoces del río Duratón, Segovia Province, central Spain.

Southeast Madrid

The number of breeding pairs showed a fluctuating pattern with lower values in the early 1990s and highest numbers in recent years (Figure 8). During this time period, the area occupied by breeding pairs decreased in size in the north of the study area (riparian forest of the Jarama River before the confluence with the Manazanares River). The number of breeding pairs in the North Jarama River showed a declining trend until no pair occupied the area since 2000 (Figure 8). No data on the number of pairs were recorded in Manazanares-South Jarama rivers from 1998 to 2000 (Figure 8), although the number of breeding pairs has further concentrated in the Manzanares Basin, close to the contamination trap since then (Figure 8).

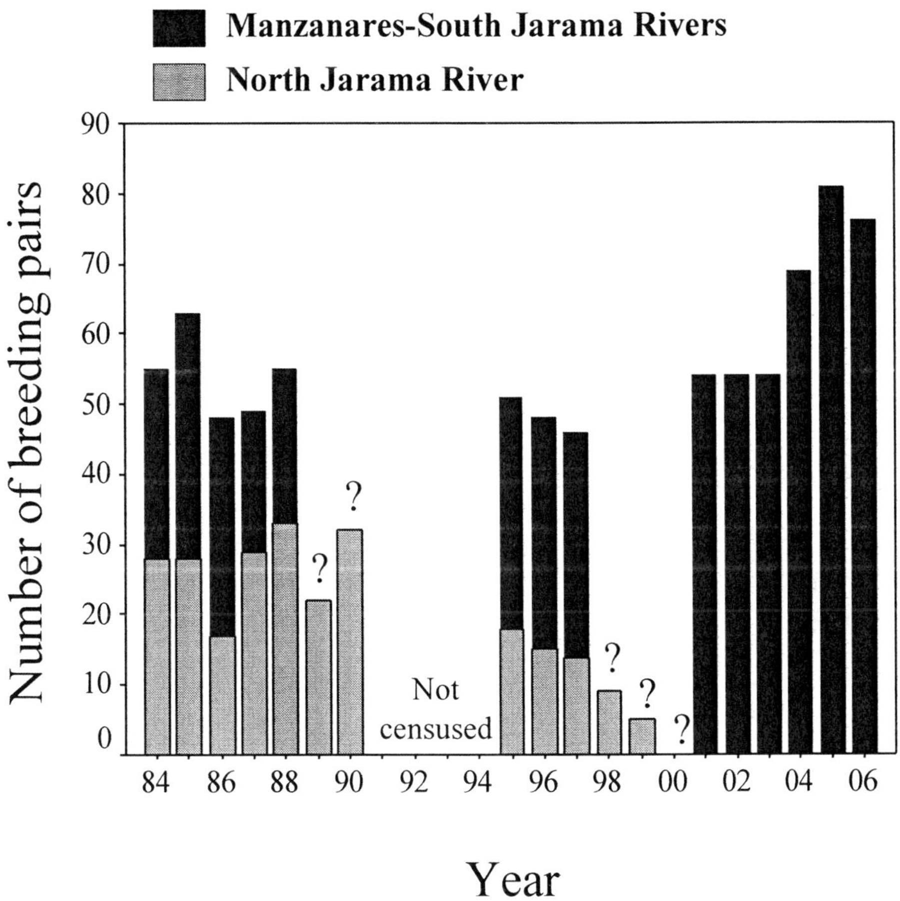

Figure 8. Annual trends in the number of black kite breeding pairs in the Regional Park of southeastern Madrid, central Spain. Two areas are distinguished corresponding to the northern (North Jarama river) and to the southern (Manazanares-south Jarama rivers) parts of this study area. Question marks indicate an unknown number of breeding pairs in the area corresponding to the Manazanares-south Jarama rivers.

Attraction Effect of Contamination Traps for Migrating and Floating Black Kites

The number of floaters associated with municipal dumps in Southeast Madrid showed a stable trend during the first decade of the study period (1990-early 2000s, Figure 9a). During this period the number of floaters was estimated at an average of about 200-400 individuals present throughout the breeding season, with fluctuations within years probably due to movements between nearby temporary settling areas. The number of migrating black kites showed a similar pattern during this time period, with a relatively stable number of about 1000 individuals at the migratory peak (Figure 9b). Afterwards, beginning in 2002, the number of both floaters and migrating individuals increased sharply to reach an average of about 800 floaters and 9229 migrating individuals during the migratory peak in 2005 (Figure a,b). These individuals, both floaters and migrants, primarily forage at the refuse dumps supplied with domestic refuse from Madrid and the surroundings cities, which were partially incinerated in the associated solid waste incinerator creating the potential contamination trap. Therefore, both the number of floaters and migrant black kites have increased simultaneously during recent years attracted by the huge quantity of food available in the large refuse dumps supplied by Madrid and surrounding cities. As a consequence, an increasing number of floaters and migrants may be more at risk from exposure to air-borne contaminants from the associated solid waste incinerator.

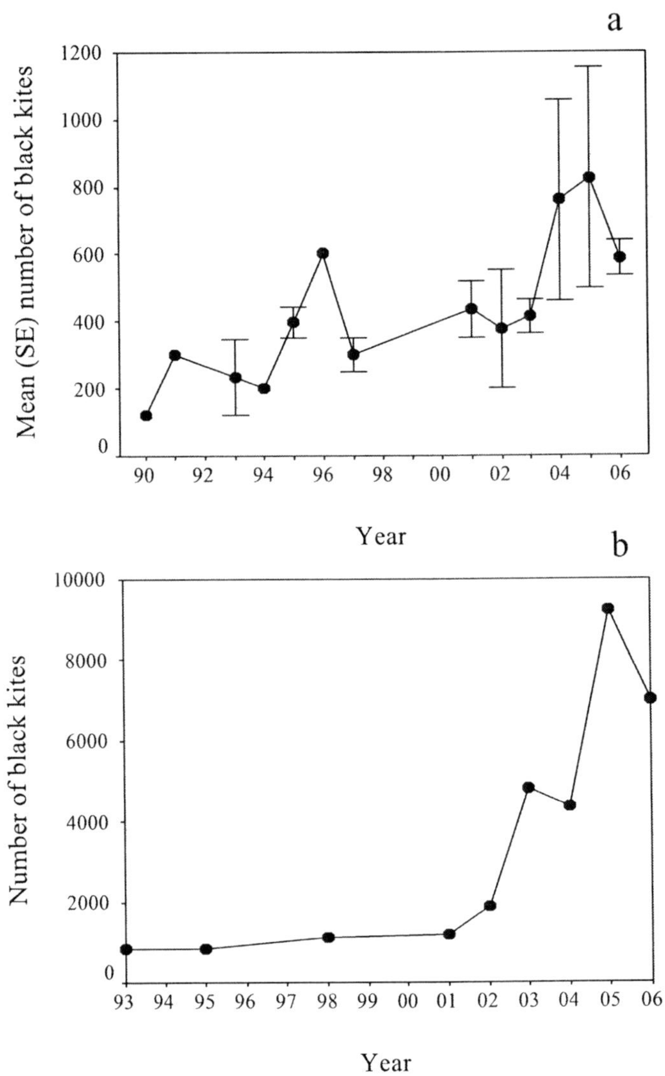

Figure 9. Trends in (a) the mean number (± SE) of floater black kites counted at a roosting site in the Regional Park of southeastern Madrid, central Spain, and (b) in the number of migrant black kites counted in the same area during the post-nuptial migratory peak (second week of August).

Dumps of Livestock Carcasses as a Source of Pathogens

Wintering Red Kites

Cloacal microflora of wintering red kites were composed of cosmopolitan intestinal bacteria, bacteria commonly present in their food, and bacteria associated with sewage and sludge commonly occurring at sites for disposal of livestock carcasses and municipal solid waste (Table 2). Previously, we examined composition, species richness and prevalence of faecal microflora of wintering red kites in two areas of central Spain where kites primarily fed on carrion from stabled livestock reared in intensive conditions (Segovia) *vs.* wild rabbits *Oryctolagus cuniculus* (Madrid) (Blanco et al. 2006). Here, we extended the study to a third area (Extremadura) where faeces from four different roosting sites were sampled.

As previously stated, bacterial flora composition and the prevalence of pathogens, especially *Salmonella* sp., showed differences attributed to diet and the environmental conditions where wintering red kites obtain most of their food. Bacterial flora cultured from faeces collected in the Madrid area was characterised by the presence of typical species of the skin and intestinal flora of their main prey (wild rabbits), primarily *Staphylococcus epidermidis*, *Klebsiella sp.* and *Corynebacterium sp.* Bacterial flora from faeces collected in the Segovia area was also composed of species commonly present in their food, especially meat from pigs, poultry and other livestock carcasses (*Enterococcus suis, Enterococcus bovis*), as well as in sewage and sludge (*Citrobacter freundii, Proteus vulgaris, Alcaligenes faecalis, Hafnia alvei, Providencia rottgeri*) commonly occurring at sites used for disposal of livestock carcasses and municipal solid waste (Blanco et al. 2006).

Wintering red kites sampled in Extremadura showed a bacterial flora that may be considered intermediate between those isolated from faeces in Madrid and Segovia (Table 2). Bacterial flora differed between roosting sites within the Extremadura region, probably because of local variations in diet. Food consumed by wintering red kites in Extremadura ranged from a diet primarily composed of wild prey in Torrecillas de la Tiesa to a diet mostly based on refuse from a municipal dump in Navalmoral de la Mata (see methods). These diet differences may be translated to differences in bacterial flora composition and prevalence. Thus, flora composition in areas where kites fed mostly or partially on refuse and livestock remains was composed of bacteria associated with those food sources (see above and Table 2). In Extremadura, unlike other sampled areas in Spain, red kites fed on sheep carrion, which transmitted its typical bacteria (*Enterococcus ovis*) to kite flora (Table 2). Similarly, as already shown in the comparison of flora composition and prevalence between Segovia and Madrid (Blanco et al. 2006), differences in diet and bacterial flora composition may influence faecal microflora species richness and prevalence in samples from different areas in Extremadura. Specifically, flora composition, species richness and high prevalence of opportunistic pathogenic and commensal bacteria typically associated with unsanitary conditions may out-compete the normal development of protective microflora. Competition between bacteria may cause alteration of normal flora by inhibiting proliferation of some bacterial species compared with others. This was previously illustrated by the occurrence of *Salmonella* sp. in pure cultures indicating competitive exclusion in wintering red kites in Segovia (Blanco et al. 2006). Although we found no pure cultures of *Salmonella* sp. in samples from Extremadura, we further explored the relationships between the prevalence of this opportunistic pathogen and features indicating the alteration of normal flora. We tested the hypothesis that multiple competitive interactions between bacteria may determine a heterogeneous pattern of flora composition among individuals, which may be reflected in a low level of evenness of bacterial species or, in other words, a high richness of species with a low prevalence. We found a significant positive relationship between the prevalence of *Salmonella* sp. and the number of bacterial species isolated in each roosting site (Spearman correlation rank coefficient, $rs = 0.928$, $P = 0.008$, $n = 6$, Figure 10). To avoid pseudoreplication, we pooled data from the two sampled roosts in Segovia because of the similar proportion of *Salmonella* sp. prevalence (Blanco et al. 2006) and the similar diet of kites using those roosts. In any case, the relationship remained significant when the two roosts of Segovia were included separately ($rs = 0.847$, $P = 0.016$, $n = 7$). This suggests that a rich heterogeneous bacterial flora was associated positively with the prevalence of pathogens, such as *Salmonella* sp., acquired by the consumption of carrion from livestock under unsanitary conditions in carcass dumps.

Table 2. Prevalence (% of faecal samples) of bacterial species present in faeces of communally roosting red kites wintering in three different areas of Spain

Species	*Segovia*	*Madrid*	*Extremadura*			
			Torrecillas de la Tiesa	Escurial	Navalmoral de la Mata	Valdesalor
	$n = 80$	$n = 33$	$n = 12$	$n = 21$	n = 18	n =21
Gram-negative bacteria						
Escherichia coli	73.8	75.8	58.3	66.7	72.2	61.9
Salmonella sp.	25.0	3.0	0.0	4.8	22.2	4.8
Klebsiella sp.	3.8	18.2	16.7	9.5	5.6	9.5
Citrobacter freundii	35.0	0.0	16.7	9.5	16.7	14.3
Proteus vulgaris	50.0	0.0	8.3	14.3	5.6	0.0
Proteus miriabilis	8.8	0.0	0.0	14.3	22.2	0.0
Alcaligenes faecalis	2.5	0.0	0.0	19.0	16.7	19.0
Alcaligenes sp.	2.5	0.0	0.0	0.0	27.8	14.3
Hafnia alvei	2.5	0.0	0.0	23.8	16.7	23.8
Serratia marcescens	7.5	0.0	0.0	14.3	11.1	4.8
Serratia sp.	5.0	0.0	0.0	0.0	0.0	0.0
Yersinia sp.	2.5	0.0	25.0	0.0	0.0	0.0
Providencia rettgeri	5.0	0.0	0.0	0.0	0.0	0.0
Enterobacter cloacae	18.8	42.4	25.0	38.1	55.6	66.7
Pseudomonas aeruginosa	3.8	3.0	0.0	4.8	11.1	0.0
Gram-positive bacteria						
Enterococcus avium	0.0	3.0	25.0	19.0	11.1	14.3
Enterococcus suis	23.8	0.0	0.0	14.3	22.2	4.8
Enterococcus faecalis	56.3	48.5	33.3	66.7	33.3	33.3
Enterococcus faecium	46.3	42.4	16.7	14.3	5.6	14.3
Enterococcus bovis	6.3	0.0	8.3	14.3	11.1	9.5
Enterococcus ovis	0.0	0.0	25.0	14.3	11.1	9.5
Staphylococcus epidermidis	7.5	54.5	25.0	23.8	16.7	28.6
Staphylococcus aureus	3.8	0.0	0.0	0.0	5.6	0.0
Bacillus sp.	6.3	33.3	0.0	9.5	11.1	9.5
Corynebacterium sp.	7.5	36.4	16.7	9.5	5.6	28.6
Aerococcus viridans	1.3	0.0	0.0	0.0	0.0	0.0
Aerococcus sp.	1.3	0.0	0.0	0.0	0.0	0.0

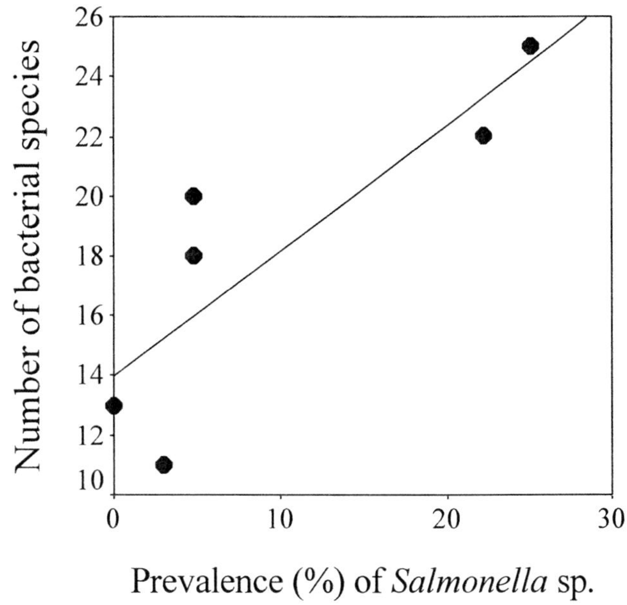

Figure 10. Relationship between the proportion of faecal isolates with *Salmonella* sp. and the number of faecal bacterial species isolated in six communal roosts of wintering red kites in Spain.

Floater and Wintering Black Kites

The intestinal microflora of floater black kites included normal faecal bacteria as well as bacteria typical of the skin and intestinal flora of their wild prey (Table 3). Microflora of kites from the Southeast Madrid showed a combination of bacteria associated with scavenging and foraging in refuse dumps (e.g. *Salmonella* sp., *Enterobacter cloacae*) and those acquired from wild prey, especially wild rabbits and birds (*Staphylococcus epidermidis* and *Enterococcus avium*, respectively). A similar pattern was found in samples from local and wintering kites in the African Sahel (Table 3) where the prevalence of particular bacteria was associated with the different components of their eclectic diet, including livestock remains and refuse (*Salmonella* sp., *Enterococcus bovis*, *Enterobacter cloacae*) and wild prey (*Staphylococcus epidermidis*, *Corynebacterium sp.*). Floater black kites from Extremadura showed a bacterial microflora associated with a diet primarily composed of wild prey (Table 3). The prevalence of the opportunistic pathogens, primarily *Salmonella* sp., was thus associated with foraging in refuse dumps in South-east Madrid (domestic refuse from municipal dumps) and the Sahel (livestock carcass dumps). No *Salmonella* sp. isolate was found in pure cultures. Other potentially pathogenic thermolabile bacteria such as *Campylobacter jejuni* and *Clostridun* sp. could not be isolated by sampling faeces, although they may be present in the intestinal flora.

Nestling Red and Black Kites

A variety of bacterial species were isolated from cloacal samples of nestlings of both kite species in several areas (Table 4). Most of these bacteria have previously been isolated from faeces of wintering and floating kites (Table 2 and 3).

Table 3. Prevalence (% of faecal samples) of bacterial species present in faeces of communally roosting black kites (*Milvus migrans*) belonging to the floating population of Southeast Madrid and Extremadura (sampled in June) and the wintering (*M. migrans*) and local population (*M. aegyptius*) in the Sahel (Mauritania and Senegal), sampled in October

Species	SE Madrid $n = 57$	Extremadura $n = 28$	Sahel $n = 20$
Gram-negative bacteria			
Escherichia coli	66.7	78.6	70.0
Salmonella sp.	21.1	7.1	25.0
Klebsiella sp.	22.8	10.7	15.0
Citrobacter freundii	17.5	21.4	10.0
Serratia marcescens	17.5	10.7	0.0
Serratia sp.	0.0	0.0	20.0
Yersinia sp.	19.3	25.0	30.0
Enterobacter cloacae	22.8	0.0	50.0
Gram-positive bacteria			
Enterococcus avium	26.3	28.6	20.0
Enterococcus faecalis	19.3	0.0	35.0
Enterococcus faecium	17.5	0.0	25.0
Enterococcus bovis	0.0	0.0	15.0
Staphylococcus epidermidis	28.1	10.7	60.0
Staphylococcus aureus	14.0	0.0	0.0
Bacillus sp.	0.0	14.3	0.0
Corynebacterium sp.	24.6	7.1	70.0

In addition, we found thermolabile bacteria that could not be isolated by sampling faeces (see above). In general, bacterial flora was similar to that of adults, although the number of species was higher in nestlings than in floating and wintering black kites and similar between nestlings and wintering red kites in Segovia. We isolated a single pure culture of *Salmonella* sp. from a nestling red kite from Segovia. Pathogenic bacteria associated with livestock refuse and places containing sludge and sewage (*Salmonella* sp.) showed the highest prevalence in black kites from Segovia (Table 4). Other pathogenic bacteria, such as *Campylobacter jejuni* was found in red kites as well as in black kites except in those from Segovia. This bacteria, as well as *Clostridium* sp., showed a higher prevalence in areas where nestling kites were fed both with livestock remains (Segovia) and wild rabbits (southeast Madrid) as compared to Extremadura, where a diet predominantly composed of wild prey did not include wild rabbits, a common source of *Clostridium* sp.

Geographic Variations in Bacterial Resistance to Antibiotics

Wintering Red Kites

We found bacterial resistance to all tested antibiotics in Segovia and central Madrid, while bacterial isolates did not present resistance to most antibiotics in Extremadura (Figure

11). The pattern of resistance to the tested antibiotics showed a similar profile between Segovia and central Madrid (Spearman correlation rank coefficient $r_s = 0.872$, $P < 0.0001$, $n = 12$, Figure 11), although the proportion of resistant bacterial isolates was higher in Segovia than in central Madrid (Wilcoxon matched-pairs test, $z = 3.059$, $P = 0.002$, $n = 12$) The highest bacterial resistance was found for semisynthetic penicillins (AX, AP) and EN in all areas. Lincomycin, NE, OT and PG showed variable resistance ranging from about 50-70% of resistant isolates in Segovia to about 20-40% in central Madrid (Figure 11). The lowest proportion of resistant bacterial isolates was found in Extremadura (pooling samples from all roosts), where only amoxocillin exceeded a level of resistance of a 20% (Figure 11).

Table 4. Prevalence (% of faecal samples) of bacterial species sampled from the cloaca of nestling red and black kites in different areas in Spain

Species	*Red kites*	*Black kites*		
	Segovia $n = 81$	Segovia $n = 31$	SE Madrid $n = 73$	Extremadura $n = 37$
Gram-negative bacteria				
Escherichia coli	59.3	71.0	60.3	48.6
Salmonella sp.	12.3	25.8	13.7	16.2
Campylobacter jejuni	8.6	0.0	23.3	8.1
Clostridium sp.	22.2	29.0	37.0	13.5
Klebsiella oxytoca	12.3	29.0	11.0	0.0
Citrobacter freundii	6.2	0.0	1.4	18.9
Citrobacter sp.	4.9	3.2	0.0	0.0
Proteus vulgaris	18.5	0.0	4.1	0.0
Proteus miriabilis	8.6	2.6	23.3	5.4
Proteus sp.	0.0	19.4	8.2	0.0
Hafnia alvei	7.4	0.0	1.4	10.8
Serratia marcescens	6.2	25.8	16.4	5.4
Serratia sp.	16.0	0.0	1.4	0.0
Yersinia sp.	4.9	22.6	28.8	21.6
Morganella morganii	8.6	22.6	19.2	27.0
Providencia rettgeri	9.9	19.4	23.3	0.0
Providencia sp.	3.7	0.0	0.0	0.0
Enterobacter cloacae	38.3	29.0	38.4	16.2
Enterobacter sp.	19.8	0.0	2.7	0.0
Pseudomonas aeruginosa	0.0	0.0	12.3	0.0
Gram-positive bacteria				
Enterococcus faecalis	29.6	51.6	42.5	48.6
Enterococcus faecium	19.8	38.7	31.5	29.7
Staphylococcus epidermidis	14.8	19.4	31.5	27.0
Staphylococcus aureus	1.2	6.5	9.6	0.0
Staphylococcus simulans	0.0	0.0	11.0	0.0
Streptococcus faecalis	0.0	16.1	45.2	27.0
Corynebacterium sp.	21.0	19.4	16.4	8.1

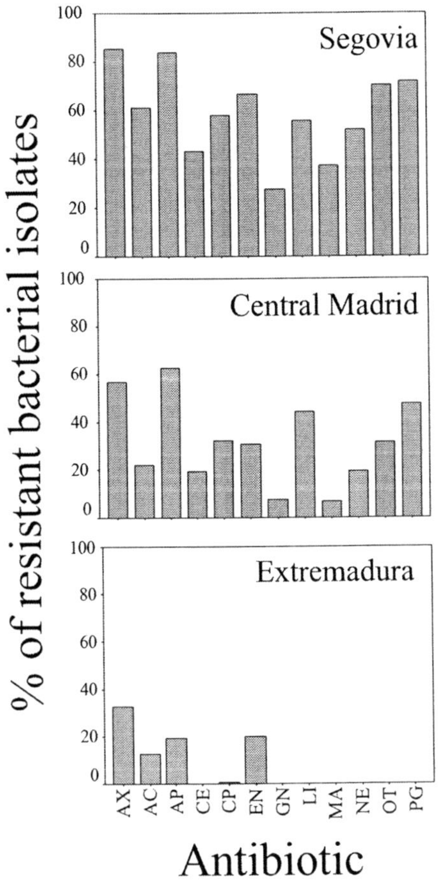

Figure 11. Profile of bacterial resistance to twelve antibiotics in wintering red kites from different regions in Spain. Ciprofloxacin (CP), gentamicin (GN), enrofloxacin (EN), marbofloxacin (MA), amoxicillin (AX), amoxicillin-clavulanic (AC), ampicillin (AP), cephalexin (CE), lincomycin (LI), neomycin (NE), oxitetracycline (OT), penicillin G procain (PG).

Floater and Wintering Black Kites

We did not found resistance to all tested antibiotics in any of the sampled areas (Figure 12). There were clear differences between the resistance profile of southeast Madrid, where we found high levels of resistance to most antibiotics, and the remaining two areas, where resistance was low for the few antibiotics towards which resistance was developed (Figure 12). Values of antibiotic resistance were not correlated between southeast Madrid and Extremadura (r_s = 0.356, P = 0.256, n = 12) due to the comparatively higher bacterial resistance to OT and PG, two antibiotics specifically used in livestock, in Extremadura (Figure 12). Bacterial resistance in wintering and local black kites in the Sahel were the lowest found in this study, with a low susceptibility to only four antibiotics (Figure 12).

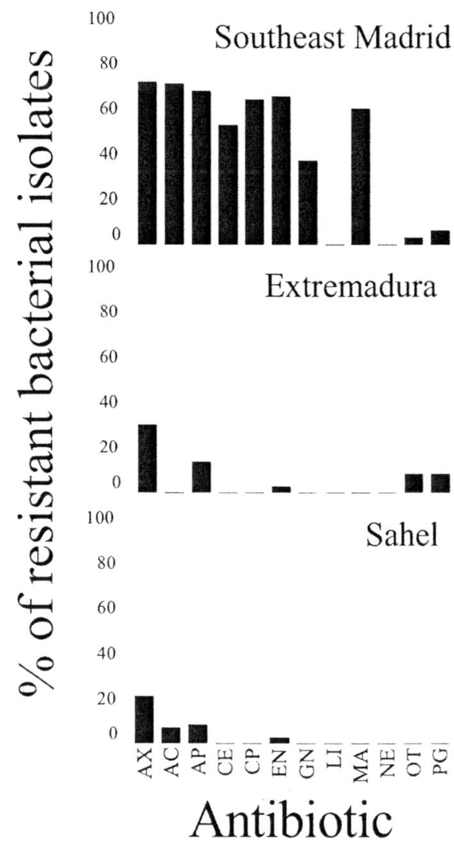

Figure 12. Profile of bacterial resistance to twelve antibiotics in floater black kites from different regions in Spain and wintering and local kites in the African Sahel (pooling black and yelow-billed kites). Antibiotic abbrevations as in Figure 11.

Nestling Red and Black Kites

Nestlings of both species showed bacterial resistance to all antibiotics in Segovia, with a high proportion of resistant bacterial isolates to most antibiotics (Figure 13). The profile of resistance between the two kite species in Segovia was very similar as indicated by the correlation between the proportion of resistant isolates to each antibiotic in each species (r_s = 0.837, P = 0.001, n = 12) and the lack of differences in the proportion of resistance to particular antibiotics between species (Wilcoxon matched-pair test, z = 1.022, P = 0.307, n = 12). Black kite nestlings from Extremadura showed a similar profile of antibiotic resistance to Segovia (r_s = 0.663, P = 0.019, n = 12), but a lower proportion of resistant bacterial isolates for each antibiotic (Wilcoxon matched-pair test, z = 3.059, P = 0.002, n = 12, Figure 13). However, red kites from Segovia and Extremadura showed a similar profile of antibiotic resistance (r_s = 0.878, $P < 0.0001$, n = 12) and similar proportions of resistant bacterial isolates (Wilcoxon matched-pair test, z = 1.414, P = 0.157, n = 12). Black kite nestlings in Southeast Madrid showed a resistance profile more similar to Extremadura (r_s = 0.908, $P <$ 0.0001, n = 12) than to Segovia (r_s = 0.662, P = 0.019, n = 12) and a proportion of resistant isolates intermediate between Segovia and Extremadura (Figure 13).

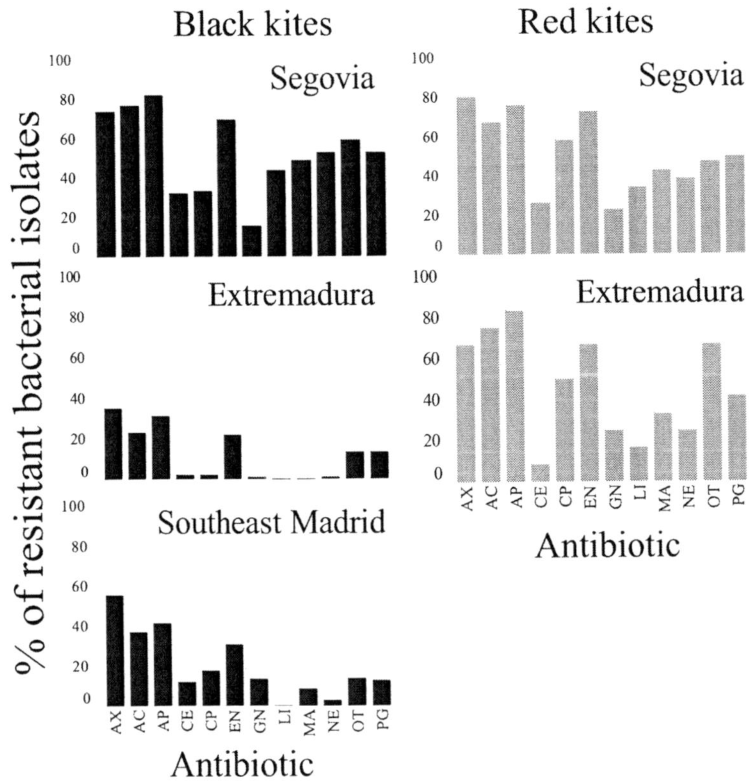

Figure 13. Profile of bacterial resistance to twelve antibiotics in nestling red (grey bars) and black (black bars) kites in several regions of Spain. Antibiotic abbrevations as in Figure 11.

Determinants of the Antibiotic Resistance Profile: Human Versus Livestock Medication

Tested sources of variation for bacterial antibiotic resistance included season, species, area and age (see above). This may lead to several resistance patterns and antibiotic profiles derived from the multiple relationships between these factors. In addition, testing of twelve different antibiotics generally used at different levels in human or veterinary medicine, or both, may complicate the interpretation of results. Therefore, in order to attempt to obtain a general objective pattern making biological sense, we conducted principal components factor analysis using the correlation matrix of antibiotics and groups of kites pooled according to species, age (nestling or full-grown) and geographical area. Thus, we attempt to reduce objectively the original variable set to smaller, mutually uncorrelated composite variables or factors. We obtained two axes with an eigenvalue greater than 1 (PC1 = 9.429, PC2 = 1.734) that accounted, respectively, for 78.58 and 14.45% of the variance. The first component (PC1) revealed high loading for all antibiotics and hence, defines a gradient of general bacterial resistance to antibiotics (Table 5). The second component (PC2) showed high positive loading for lincomycin, neomycin, oxitetracycline and penicillin G procaine and high negative loading for cephalexin, ciprofloxacin, gentamicin and marbofloxacin (Table 5). Therefore, it defined a gradient of antibiotic resistance profile determined by the resistance to antibiotics generally used in human (negative loadings) and veterinary medicine (positive loadings).

Table 5. Loadings of the principal components factor analysis using the proportion of antibiotic resistant bacterial isolates in each group of samples from kites pooled according to species, age and geographical area.

Antibiotic	*PC1*	*PC2*
Amoxicillin	0.980	-0.063
Amoxicillin-clavulanic	0.924	-0.235
Ampicillin	0.979	0.029
Cephalexin	0.819	-0.355
Ciprofloxacin	0.929	-0.266
Enrofloxacin	0.966	-0.154
Gentamicin	0.857	-0.474
Lincomycin	0.760	0.577
Marbofloxacin	0.886	-0.388
Neomycin	0.840	0.488
Oxitetracycline	0.830	0.471
Penicillin G procaine	0.836	0.534

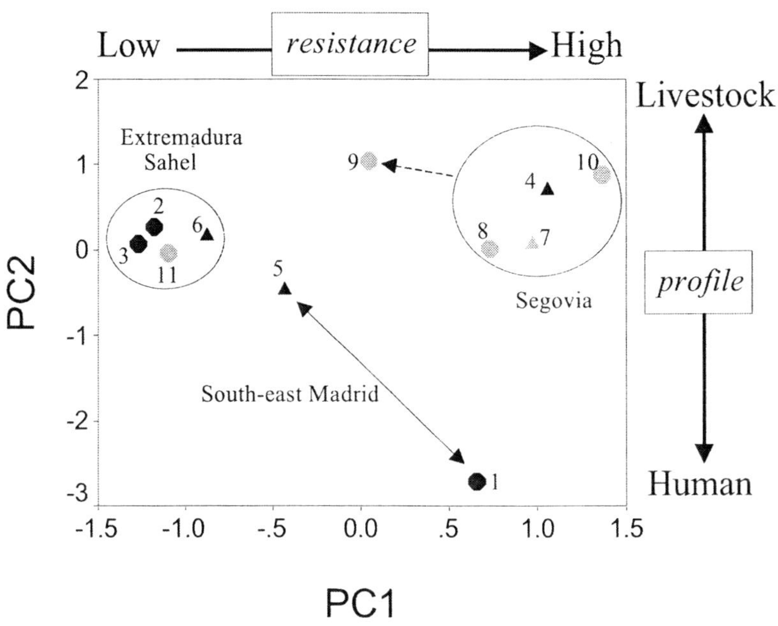

Figure 14. Location of the groups of study areas/species on the plane generated by principal components (PC1 and PC2) of a factorial analysis on the proportion of bacterial antibiotic resistance. Black and grey symbols indicate black and red kites respectively, while circles and triangles correspond to full-grown individuals and nestling respectively. 1: floater black kites in southeast Madrid, 2: floater black kites in Extremadura, 3: Sahel kites (wintering black and local yelow-billed kites), 4: nestling black kites in Segovia, 5: nestling black kites in southeast Madrid, 6: nestling black kites in Extremadura, 7: nestling red kites in Segovia, 8: nestling red kites in Extremadura, 9: wintering red kites in Central Madrid Province, 10: wintering red kites in Segovia, 11: wintering red kites in Extremadura. The continuos arrow connected groups from southeasterm Madrid while the discontinuous arrow indicates possible movements of wintering red kites from Segovia (10) towards Central Madrid Province (9).

Graphical representation of factor scores of principal components showed association between groups of samples from areas where kites showed a high (Segovia) or low (Extremadura, Sahel) proportion of bacterial resistance to antibiotics used in livestock operations (Figure 14).

Nestling black kites from southeastern Madrid showed intermediate values of bacterial antibiotic resistance (PC1) and antibiotic profiles (PC2) influenced by resistance to antibiotics used in veterinary and human medicine. This pattern was more clearly revealed in floater black kites from the same area, where a high proportion of bacterial resistance to antibiotics typically used in human medicine was found (Figure 14). Wintering red kites from central Madrid Province showed intermediate values of bacterial resistance to antibiotic used in livestock (Figure 14), probably because they also used near geographic areas (e.g. Segovia Province) in which they acquired bacterial antibiotic resistance.

Contamination with Lead

The concentration of lead in black kite nestlings from Southeast Madrid differed between years (ANOVA, $F_{3,152} = 3.901$, $P = 0.010$), primarily due to differences between 2001 and 2006 (Tukey post hoc test, $P = 0.004$, Figure 15), when sample sizes were higher. Pooling years, the average concentration of lead in nestlings was similar to that found in a small sample of floaters from the same area ($F_{1,159} = 0.043$, $P = 0.836$, Figure 15).

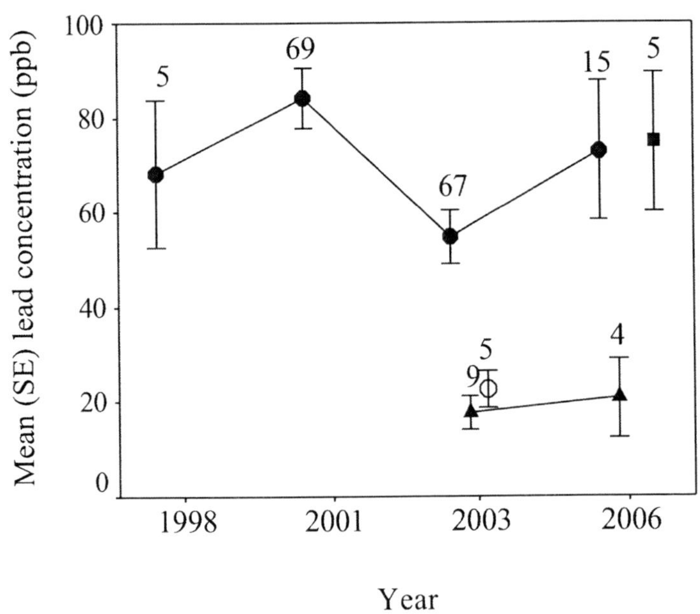

Figure 15. Annual variation in the mean (± SE) lead concentration (ppb) in nestling black kites from the Regional Park of southeastern Madrid, central Spain (black points), floaters from the same area (black squares), and nestling black (black triangles) and red (white points) kites from the Natural Park de las Hoces del río Duratón, Segovia Province. Sample size is shown above bars.

Considering only 2006, when floaters were sampled, the differences between nestlings and floaters remained non-significant ($F_{1,18} = 0.005$, $P = 0.945$). Lead levels in black kite nestlings from Segovia (mostly sampled in the Natural Park of Las Hoces del Rio Duratón) did not differ between years ($F_{1,11} = 0.168$, $P = 0.690$, Figure 15), but were significantly lower than in Southeast Madrid (pooling years, $F_{1,167} = 12.945$, $P < 0.0001$, Figure 15). A small sample of red kite nestlings from Segovia showed similar lead concentrations to black kite nestlings from the same area (pooling years, $F_{1,16} = 0.431$, $P = 0.521$, Figure 15).

CONCLUSION

Attraction and Co-Occurrence of Local and Foreign Kites in Refuse Dumps

Individuals from demographically and spatially independent subpopulations often occur together in geographically distant areas during some period of the annual cycle. This has multiple implications for the application of metapopulation theory to conservation of migratory birds (Esler 2000, Trakhtenbrot et al. 2005). However, due to practical and logistic reasons most population studies have neglected long-distance movements when focused at local scales and short-distance immigration and emigration, which could be insufficient to understand population dynamics, even within those local scales (Baillie et al. 2000). This may be especially marked when different population limiting factors have variable impacts at different temporal and spatial scales and at different areas used throughout the annual cycle. This work supports previous evidence that dumps of livestock carcasses and domestic refuse located in Spain potentially attract a considerable fraction of red and black kites from central and northern Europe, as well as local breeding populations and floaters. Therefore, understanding the dynamics of use of these foraging locales and their associated costs and benefits may be crucial to the development of successful management strategies for large-scale conservation.

Red kites have suffered a dramatic decline throughout their breeding and wintering range during recent decades (Thiollay 2001, Mammen and Stubbe 2002, Organbidexka 2003, Blanco and Montoya 2004). A major fraction of red kites from central and northern Europe, primarily from Germany, spend the winter in Spain (De Juana et al. 1988, this chapter), while local breeding individuals are sedentary. In addition, a small fraction of sub-adult and adult individuals from central Europe spend the breeding season as floaters in Spain. Therefore, knowledge of the threats incurred by both local and wintering individuals in Spain should help to understand factors associated with decline in northern breeding populations. Most studies of population dynamics have rarely considered that populations connected as metapopulations during the breeding season may be primarily influenced by limiting factors during other parts of the annual cycle (Rappole et al. 1993, Esler 2000), when the population may remain spatially and temporarily concentrated. Data of recoveries of foreign red kites in Spain suggest that individuals from a wide breeding range co-occurred in wintering quarters, primarily associated with large and predictable food sources such as refuse and livestock carcass dumps (Sunyer and Viñuela 1994, Viñuela et al. 1999, Garcia et al. 1998, Serrano 1999, Blanco et al. 2006). Therefore, factors associated with mortality during potential winter panmixia can similarly affect most breeding subpopulations in central and northern Europe.

The generalized declining trend in continental European populations wintering in Spain, where the local breeding population has suffered a concomitant sharp decline, clearly illustrated this point when compared to the recovering sedentary population in Great Britain (Wotton et al. 2002). Behavioural mechanisms such as natal philopatry common to both kite species (Carter 2001, Forero et al. 2002) may promote demographically independent subpopulations during the breeding season. However, the co-occurrence during winter of most northern and central populations of red kites in Spain, intensified by attraction to contamination traps, could prevent demographic independence of breeding subpopulations. Contamination traps influencing the state of health and mortality rates through infection with pathogens and contamination with pollutants may thus indirectly contribute to demographic panmixia due to a lack of independence of extinction probabilities between geographical areas.

The decline of wintering red kite populations appears to be associated with a constriction of the wintering range during the last century. This change of spatial distribution seems to be linked to an increasing use of livestock refuse dumps as food sources, especially from 1970 to 1990 in the Northern Plateau (De Juana 1998, Sunyer and Viñuela 1994) and, after 1990, in the Ebro Valley. On the contrary, these changes did not support the hypothesis suggesting that the increasing proportion of recoveries in the Northern Plateau corresponds to the opportunistic exploitation of common voles since the late seventies (Sunyer and Viñuela 1994). This is further supported by the increasing supply of carrion from growing livestock operations, primarily fattening pigs, in the Northern Plateau since the sixties (Sunyer and Viñuela 1994, Contreras 2001, Soto-Largo 2002, authors' unpublished data) and especially during recent decades in the Ebro Valley (Dean and Fernández 2006, authors' unpublished data), where there is no population of common voles (González-Esteban and Villate 2002). In addition, in other areas within the distribution range of common voles (e.g. Teruel and Guadalajara provinces in the Northern Plateau, Palomo and Gisbert 2002), there are neither wintering nor breeding populations of red kites (Viñuela 2003). Therefore, we suggest that historic and recent changes in the distribution of wintering red kites in Spain were primarily determined by their increasing attraction to livestock carcass dumps due to growing intensive livestock operations (Blanco et al. 2006). This distribution shift may thus reflect a preferential spatial track of this type of food resource, likely due to their abundance and temporal predictability once detected (Hiraldo et al. 1991, 1993). This increasing use and probable preference of refuse dumps as food sources also appear temporarily associated with the major decline of breeding populations to which wintering individuals belong, suggesting that dumps of livestock carcasses may entail risks leading to the formation of an ecological trap (Schlaepfer et al. 2002, Battin 2004).

Black kites migrating to Africa and recovered in Spain were mostly ringed in central Europe. A high proportion of two-year old sub-adults were recovered as floaters during the breeding season in Spain. These floating populations were associated with large concentrations of predictable food found in refuse dumps (Blanco 1997, Junta de Extremadura 2005). Therefore, the impact of these foraging places in the demography of populations of central Europe may be crucial for their conservation, as survival probabilities of pre-breeding individuals greatly depend on conditions outside of their natal areas. The use of distant areas by floaters during the breeding season, both for kites from central Europe in Spain and for Spanish kites in northern Africa, and their mobile nature unrestrained by breeding activities, may promote the typical fluctuating trend of breeding populations (this chapter, Sergio et al. 2005). Individuals use public information such as the presence of conspecifics and other cues as indicators of habitat quality (Sergio and Pentiriani 2005). Thus, the role of attraction of refuse dumps may further result in increased conspecific attraction with detrimental demographic consequences when the use of these areas is based on a maladaptive selection of foraging habitat. This social attraction can also influence the probability of breeding recruitment in a given area depending on the abundance of conspecifics (Serrano et al. 2001). Therefore, local breeders may disperse to particular nearby areas and eventually recruit in these areas due to the social attraction effects of conspecifics. The role of conspecific cues in the settlement of floaters has been previously established (Reed and Dobson 1993, Serrano et al. 2003), but the role of breeder cueing on the abundance of floaters and migrants has been largely neglected despite their potential implications in population dynamics and large-scale conservation of migratory species. Although black kites showed high breeding philopatry to birth areas (Forero et al. 2002, authors' unpublished data), breeding and natal dispersal potentially vary between populations depending on population size and degree of isolation among breeding patches or colonies (Serrano and Tella 2003). This may further explain the fluctuating population trends in spatial fragments of widely distributed populations as clearly illustrated by the population trend in southeastern Madrid. In this area, a concentration of local pairs has progressively occurred in the proximity of the refuse dumps of Madrid from 1995 onwards until the complete abandonment of the northern half of the local breeding range (Figure 3). In addition, the number of breeding pairs has greatly fluctuated, for instance increasing more than 30% in a period of two years (2003-2005) despite the relatively isolated location of this loose aggregation of breeding pairs (Blanco and Viñuela 2003). Overall, the increase in the number of breeding pairs was temporally associated to a concomitant increase in the number of migrating kites and floaters settled in the area attracted by the huge abundance of food found in the refuse dumps of the city of Madrid (Blanco 1993, 1997, Figure 9). Thus, the recent apparent increasing trend in this area may be a population mirage, because it is based on individuals from other populations rather than on the breeding recruitment of local individuals in their natal area. This area may hence be considered a contamination trap determining a demographic sink primaririly due to the low survival prospects of nestlings in a very poor health state during development in the nest due to contaminants (Blanco 2004, 2005, authors' unpublished data). The apparent recent increase in the number of wintering rek kites in the Ebro Valley (Dean and Fernández 2006) represents another clear example of populational mirage created by the atraction exerted by contamination traps on the spatial concentration of a globally declining population involved in a severe distribution range constriction.

Health Risks in Contamination Traps

Dumps of livestock carcasses and domestic refuse are superabundant and predictable sources of food, primarily meat remains from livestock operations, which may be associated with increased mortality and health risks. Risks directly linked to this food include its suitability in terms of nutritional quality (Annett and Pierotti 1999, Veiga and Hiraldo et al. 1989), pathogen acquisition and alteration of intestinal flora (Blanco et al. 2006), acquisition of bacterial resistance to veterinary and human drugs (this chapter) and direct ingestion of residues of these drugs (Langelier 1993, Thomas 1999, Shultz et al. 2004). In addition, these places are frequently associated with other sources of urban and industrial contamination, including heavy metals and organic pollutants (Blanco et al. 2003). All of these potential risks for health of red and black kites were found in different degrees in refuse dumps from different areas in Spain. These risks may be influencing mortality rates and health status of large numbers of foreign individuals from central and northern Europe, as well as local breeding populations in Spain. Therefore, the responsibility of Spain in the large-scale conservation of these species cannot be neglected.

This work supports the increasing evidence that dumps of livestock carcasses from intensive farming operations are a primary source of pathogens, especially food-borne pathogenic bacteria, such as *Salmonella, Campylobacter* and *Clostridium*. These bacterial species showed a relatively high prevalence in kites feeding upon livestock carcasses when compared with other wild non-passerines (Bangert et al. 1988, Brittingham et al. 1988, Singh and Sherikar 1999, Fallacara et al. 2001, Reche et al. 2003, Refsum et al. 2003, Waldenström et al. 2002). On the contrary, the consumption of carrion from unstabled livestock grazing freely and wild prey was linked to a lesser prevalence of detrimental pathogens. This kind of food has become progressively less available to scavengers due to legal restrictions on the abandonment of livestock carcasses in the countryside (Tella 2001, Blanco et al. 2006) and the decline of wild rabbits (Calvete 2006), the major wild prey for kites in Spain. Both nestlings and full-grown red and black kites were exposed to different degrees to these pathogens depending on their variable use of livestock carcass dumps as food sources.

Livestock carcass dumps supplied from intensive livestock farming operations, primarily fattening pigs and poultry, greatly increased in number and hence, as a food supply available to scavengers from the sixties onwards in the Northern Plateau (especially Segovia Province in Castilla-León) and later in the Ebro Valley. As a consequence, foreign and local kites may be at an increasing risk of pathogen acquisition with detrimental consequences on health that ultimately may be related to the decline of their populations. This may be especially salient given the strong attraction that these kind of dumps seem to exert on wintering kites, which likely have no alternative to the stimuli of a large concentration of predictable food during winter. Current efforts to identify causes of population decline of kites, especially red kites, in central and northern Europe should consider the potential impact of pathogens acquired in Spain. Effects on health after flora alteration and pathogen acquisition may not be immediate, but may surface after some time, often months later, for instance during migration or during periods of starvation and crowding, even when kites are feeding upon healthy food. Indeed, pathogens such as *Salmonella* may cause embryonic and neonatal mortality and even infertility (Battisti et al. 1998, Tizard 2004) contributing to population decline.

High prevalence of pathogenic bacteria, specifically *Salmonella*, and alteration of intestinal microflora were found associated with high levels of bacterial resistance to antibiotics used in intensive livestock operations. The use of huge quantities of antibiotics in stabled livestock operations was spatially associated with the larger concentration and quantity of carrion available for kites. Therefore, kites may be at an increasing risk of simultaneously acquiring pathogens and bacterial resistance to antibiotics used to combat these pathogens. Pathogens and commensal bacteria resistant to antibiotics may have fitness advantages that can persist over time and invade new strains even under conditions of low antibiotic selection (Martinez and Baquero 2002, Dionisio et al. 2005). The effects of pathogens resistant to antibiotics compared with those susceptible to antibiotics remains unexplored for wild birds. However, recent epidemiological studies in humans have highlighted the consequences of antibiotic resistance in terms of excess mortality, increased duration of illness, increased risk of invasive illness following infections, and ultimately a possible co-selection of virulence traits (Barza and Travers 2002, Barza 2002), especially with resistant *Salmonella* (Mølback 2004). Therefore, high levels of antibiotic resistance in detrimental pathogens acquired as a consequence of feeding on intensive stabled livestock carrion may make the contamination trap more dangerous to kites. In this sense, it is remarkable that a comparatively low prevalence of pathogens and bacterial resistance due to a healthier diet (wild prey and carrion from un-stabled livestock) occurs in areas apparently being progressively abandoned as major wintering quarters (i.e. Southern Plateau, especially Extremadura and Andalucia). These risks are hypothesized as being stronger for red than for black kites, because the wintering in Africa of the latter species seems not to be associated with a higher probability of infection with resistant pathogenic bacteria associated to food sources, but rather the contrary. In fact, overwintering in Africa could be viewed as a form of escape from the contamination trap linked to intensive practices in livestock operations which are less common but increasing in developing countries of Africa (Okeke et al. 1999).

Insufficient information about the effects of resistant pathogens exists to assess their influence on the conservation status of kites and other scavengers. Further investigation is required as these bacterial species have potential negative effects for wildlife and human health (Friend and Franson 1999), especially when resistance has been acquired by pathogens towards multiple antibiotics (Mølback 2004). We found bacterial resistance to multiple antibiotics in wintering, floating and nestling kites of both species in areas in which livestock remains and domestic refuse constitute an important proportion of the diet. In addition, high levels of bacterial resistance were also found in nestling black kites in southeastern Madrid despite the low consumption of food obtained in refuse dumps compared to floaters (Blanco 1997). Acquisition of resistant bacteria may be due to the direct ingestion of resistant strains already present in livestock carrion, but also through transference of mobile genetic elements from bacteria to bacteria in or across bacterial species and genera (Henriques Normark and Normark 2002, McEwen and Fedorka-Cray 2002; Mølback 2004), and by transference of resistant bacteria from parents to nestlings. The latter manner of acquisition may also occur as suggested by the fact that nestling red kites in Extremadura showed a high proportion of bacterial resistance to antibiotics. This high level of antibiotic resistance is hypothesized as coming from parents who acquired resistance from their more frequent contact with resistant strains present in livestock carrion in Extremadura or even in other areas. For instance, parental red kites may acquire bacterial antibiotic resistance during temporary settlement in risky areas (e.g. Northern Plateau) during the non-breeding season or during the pre-breeding stage as mobile floaters. This contrasts with the lower proportion of bacterial resistance in wintering red and floater black kites in the same area, probably due to the greater use of wild prey and sheep carrion relatively free of bacterial resistance to antibiotics. Based on these results, it might be argued that the impact of resistant pathogenic and commensal bacteria are likely variable depending on the area selected by foreign red and black kites. Apparently, there is no segregation of wintering red kites in Spain according to their geographical origin in central or northern Europe. On the contrary, this work provides evidence about the progressive abandonment of areas relatively free of bacterial antibiotic resistance (e.g. Extremadura) towards contamination traps in the Northern Plateau and Ebro Valley. In these areas, wintering remained more stable and even increased in recent years despite the general decline presumably promoted by the health problems caused by foraging in livestock carcass dumps.

The potential negative health effects of direct antibiotic ingestion with livestock meat are also possible but undocumented in wildlife. For instance, in addition to diclofenac (Shultz et al. 2004, Green et al. 2004) and other veterinary drugs (Langelier 1993, Thomas 1999), the recent and widespread increase in the use of antibiotics in intensive livestock farming operations in developing countries may be involved in the massive mortality of vultures on the Indian subcontinent, which should be promptly evaluated. Potential detrimental effects of antibiotics in wildlife may include those reported as being associated with their overuse or misuse in humans and domestic animals, including a transient decrease in host resistance upon exposure to pathogens (Barza 2002, Mølback 2004). Given the intensive use of antibiotics worldwide, their presence in wild birds may have unknown global implications in the uncontrolled generation of resistant pathogens and their spread through the environment. Nowadays, only bacterial resistance to antibiotics can be contemplated as a potential problem in human health through their effects on the transmission of resistant strains whose hosts may act as reservoirs of infection. In this sense, our results suggest that a resistance profile to antibiotics used in human medicine was typically represented in floater black kites foraging in the large dumps of the city of Madrid, clearly influenced by the strong impact of antibiotic use in this area of high concentrations of humans. On the other hand, the potential spread and transmission by migratory species of resistant strains between distant geographic areas and even different species cannot be ruled out. Results from this study are not enough to reject the hypothesis that wintering red and migrating black kites can be dispersing resistant strains to their breeding and wintering quarters, respectively. Sampling of red kites in breeding areas in central Europe (e.g. Germany) after their wintering in Spain may help understanding the real impact of pathogens and bacterial resistance on the conservation of this endangered species. Our sampling of European black kites overwintering in the African Sahel appears clearly insufficient to reject the hypothesis of a trans-continental dispersion of resistant bacterial strains, primarily because of the predominant presence of yellow-billed kites in the sampled areas. Nevertheless, the low proportion of resistant bacteria recorded in the Sahel kites suggests that the potential impact of a widespread transmission of resistant strains between species from developed to developing countries is currently not very strong, at least in the case of black kites.

The impact of detrimental pollutants from industrial activities represents an increasing threat of universal concern for wildlife conservation. Nowadays, developed countries have a major role in contamination with negative implications on the health of individuals even from very distant areas. Heavy metals are usually associated with dumps of urban and industrial residues, although their impact on wildlife has been mainly reported as derived from the ingestion of lead shots from hunting activities (Pain 1996). This work demonstrated that the impact of contamination with lead previously reported in nestling black kites from southeastern Madrid (Blanco et al. 2004) continues today at similar degrees as in past years. In addition, floaters from the same area also showed high levels of lead, similar to those found in nestlings, which indicate a more widespread impact than previously reported because of the large numbers of floaters using the area throughout the breeding season (this chapter). More importantly, the same area is being increasingly used by huge numbers of migrants (up to more than 9000 individuals roosting in the area in a single day) representing a major proportion of the European population crossing the Strait of Gibraltar (about 160,000 individuals in recent years, MIGRES Program, SEO/BirdLife). Therefore, the impact of this contamination trap may be considered of global importance for the large-scale conservation of black kites if migrants become contaminated in this area with similar levels of lead as nestlings and floaters.

Refuse Dumps for the Conservation of Scavengers: A Trans-Frontier Challenge for the XXI Century

Refuse and livestock carcass dumps may be dangerous places for avian scavengers in the modern world because production intensification and large concentration of exploitable food might cause direct and indirect health problems. In the past, traditional livestock carcass dumps were attractive and appropriate foraging places that might have certainly played an important role in the growth and maintenance of avian scavenger populations (Donázar et al. 1997). Even today, many of these places represent important appropriate sources of suitable food, especially those small dumps associated with traditional and family livestock operations, as well as those spatially and temporally limited small dumps originating from free-range livestock operations (Donázar et al. 1997, Blanco et al. 2006, Carrete et al. 2007). Intentionally created dumps for specific conservation programs of avian scavengers such as 'vulture restaurants' may have played a crucial role for population recovery in the past (Donázar et al. 1997, Carrete et al. 2006). However, the changes that have taken place in the livestock operations, especially the intensification of production and the increase of disposable refuse with the commercialisation process, have lead to a large accumulation of refuse in fewer dumping locations. This fact, together with the dangerous nature of the available meat waste from stabled livestock reared under intensive conditions, have favoured the creation of ecological traps for scavenger bird species, especially contamination and infection traps. The knowledge of the circumstances related to the creation of these ecological traps, as well as the costs and benefits of their use, should be evaluated in order to attempt to revert the dangerous use of these resources into a beneficial tool for conservation. This challenge should be considered on a long term and large scale, with the participation of all the countries and continents involved. Taking into account Spain's crucial responsibility in the case of migratory kite species, a first step in utilizing the opportunities that dumps provide for conservation would be to increase the state of knowledge about the costs and benefits at larger scales of the use of dumps in Spain. This work represents a first attempt at reaching this goal. Researchers, and especially wildlife managers who bear the ultimate responsibility of implementing conservation measures, should consider global rather than local implications of the different threats for wildlife (Sutherland 1998). Communication and the transfer of information should be coordinated by government agencies of the different countries, especially regarding global conservation guidelines and international regulations.

The elimination of livestock dumps should be considered as a conservation tool that may provide fruitful results if managed adequately based upon scientific evidence. Accordingly, this work provides insights about the risks that refuse dumps entail and their potential demographic consequences. These issues should be discussed in forums involving government agencies responsible for the organization and regulation of livestock operations, environmental managers, and public health agencies with the goal of identifying the best measures to eliminate refuse in a way that minimizes the risk for endangered scavengers.

Finally, this work warns of the potential risks of simplifying or undervaluing the natural processes associated with human activities, especially as they are increasingly becoming globalized in ways that may impose universal risks on wildlife but that, with suitable knowledge and action, may be turned into global conservation measures. We want to specifically underline that narrowly focused appreciation of local problems, which are only a part of the total process, might lead to partial or inappropriate management and conservation measures, which can be avoided if critical scientific evidence is taken into account (Pullin and Knight 2001, Ashely et al., 2003, Pullin et al., 2004, Sutherland et al., 2004).

ACKNOWLEDGEMENTS

We thank A. Tejedor, S. Sacristan, F. Sánchez-Aguado, E. Soto-Largo, J. Zalba, A. Ortega, I. Cardiel, L. Bolonio, R. Barrientos and L. Matheus and Naumanni group, for their help in fieldwork, and C. Lakunza for improving the English. Data and samples of kites from the Sahel were kindly provided by F. Hiraldo and M. de la Riva (Estación Biológica de Doñana, CSIC). Fieldwork was partially supported financially by Seo/BirdLife, Junta de Extremadura, Junta de Castilla y León and the projects BOS2003-05066 and PAI05-051. Our special acknowledgements to J.C. Rincón for his valuable help and friendship across the years.

REFERENCES

Angulo, F.J., Nunnery, J.A. and Bair, H.D. (2004) Antimicrobial resistence in zoonotic enteric pathogens. *Revue scientifique et technique de la Office international des Epizooties*, 23, 467-496.

Annett, C., and Pierotti, R. (1999). Long-term reproductive output in Western Gulls: consequences of alternate tactics in diet choice. *Ecology*, 80, 288-297.

Ashley, M. V., Willson, M.F., Pergams, O.R.W., O'Dowd, D.J., Gende, S.M. and Brown, J.S. (2003). Evolutionary enlightened management. *Biological Conservation*, 111, 115-123.

Baillie, S. R., Sutherland, W.J., Freeman, S. N., Gregory, R.D. and Paradis, E. (2000). Consequences of large-scake processes for the conservation of bird populations. *Journal of Applied Ecology*, 37, 88-102.

Bangert, R.L., Ward, A.C.S., Stauber, E.H., Cho, B.R. and Widders, P.R. (1988) A survey of the aerobic bacteria in the feces of captive raptors. *Avian Diseases*, 32, 53-62.

Barza, M. (2002). Potential mechanisms of increased disease in humans from antimicrobial resistance in food animals. *Clinical Infectious Diseases*, 34 (Suppl. 3), S123-S125.

Barza, M., and Travers, K. (2002). Excess infections due to antimicrobial resistance, the 'Attributable Fraction'. *Clinical Infectious Diseases*, 34 (Suppl. 3), S126-S130.

Battin, J. (2004). When good animals love bad habitats: ecological traps and the conservation of animal populations. *Conservation Biology*, 18, 1482-1491.

Battisti, A., Di Guardo, G., Agrimi, U. and Bozzano, A. (1998) Embryonic and neonatal mortality from salmonellosis in captive bred raptors. *Journal of Wildlife Diseases*, 34, 64-72.

Bengis, R.G., Leighton, F.A., Fischer, J.R., Artois, M., Mörner, T. and Tate, C.M. (2004). The role of wildlife in emerging and re-emerging zoonoses. *Revue scientifique et technique de la Office international des Epizooties*, 23, 497-511.

Bernis, F. (1980). *La Migración de las aves en el Estrecho de Gibraltar. I. Aves Planeadoras*. Madrid: Universidad Complutense.

Blanco, J.C., Hiraldo, F. and Heredia, B. (1990) Variations in the diet and foraging behavior of a wintering red kite (*Milvus milvus*) population in response to changes in food availability. *Ardeola*, 37, 267-278.

Blanco G .(1994). Seasonal abundance of black kites associated with the rubbish dump of Madrid. *Journal of Raptor Research*, 28, 242-245.

Blanco, G. (1996). Population dynamics and communal roosting of white storks foraging at a Spanish refuse dump. *Colonial Waterbirds*, 19, 273-276.

Blanco G .(1997). Role of refuse as food for migrant, floater and breeding Black Kites (*Milvus migrans*). *Journal of Raptor Research*, 31, 71-76.

Blanco, G., Frías, O., Jiménez, B. and Gómez, G. (2003). Factors affecting variability and potential uptake routes of heavy metals in black kites exposed to emissions from a solid-waste incinerator. *Environmental Toxicology and Chemistry*, 22, 2711-2718.

Blanco, G. and Viñuela, J. (2003). Milano Negro, *Milvus migrans*. In: *Atlas de las Aves Reproductoras de España* (pp. 160-161), Madrid: Dirección General de Conservación de la Naturaleza-SEO/BirdLife.

Blanco, G., Jiménez, B., Frías, O., Millan, J. and Dávila, J. A. (2004). Contamination with non-essential metals from a solid-waste incinerator correlate with nutritional and immunological stress in prefledgling black kites (*Milvus migrans*). *Environmental Research* 94: 94-101.

Blanco, G. and Viñuela, J. 2004. Milano Negro, *Milvus migrans*. Libro Rojo de las Aves de España, pp. 116-120 Dirección General para la Biodiversidad-SEO/BirdLife. Madrid.

Blanco, G. and Montoya, R. (2004) Landscape, "muladares" or poisoning?: using GIS to assess factors related to the decline of breeding red kites (*Milvus milvus*) in Spain. In: R. Rodríguez-Estrella and L.A. Bojórquez-Tapia (Eds.), *Spatial Analysis in Raptor Ecology and Conservation* (pp. 133-152). Mexico:CIBNOR and CONABIO.

Blanco, G., Lemus, J.A. and Grande, J. (2006). Faecal bacteria associated with different diets of wintering red kites: influence of livestock carcass dumps in microflora alteration and pathogen acquisition. *Journal of Applied Ecology*, 43, 990-998.

Blount, J.D., Houston, D. C, Møller, A. P. and Wright, J. (2003). Do individual branches of immune defence correlate? A comparative case study of scavenging and non-scavenging birds. *Oikos*, 102, 340-350.

Brittingham, M.C., Temple, S.A. and Duncan, R.M. (1988) A survey of the prevalence of selected bacteria in wild birds. *Journal of Wildlife Diseases*, 24, 299-307.

Burger, J. and Gochfeld, M. (1997). Lead and neurobehavioural development in gulls: A model for understanding effects in the laboratory and the field. *Neurotoxicology*, 18, 486-506.

Calvete, C. (2006). Modelling the effect of population dynamics on the impact of rabbit hemorrhagic disease. *Conservation Biology*, 20, 1232-1241.

Carrete, M., Donázar, J.A., and Margalida, A. (2006). Population increase causes productivity depression in Pyrenean bearded vultures: the need of considering density-dependence processes in conservation plans. *Ecological Applications*, 16: 1674-1682.

Carrete, M., Grande, J.M., Tella, J.L., Sánchez-Zapata, J.A., Donázar, J.A., Díaz-Delgado, R. and Romo, A. (2007). Habitat, human pressure, and social behavior: partialling out factors affecting large-scale territory extinction in an endangered vulture. *Biological Conservation*, 136: 143-154.

Carter, I. (2001). *The Red Kite*. Chelmsford: Arlequin Press, UK.

Caughley, G. and Gunn, A. (1995). *Conservation Biology in Theory and Practice*. Oxford: Blackwell Science, UK.

Contreras, A. (2001). *Impacto sobre la avifauna de la implantación del Plan de Gestión de los Resíduos Sólidos Urbanos en Segovia*. Segovia: Caja Segovia, Spain.

Daszak, P., Cunningham, A.A. and Hyatt, A.D. (2000) Emerging infecting diseases of wildlife: threats to biodiversity and human health. *Science*, 287, 443-449.

Dean, J. I., $ Fernández, A. (2006). La invernada del milano real en Navarra. *Quercus* 248, 10-16.

De Juana, E., De Juana, F., and Calvo, S. (1988). La invernada de aves de presa (O. Falconiformes) en la Península Ibérica. In: J.L. Tellería (Ed.), *Invernada de aves en la Peninsula Iberica* (pp. 97-123), Madrid: Monografias SEO 1.

Delibes, M., Gaona, P. and Ferreras, P. (2001). Effects of an attractive sink leading into maladaptive habitat selection. *American Naturalist*, 158, 277-285.

DeVault, T. L., Rhodes, O. E., and Shivik, J.A. (2003). Scavenging by vertebrates: behavioral, ecological, and evolutionary perspectives on an important energy transfer pathway in terrestrial ecosystems. *Oikos*, 102, 225-234.

Diamond, J. (1989). Overview of recent extinctions. In: D. Western and M. Pearl (Eds.), *Conservation for the Twenty-first century*, Oxford University Press: New York.

Díaz, M., Asensio, B. and Tellería, J.L. (1996). Aves Ibéricas. I. No Paseriformes. J.M. Reyero, Ed: Madrid.

Dionisio, F., Conceicao, I.C., Marques, A.C.R., Fernandes, L. and Gordo, I. (2005). The evolution of conjugative plasmid and its ability to increase bacterial fitness. *Biology Letters*, 1, 250-252.

Donázar, J.A., Naveso, M.A., Tella, J.L. and Campión, D. (1997) Extensive grazing and raptors in Spain. In: D. Pain and M.W. Pienkowsky (Eds.), *Farming and Birds in Europe* (pp. 117-149). London: Academic Press.

Donázar, J.A., Blanco, G., Hiraldo, F., Soto-Largo, E. and Oria, J. (2002). Effects of forestry and other land-use practices on the conservation of cinereous vultures. *Ecological Applications,* 12, 1445-1456.

Esler, D. (1999). Applying metapopulation theory to conservation of migratory birds. *Conservation Biology*, 14, 366-372.

Fallacara, D.M., Monahan, C.M., Morishita, T.Y., and Wack, R.F. (2001). Fecal shedding and antimicrobial susceptibility of selected bacterial pathogens and a survey of intestinal parasites in free-living waterfowl. *Avian Diseases*, 45,128-135.

Ferrer, M., García, L. and Cadenas, R. (1990). Long-tern changes in nest defence intensity of the Spanish Imperial Eagle *Aquila adalberti*. *Ardea*, 78, 395-398.

Forero, M.G., Donázar, J.A., and Hiraldo, F. (2002). Causes and fitness consequences of natal dispersal in a population of black kites. *Ecology*, 83, 858-872.

Fox, A.D. and Madsen, J. (1997). Behavioural and distributional effects of hunting disturbance on waterbirds in Europe: implications for refuge design. *Journal of Applied Ecology*, 34, 1-13.

Fretwell, S. D. (1972). *Populations in a Seasonal Environment.* Princeton, NJ: Princeton University Press.

Friend, M., and Franson, C.J. (1999). *Field Manual of Wildlife Diseases*, Washington, D.C.: U.S. Goverment Printing Office and USGS.

Friend, M., McLean, R.G., and Dein, F.J. (2001). Disease emergence in birds, challenges for the twenty-first century. *The Auk*, 118, 290-303.

Fry, D. M. (1995). Reproductive effects in birds exposed to pesticides and industrial chemicals. *Ennviromental Health Perspectives,* 103, 165-171.

García, J.T., Viñuela, J. and Sunyer, C. (1998) Geographic variation of the winter diet of the Red kite *Milvus milvus* in the Iberian Peninsula. *Ibis*, 140, 302-309.

Gaston, K. J., R. L. Pressey, and C. R. Margules. (2002). Persistence and vulnerability: retaining biodiversity in the landscape and in protected areas. *Journal of Biosciences*, 27, 361–384.

González-Esteban, J., and Villate, I. (2002). Topillo Campesino, *Microtus arvalis*, In: *Atlas de los Mamiferos Terrestres de España.* Madrid:vDirección General de Conservación de la Naturaleza-SECEM-SECEMU.

Green, R.E., Newton, I, Shultz, S., Cunningham, A.A., Gilbert, M., Pain, D.J. and Prakash, V. (2004) Diclofenec poisoning as a cause of vulture population declines across the Indian subcontinent. *Journal of Applied Ecology*, 41, 793-800.

Hanski, J. and Gilpin, M. (1991). Metapopulation dynamics-brief history and conceptual domain. *Biological Journal of the Linnean Society*, 42, 3-16.

Henriques Normark, B., and Normark, S. (2002). Evolution and spread of antibiotic resistance. *J.Int. Med.*, 252, 91–106.

Hiraldo, F., Blanco, J.C. and Bustamante, J. (1991) Unspecialized exploitation of small carcasses by birds. *Bird Study*, 38, 200-207.

Hiraldo, F., Heredia, B. and Alonso, J.C. (1993). Communal roosting of wintering red kites Milvus milvus (Aves. Acipitridae): social feeding strategies for the exploitation of food resources. *Ethology,* 93, 117-124.

Houston, D.C. and Cooper, J.E. (1975) The digestive tract of the white-back griffon vulture and its role in disease transmission among wild ungulates. *Journal of Wildlife Diseases*, 11, 306-313.

Hubálek, Z. (1994) Pathogenic microorganisms associated with free-living birds (a review). Acta *scientiarum naturalium Academiae scientiarum Bohemicae-Brno*, 28, 1-74.

Iwata, H., Tanabe, S., Sakai, N., and Tatsukawa, R. (1993). Distribution of persistent organochlorines in oceanic air and surface seawater and the role of ocean in their global transport and fate. *Environmental Science and Technology*, 27, 1080-1098.

Janzen, D. H. (1986). The eternal external threat. In: M.E. Soulé (Ed.), *Conservation Biology* (pp. 286-303), Sunderland, USA: Sinauer.

Jha, S. and Bawa, K.S. (2006). Population Growth, Human Development, and Deforestation in Biodiversity Hotspots. *Conservation Biology*, 20, 906–912.

Jiménez, B., Hernández, L.M., González, M.J., Eljarrat, E., and Rivera, J. (1996). PCDDs and PCDFs in soils near a clinical waste incinerator in Madrid, Spain. Chemometric comparison with other pollution sources and soils. *Chemosphere*, 32, 1327-1348.

Jiménez, M., Cuesta, S., Jimenez, O., Garcia, M. A., Hernandez, L.M., Marina, M. L., and González M. J. (2000). Organochlorine and heavy metal residues in the water/sediments system of the Southeast Regional Park in Madrid, Spain. *Chemosphere,* 4, 801-812.

Johnson, J. E., Watson, R. T., and Mindell, D. P. (2005). Prioritizing species conservation: does the Cape Verde kite exist?. *Proceedings of the Royal Society B, London*, 272, 1365-1371.

Junta de Extremadura (2005). *Asistencia técnica para el estudio de determinación del tamaño, distribución, problematica y propuestas de conservación de las poblaciones de algunas especies amenazadas en Extremadura.* Cáceres: Dirección General de Medio Ambiente-Consultores en Biología de la Conservación.

Kenney, J.S., Smith, J.L.D., Starfield, A.M., and Mcdougal, C.W. (1994). The long-term effects of tiger poaching on population viability. *Conservation Biology*, 9, 1127-1133.

Lamberski, N., Hull, A.C., Beckmen, K., and Morishita, T.Y. (2003) A survey of the choanal and cloacal aerobic bacterial flora in free-living and captive red-tailed hawks (*Buteo jamaicensis*) and Cooper's hawks (*Accipiter cooperii*). *Journal of Avian Medicine and Surgery*, 17, 131-135.

Langelier, K.M. (1993) Barbituate poisoning in twenty-nine bald eagles. In: P.T. Redig, J.E. Cooper, and J.D. Remple (Eds.), *Raptor Biomedicine* (pp. 231-232). Minneapolis: University of Minnesota Press.

Mammen, U., and Stubbe, M. (2002). Jahresbericht 2001 zum Monitoring Greifvögel und Eulen Europas. Jahresber Monit. *Greifvögel Eulenarten*, 14, 1-11.

Martínez, J.L., and Baquero, F. (2002). Interactions among strategies associated with bacterial infection: pathogenicity, epidemicity, and antibiotic resistance. *Clinical Microbiology Review*, 15, 647-679.

Martínez, F., Arroyo, B., Lemus, J.A. and Blanco, G. (2004) El declive del alimoche en Segovia da la pista sobre la situación actual de la especie. *Quercus*, 227, 69.

Mateu, E., and Martin,M. (2001). Why is anti-microbial resistance a veterinary problem as well?. *Journal of Veterinary Medicine B*, 48, 569-581.

McCullough, D. R. (1996). Spatially structured populations and harvest theory. *Journal of Wildlife Management*, 60, 1-9.

McEwen, S.A. and Fedorka-Cray, P.J. (2002). Antimicrobial Use and Resistance in Animals. *Clinical Infectious Diseases*, 34 (Suppl 3),S93–106.

Meffe, G. and Carroll, C. (1997). *Principles of Conservation Biology*, 2nd ed. Sinauer Associates, Inc.

Mølbak, K. (2004). Spread of resistant bacteria and resistance genes from animals to humans-the public health consequences. *Journal of Veterinary Medicine B*, 51, 364-369.

Newton, I. (1998). *Population Limitation in Birds*. London: Academic Press.

O'Brien, T. F. (2002). Emergence, spread and environmental effect of antimicrobial resistance, how use of an antimicrobial anywhere can increase resistance to any antimicrobial anywhere else. *Clinical Infectious Diseases*, 34 (Suppl. 3), S78-S84.

Oehme, M. (1991). Further evidence of long-range air transport of polychlorinated aromated and pesticides: North America and Eurasia to the Arctic. *Ambio*, 20, 293-297.

Okeke, I.N., Lamikanra, A., and Edelman, R. (1999). Socioeconomic and behavioral factors leading to acquired bacterial resistance to antibiotics in developing countries. *Emerging Infectious Diseases*, 5, 18-27.

Opdam, P. (1991). Metapopulation thery and habitat fragmentation: a review of holarctic breeding bird studies. *Landscape Ecology*, 5, 93-106.

Organbidexka Col Libre. (2003). Suivi des populations migatrices post nuptiales transpyreneenes de Milans Royaux?. *OCL*, 42:6.

Ortega, A. and Casado, S. (1991) Alimentación del milano real en la provincia de Madrid. *Doñana Acta Vertebrata*, 18, 195-204.

Pain, D.J. (1996). Lead in Waterfowl. In: Beyer, W.N., Heinz, G.H., Redmon-Norwood, A. W. (Eds.), *Environmental Contaminants in Wildlife: Interpreting Tissue Concentrations* (pp. 251-264), Boca Ratón, FL: CRC Press.

Pressey, R. L., H. P. Possingham, and C. R. Margules. (1996). Optimality in reserve selection algorithms: when does it matter and how much? *Biological Conservation,* 76, 259–267.

Pulliam, H.R. (1988). Sources, sinks, and population regulation. *American Naturalist*, 132, 652-661.

Pulliam, H. R., Liu, J., Dunning, J.B., Stewart, D.J., and Bishop, T. D. (1994). Modelling animal populations in changing landscapes. *Ibis,* 137, 120-126.

Pullin, A. S., and Knight, T.M. (2001). Effectiveness in conservation practice: pointers from medicine and public health. *Conservation Biology*, 15, 50-54.

Pullin, A. S., Knight, T.M., Stone, D.A. and Charman, K. (2004). Do conservation managers use scientific evidence to support their decision-making? *Biological Conservation*, 119, 245–252.

Rappole, J.H., Mortin, E. S., Lovejoy III, T.E., and Ruos, J.R. (1993). *Aves migradoras Nearticas en los Neotrópicos*. Smithsonian Institution.

Reche, M.P., Jiménez, P.A., Alvárez, F., García de los Ríos, J.E., Rojas, A.M. and De Pedro, P. (2003) Incidence of Salmonellae in captive and wild free-living raptorial birds in central Spain. *Journal of Veterinary Medicine B*, 50, 42-44.

Redford, K.H. and Richter, B.D. (1999). Conservation in a world of use. *Conservation Biology*, 13, 1246-1256.

Reed, J. M. and Dobson, A. P. (1993). Behavioural constraints and conservation biology: conspecific attraction and recruitment. *Trends in Ecology and Evolution*, 8, 353-356.

Reed, J.M. (1999). The role of behaviour in recent avian extinction and endangerment. *Conservation Biology,*: 13, 232-241.

Refsum, T., Vikoren, T., Handeland, K., Kapperud, G. and Holstad, G. (2003) Epidemiologic and pathologic aspects of *Salmonella typhimurium* infection in passerine birds in Norway. *Journal of Wildlife Diseases*, 39, 64-72.

Rodrigues, L., Macedo, L., Robert, J., Fernandes, M.C., Santos Meira, A.T., de Lima, L.A., Fernandes, R.A:, Peregrino, P.C. and Bernardes, M.E. (2003) Dominant culturable bacterial microbiota in the digestive tract of the american black vulture (*Coragyps atratus* Bechstein 1793) and search for antagonistic substances. *Brazilian Journal of Microbiology*, 34, 218-224.

Schlaepfer, M.A., Runge, M.C. and Sherman, P. W. (2002). Ecological and evolutionary traps. *Trends in Ecology and Evolution,* 17, 474-480.

Schneitz, C. (2005) Competitive exclusion in poultry—30 years of research. *Food Control*, 16, 657-667.

Sergio, F. and Penteriani, V. (2005). Public information and territory establishment in a loosely colonial raptor. *Ecology,* 86, 340-346.

Sergio, F., Blas, J., Forero, M.G., Fernández, N., Donázar, J.A. and Hiraldo, F. (2005). Preservation of wide-ranging top predators by site protection: black and red kites in Doñana National Park. *Biological Conservation*, 125, 11-21.

Serrano, D. (1999) Dumps for dead livestock and the conservation of wintering red kites (*Milvus milvus*). *Journal of Raptor Research*, 33, 338-340.

Serrano, D., Tella, J.L., Forero, M., and Donázar, J.A. (2001). Factors affecting breeding dispersal in the facultatively colonial lesser kestrel: individual experience vs. conspecific cues. *Journal of Animal Ecology,* 70, 568-578.

Serrano, D. and Tella, J.L. (2003). Dispersal within a spatially structured population of lesser kestrels: the role of spatial isolation and conspecific attraction. *Journal of Animal Ecology,* 72, 400-410.

Shultz, S., Sagar Baral, H., Charman, S., Cunningham, A.A., Das, D., Ghalsasi, G.R., Goudar, M.S., Green, R. E., Jones, A., Nighot, P., Pain, D.J. and Prakash, V. (2004) Diclofenac poisoning is widespread in declining vulture populations across the Indian subcontinent. *Proceedings of the Royal Society of London B (Supplement),* DOI 10, 1098/rsbl. 2004. 0223.

Singh, R.B., and Sherikar, A.A. (1999). An experimental study on the normal gut microflora and the survival of known bacterial pathogens through the intestinal tract of indian whitebacked vultures *Gyps bengalensis*. *Vulture News,* 40, 20-25.

Spahn, S.A., and Sherry, T. W. (1999). Cadmium and lead exposure associated with reduced growth rates, poorer fledgling success of little blue herons chicks (*Egretta caerulea*) in south Louisiana wtlands. *Archives in Environmental Contamination and Toxicology,* 37, 377-384.

Soto-Largo, E. (2002). *Tipificación de muladares y control de los cadáveres depositados. Utilización como comederos por especies protegidas de fauna silvestre en la provincia de Segovia.* Informe inédito. Boscaje S.L. Consejería de Medio Ambiente. Junta de Castilla-León.

Sutherland, W. (1998). *Conservation Science and Action.* Oxford: Blackwell Science.

Sutherland, W., Pullin, A.S., Dolman, P.M. and Knight, T.M. (2004). The need for evidence-based conservation. *Trends in Ecology and Evolution,* 19, 305-308.

Sunyer, C. and Viñuela, J. (1994). Variación temporal en los habitos alimentarios del milano real durante la invernada en la Meseta Norte. *Ardeola,* 41, 161-167.

Tella, J.L. (2001). Action is needed now, or BSE crisis could wipe out endangered birds of prey. *Nature,* 410, 408.

Thiollay, J.M. (2001). The red kite in Europe: unprecedent decline and a call for action. *Raptor News,* 2, 29-32.

Thomas, N. J. (1999). Barbituates. In: M. Friend and J. C. Franson (Eds.), *Field Manual of Wildlife Diseases* (pp. 349-350), Washington, D. C: U.S. Goverment Printing Office and USGS.

Thorns, C.J. (2000) Bacterial food-borne zoonoses. *Review of Science and Technology,* 19, 226-239.

Tizard, I, (2004) Salmonellosis in wild birds. *Seminars in Avian and Exotic Pet Medicine,* 13, 50-66.

Trakhtenbrot, A., Nathan, R., Perry, G., and Richardson, D.M. (2005). The importance of long-distance dispersal in biodiversity conservation. *Diversity and Distributions,* 11, 173-181.

Tucker, G.M. and Heath, M.F. (1994). *Birds in Europe: their Conservation Status.* Cambridge: Birdlife International.

Veiga, J.P., and Hiraldo, F. (1989). Food habits and the survival and growth of nestlings in two sympatric kites (*Milvus milvus* and *Milvus migrans*). *Holarctic Ecology,* 13, 62-71.

Viñuela, J., Martí, R. and Ruiz, A. (1999) *El Milano Real en España.* Madrid: SEO/Birdlife.

Viñuela, J. 2003. Milano Real, *Milvus milvus*. In: *Atlas de las Aves Reproductoras de España* (pp. 162-163), Madrid: Dirección General de Conservación de la Naturaleza-Sociedad Española de Ornitología.

Waldenström, J., Broman, T., Carlsson, I., Hasselquist, D., Achterberg, R.,. Wagenaar, J., and Olsen, B. 2002. Prevalence of *Campylobacter jejuni, Campylobacter lari,* and *Campylobacter coli* in different ecological guilds and taxa of migrating birds. *Applied and Environmental Microbiolog,* 68, 5911-5917.

Watkinson, A.R. and Sutherland, W.J. (1995). Sources, sinks and pseudosinks. *Journal of Animal Ecology*, 64, 126-130.

Whitfield, D.P., Fielding, A.H., Mcleod, D.R.A., and Haworth, P.F. (2004). The effects of persecution on age of breeding and territory occupation in golden eagles in Scotland. *Biological Conservation*, 118, 249-259.

Winsor, D.K., Bloebaum, A.P. and Mathewson, J.J. (1981) Gram-negative, aerobic, enteric pathogens among intestinal microflora of wild turkey vultures in west central Texas. *Applied and Environmental Microbiology*, 42, 1123-1124.

Wotton, S.R., Carter, I., Cross, A.V., Etheridge, B., Snell, N., Duffy, K., Thorpe, R. and Gregory, R.D. (2002). Breeding status of the Red kite Milvus milvus in Britain in 2000. *Bird Study,* 49, 278-286.

Wright, R. G. (1996). *National parks and protected areas: their role in environmental protection.* Oxford: Blackwell Science.

In: Environmental Research Trends
Editor: Marvin A. Cato, pp. 205-230
ISBN: 978-1-60021-556-8

Chapter 6

INFLUENCE OF LEAD AND CADMIUM ON SOME GENETIC AND PHYSIOLOGICAL PARAMETERS OF LABORATORY MICE

R. Metcheva[1], M. Topashka-Ancheva[1] and S. E. Teodorova[2]

[1] Institute of Zoology, Bulgarian Academy of Sciences,
1 Tzar Osvoboditel boul. 1000 Sofia, Bulgaria
[2]Institute for Nuclear Research and Nuclear Energy,
Bulgarian Academy of Sciences,
72 Tzarigradsko chaussee, 1784 Sofia, Bulgaria

ABSTRACT

Some aspects of influence of industrial pollutant on small mammals were studied in an ecologo-toxicological experiment. Polymetal dust, containing Pb and Cd, mixed at 1 % concentration with conventional animal food, was administrated to laboratory mice of the inbreed line BALB/c. The experiment spanned 90 days. Structural and numerical chromosome changes were studied as characteristics of genotoxicity. Resting metabolism and thermoregulation, examined over the range of ambient temperatures 0–30°C, were considered as parameters of the general physiological status. Also body weight of the mice was established in the context of metal bioaccumulation. On days 15, 40, 60, and 90 bone marrow samples were taken and oxygen consumption, body temperature and body weight were measured. Males had significantly higher chromosome aberration frequency (CAF) than females. The highest CAF observed in males was 25.7% (on day 60), while in females it was 16.3% (on day 40). Chromatide breaks were about 1.8 times more frequent than centromeric fusions. High aneuploidy was found in males (13.5%), 3 times more than in females (4%). Fragments, isochromatide breaks, c/t aberrations and polyploidy were found infrequently. No dose-effect dependence between the frequency of chromosome aberrations and duration of exposure was recorded. This suggests that up to day 90 of the exposure there was no decline in the efficacy of the repair mechanisms. The oxygen consumption, body temperature, and body weight sharply decreased on day 90; at 25 °C they were 59%, 7%, and 28%, respectively, lower compared to the control group. On day 90 of exposure the experimental mice could not thermoregulate and their rate of resting metabolism reduction was 1.35 times higher compared to that in the controls. The results confirm that chromosome alterations are a very sensitive parameter indicating heavy metal burden. Resting metabolism, body temperature, and body weight are not among the most sensitive biomarkers but they could be used for estimation of tolerable metal levels in physiological aspect.

INTRODUCTION

Two main tendencies in human society have developed in contrast in the recent decades; two types of human needs have met at a crossroad in the contemporary civilization highway. On one hand, the modern man desires to spend more time amid native nature to recover his health and power. On the other hand, today's life style requirements strongly accelerate the time course of industry, which in turn creates an adverse impact on the environment. Besides, without industry it is not possible to apply the new products of advanced scientific and technical endeavor, directed towards further progress. Luckily, the science development is becoming progressively involved in the invention of new technologies aimed at preserving an unpolluted environmental.

New sources of energy will be created and new technologies will be used but the traditional technologies remain absolutely necessary for the present and probably yet for a long time. Regardless of the existing multiple different ways to limit waste, many plants and factories are serious sources of environmental pollution. One of the most harmful industries is related to heavy metals production. Small particles of the waste can be deposited on plants and in the soil or remain in the air over a long period of time. They could influence humans by contamination of the food chain or directly by breathing. Because of that it is necessary to study the effect of pollutants on the human health. The specific hazards, the scale of their effect and the adaptive possibilities of the organism should be elucidated.

Small mammals are often explored both in the wild life and in laboratory in order to make reliable conclusion that are applicable to humans as well. If their genetic and physiological functions are closely related to those of humans then the results obtained may be reasonably extrapolated to people. Such animals, usually rodents, are selected on the basis of their position in the food chain, rapid maturity, can be studied in large populations, are wide-spread, and have biological reaction to the environmental changes similar to those in humans. Small mammals have been often used as bioindicators of pollution, with residues being determined in either the whole body or in specific organs and tissues (Martin § Coughtry 1982, Wern 1986, Talmage § Walton 1991, Pascoe et al. 1996).

In the regions with developed metallurgy the concentrations of metals such as lead, cadmium, zinc, copper, nickel, manganese etc., exceed by several fold the permissible limits in both the ecotic and abiotic components of the environment. Thus, the living nature is in danger because of the limited abilities of the ecosystems to detoxify the waste residues. It is not possible more of them to be decomposed or otherwise eliminated.

Investigation of distribution, accumulation, transfer, and toxic effects of heavy metals on the living organisms is an important topic and many studies are dedicated on these problems. Lead, cadmium and mercury are the most toxic heavy metals, affecting genetic status, hemopoesis, blood parameters, different physiological functions, and body weight. In the present chapter we focus our attention on lead and cadmium.

Lead affects a wide range of physiological systems, including the central nervous system, kidneys, hematopoetic system, cardiovascular system, and gastrointestinal tract (Ma 1996). Cadmium exposure is a chronic problem resulting in a gradual accumulation of the metal in target organs (kidney and liver) which often causes renal damages. Chronic pulmonary disease, disturbance of calcium metabolism, and hypertension and cardiovascular effects could also occur as a result of Cd intake (Goyer 1996), making the investigation of accumulation and distribution of these metals truly important. Barry (1975) has established that lead accumulates in mammalian tissues in the following order of concentrations: bone>kidneys>liver>brain>muscle. The absorption, disposition, accumulation, and release of Pb from bones have been studied in details (O'Flaherty 1995a, O'Flaherty *et al.* (1996). Fleming *et al.* (1999) have determined the relationship among blood lead concentration, plasma Pb concentration, and Pb excretion in monkeys in order to assess the comparability of Pb disposition in cynomolgus monkey to that in the humans. A significant ecotoxicological risk to a wild population of bank voles, associated with high Pb tissue concentration has been estimated (Milton *et al.* 2003). Tissue concentrations of Cd in wild species and critical tissue concentrations of this metal leading to different dysfunctions have been discussed by Cooke and Johnson (1996). Müller et al. (1986) have established that the distribution of Cd between various organs depends on the formulation of the Cd applied. A very interesting problem, connected with Cd accumulation is the decreasing of kidney/liver ratio with increasing of total body burden of Cd. This ratio has been studied by Bremner 1979; Andrews and Cooke 1984; Hunter *et al.* 1987a, 1987b; 1989; Ma *et al.* 1991; Dodds-Smith *et al.* 1992. The tendency of change in kidney/liver ratio was confirmed by our observation, which showed an inversion of cadmium kidney/liver ratio (the ratio declined bellow unity) in the conditions of very high exposure (Teodorova et al. 2003). In mice this ratio has been usually found in the interval 2/1 – 5/1 (Cooke and Johnson 1996). The order of Pb and Cd accumulation in the liver and body of rodents and insectivores has also been established (Metcheva *et al.* 2003).

A quantitative investigation of bioaccumulation of heavy metals and toxic elements could be used to determine their toxicokinetic parameters. However, not many relevant kinetic models are published. O' Flaherty (1998a) has developed a physiologically based kinetic models for arsenic, lead, chromium, and mercury. The O'Flaherty model, a physiologically based computer model of Pb disposition in humans, explaines well the trends for the accumulation of Pb in cortical bone and the release of Pb from bone stores (Fleming et al. 1999). The return of bone Pb to blood, confirmed by experimental observations, was modeled in terms of aging and osteoporotic bone loss (O' Flaherty 2000). Taking into account the important features of metal kinetics: metal-protein binding and metal-metal interactions, we proposed a mathematical model regarding Cd and Zn age-bioaccumulation in laboratory mice (Teodorova et al. 2003). Like all models, these for metals kinetics have collected a variety of observations and identified the needs for more complete experimental information.

Heavy metals toxicity affects the organism on cellular and genetic level. The mechanisms of the toxic effects of Pb have been considered at both cellular and subcellular levels (Tsuchia 1979). Venugopal and Lluckey (1978) have evaluated the carciniogenic activity of Cd. Lead is known to damage the chromosomal structure in mammalian cells (Dhir *et al.* 1990; Nayak *et al.* 1989; Hartwig *et al.* 1990). Clastogenic effects of industrial dust, containing Pb and Cd, on Guenther's vole (*Microtus guentheri*) was recorded in a 60-day ecologo-toxicological experiment (Topashka-Ancheva *et al.* 1995, 1998). Bilban and Jakopin (2005) have found that the incidence of structural chromosome aberrations in mine workers working in a lead-zinc mine is probably more strongly influenced by heavy metals than by radon. Lead has multiple hematological effects. Microcytosis and hypohromy of red blood cells appear because of disturbed heme synthesis (Goyer 1996). Reticulocytosis have been observed in adult amphibian exposed to inorganic Pb (Marcos *et al.* 2004). Significant decreases in red blood cells and mean corpuscular hemoglobin and significant increase in the frequency of micronucleated polychromatic bone marrow erythrocytes were found in Algerian mice (*Mus spretus*) exposed to Pb for 15, 45, and 90 days (Marques *et al.* 2006).

We have found severe disturbances of the blood picture in our previous studies. Micronuclei (Howell-Jolly bodies) and basophilic granulations were found in rodents inhabiting lead-polluted regions (Metcheva *et al.* 1994, 1995, 1997; Topashka-Ancheva *et al.* 2003a). All symptoms of lead-induced anemia (decrease in hemoglobin, significant reticulocytosis, microcytosis, hypohromy, and basophilic erythrocytic granulation) were recorded in laboratory mice exposed to Pb and Cd containing industrial mixture in experiment covered a period of 90 days (Topashka-Ancheva *et al.* 2003b).

The present study is focused on the genetic and physiological responses of the mammalian organism to an exposure to lead and cadmium, mixed in polymetal industrial dust. The administration of heavy metals using a procedure of supplementation of the conventional animal food by metal mixture affords to simulate toxicity, which might result from environmental pollution through contaminated food. Lead is a component of factory production. Cadmium is always associated with zinc and lead both in the lead industry and in the natural state. Lead and cadmium happen often in combination in industrial mixtures and thus, the study of toxic effects caused by such a mixture, in a long-term experimental exposure, allows for a context of more realistic environmental conditions. The metal concentrations applied were enhanced compared to those occurring in the environment but was chosen as means to assess the maximum hazard levels for the mammalian organism.

Experimental Setting

Laboratory mice of the inbreed line BALB/c were selected and treated in conditions of chronic ecologo-toxicological experiment. One hundred and eighty male and female white mice BALB/c, 8 - 10 weeks of age and weighing 25 ± 1.5g, were divided into two groups as follows: control group of 50 and experimental group of 130 animals. The experimental mice were treated with polymetal industrial dust from the electrofilters of Lead-Zinc Plant "KCM-Plovdiv", mechanically mixed in 1 % concentration with conventional, balanced animal food. The polymetal samples collected from the electrofilters represented airborne contaminants in the working zone of the Plant.

Control and experimental animals were bred in a vivarium and housed in cages constructed of solid plastic sides with stainless steel grid ceilings and floors in a room with a controlled temperatures (approximately 20°C), humidity (45-75%), and light cycle (12 h light, 12 h dark). All mice were allowed access to food and water ad libitum.

The diet of mice contained the following quantities of heavy metals, respectively in control and experimental groups: lead – 61.8 ± 20.9 mg/kg and 784 ± 243.6 mg/kg; cadmium – 3.5 ± 1.9 mg/kg and 64.1 ± 10.5 mg/kg; zinc – 90.9 ± 25.9 mg/kg and 1945.0 ± 429.2 mg/kg; and copper – 11.3 ± 1.1 mg/kg and 20.9 ± 7.3 mg/kg.

Concentrations of these metals were determined in carcass, liver, kidney, and bone of the control and experimental animals on days 15, 40, 60, and 90. To determine the bioaccumulation of heavy metals, the animals were killed and the alimentary tract was removed; tissues were dried at 60^0C to a constant weight. The dried samples were dissolved in a mixture of concentrated nitric–perhloric acid 4:1 (Sawicka-Kapusta *et al.* 1987a). The analyses were carried out by atomic absorption spectrometry using a Perkin-Elmer 2380 instrument. The data were tested using analysis of variance (Parker, 1973).

Oxygen consumption [O_2 g^{-1} h^{-1}], body temperature [^{0}C], body weight [g], and chromosome changes [%], were determined on days 15, 40, 60, and 90 of the exposure. The oxygen consumption and body temperature were measured at the following ambient temperatures: 0°, 5°, 10°, 15°, 20°, 25°, and 30°C. Bone marrow cells were taken at 20°C ambient temperature. At each time point, control and experimental study groups – subsets of 20 mice, 10 males and 10 females – were used for measurements of the O_2 consumption, body temperature, and body weight; and subsets of 10 mice, 5 males and 5 females were used for determination of chromosome aberrations.

Comparisons of the investigated characteristics were made between the experimental and control groups using one-way analysis of variance (ANOVA-SPISS 7.5).

Structural and Numerical Chromosome Changes

Chromosomal aberrations, involving gross alterations of the genetic material, have been regarded as a sensitive endpoint for detecting genotoxic effects induced by heavy metal and toxic chemicals. Many chemical clastogens induce chromatid aberrations as a result of DNA synthesis on a damaged DNA template in the S-period of cell cycle (Sorsa *et al*; 1992). According Longhese et al. (2006) DNA double-strand breaks are among the most deleterious types of damage that can occur in genome of eukaryotic cells because repair failure can lead to loss of genetic information and chromosome rearrangements. Such data are considered highly relevant in the human context (International Commission for Protection 1983, World Health Organisation 1985). Investigations of the effects of toxic metals in different concentrations and compositions will be useful to plan and undertake further experiments for explaining of mechanisms of metal-DNA interactions.

In the present experiment cytogenetical analysis was performed as described by Preston *et al.* (1987). Mitomycin C (3.5 mg/kg, in saline solution, ip ingected) was used as a positive control. The control animals were injected with only 0.2 ml 0.9% NaCl. Bone marrow chromosomal aberration assays were performed on groups of animals, each one (contol and experimental) consisting of 5 males and 5 females. The animals were injected with a dose of 0.4 mg/kg colchicine 1 h before isolation of the bone marrow cells. Bone marrow cells were flushed from the femur and hypotonized with 0.075 M KCl at 37^0C for 20 min. After that the cells were fixed in methanol-acetic acid (3:1), dropped onto cold slides and air dried. To examine chromosome aberrations the slides were stained with 5% Giemsa solution. At least 100 well-spread metaphases were analyzed per animal at random.

The frequencies of chromosome aberrations were determined for each animal. The mean ± SD for each group was calculated and the data were statistically evaluated for their significance by analysis of variance using Student t-test.

The data regarding chromosome aberrations frequency - CAF in the bone marrow cells of the mice are displayed in Figure 1. The concentrations of Pb and Cd in liver, kidney, spleen, bone, and carcass during the experiment are presented in Table 1. Lead is well known as bone-seeking element and it is expected that their concentration in the bones is 3.4 times higher than in the whole body (on day 90 of exposure).

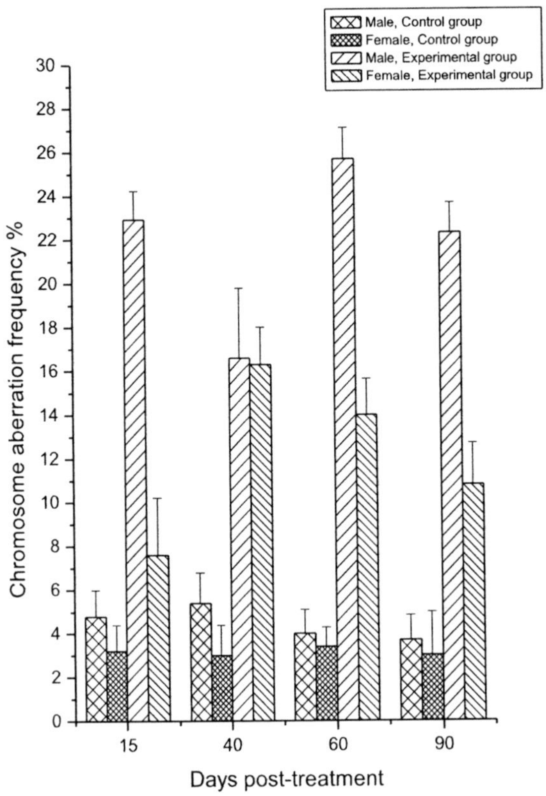

Figure 1. Frequency of chromosome aberrations (CAF) in laboratory mice BALB/c determined on days 15, 40, 60, and 90 of the exposure to Pb and Cd containing industrial dust.

The percentages of aberrant bone marrow cells of males and females from the control group were within in the boundaries of spontaneous frequencies. The differences between control and experimental animals, both males and females, are statistically significant (Table 2). The CAF was 6.4 times higher in experimental males compared to the controls on day 60. If the CAF in males and females are summarized, we obtain that on days 15, 40, 60, and 90 CAF_{M+F} is 30.54%, 32.9%, 37.9%, and 33.1%, respectively. On days 15, 60, and 90 CAFs were 3, 1.84, and 2 times, respectively, higher in male mice when compared to females ($P < 0.001$; Table 3). Only on day 40 the CAFs in males and females were not significantly different ($P > 0.1$).

Chromatide breaks, centromeric fusion (c/c), and telomere-telomer fusions (t/t) in BALB/c mice are displayed in Figure 2. The high CAF in male mice (Figure 1) was due predominantly to the high percentage of chromatide breaks. Isochromatide breaks, centromeric-telomeric fusions, and fragments were found bellow 2%, 1%, and 0.4%, respectively.

Table 1. Concentrations of heavy metals in body and tissues of laboratory mice BALB/c

Body, organs	Heavy metal	Heavy metals concentrations (m g / k g ± S D)				
		Control	Day 15	Day 40	Day 60	Day 90
Liver	Zn	114.6 ± 9.7	166.6 ± 7.9	250 ± 29.3	315 ± 24.3	195.8 ± 31.1
	Cu	20.7 ± 2.9	19 ± 1.9	15.5 ± 1	18.9 ± 1.6	12.6 ± 1.1
	Pb	-	30.3 ± 4.8	128.8 ± 28.9	192.3 ± 27.3	9.8 ± 0.7
	Cd	2.2 ± 1.4	53.2 ±13.2	149.1 ± 18.9	318.4 ± 22.9	479.1 ± 28.9
Kidney	Zn	102.6 ± 4.1	160.6 ± 19.2	241.1 ± 28.5	241.1 ± 28.5	141. 3 ± 31.7
	Cu	25.9 ± 3.8	21.5 ± 2.6	19.2 ± 1.2	23.4 ± 1.3	15.8 ± 0.7
	Pb	-	349.7 ± 37.9	357.5 ± 43.8	326.6 ± 37.6	10.56 ± 2.1
	Cd	3.8 ± 0.6	51.46 ± 4.4	206.4 ± 24.2	323.5 ± 26.3	465.7 ± 11.1
Bone	Zn	184.4 ± 26.5	145.2 ± 31.1	403.4 ± 29.9	223.3 ± 39	138 ± 23.2
	Cu	10.3 ± 1.5	7.33 ±1.4	10.4 ± 2.1	12.6 ± 1.9	6.5 ± 1.7
	Pb	-	410.4 ± 20.3	1524 ± 290	1601.1 ± 38.7	1625 ± 42
	Cd	2.51 ± 1	5.2 ± 1.8	8.7 ± 3.8	10.6 ± 1.7	20.5 ± 2.9
Body	Zn	132 ± 3.1	153.4 ± 2.8	171.2 ± 2.8	188.5 ± 4.8	108.5 ± 4.1
	Cu	9.1 ± 1.2	8.26 ± 0.22	7.4 ± 2.1	9.5 ± 1.8	5.1 ± 0.4
	Pb	-	164.4 ± 13.9	360.3 ± 17.5	484.2 ± 24.2	475 ± 17.9
	Cd	2.3 ± 0.9	14.8 ± 3.5	11.7 ± 1.4	12.6 ± 1.7	13.1 ± 2.9

Table 2. P values of a comparison of the frequency of structural and numerical chromosome aberration, and body weight in control (c) and experimental (15, 40, 60, 90) groups

P	Chromosome aberration frequency		Aneuploidy frequency		Body weight at 20°C
	Male	Female	Male	Female	
$P_{c\,15}$	< 0.001	< 0.02	< 0.001	< 0.01	<0.001
$P_{c\,40}$	< 0.01	< 0.001	< 0.001	< 0.01	> 0.1
$P_{c\,60}$	< 0.001	< 0.001	< 0.001	< 0.05	> 0.1
$P_{c\,90}$	< 0.001	< 0.01	< 0.001	> 0.1	<0.001
$P_{15\,40}$	< 0.05	< 0.001	< 0.001	> 0.1	<0.01
$P_{15\,60}$	< 0.05	< 0.02	< 0.001	< 0.1	<0.001
$P_{15\,90}$	> 0.1	< 0.1	< 0.01	< 0.05	> 0.1
$P_{40\,60}$	< 0.01	< 0.1	< 0.1	< 0.05	<0.02
$P_{40\,90}$	< 0.05	< 0.05	< 0.05	< 0.02	<0.01
$P_{60\,90}$	< 0.02	< 0.1	< 0.05	< 0.05	<0.001

Table 3. P values of a comparison of chromosome aberration frequency (%) between males and females in control and experimental groups, respectively

Aberration type	P_{MF}	Male – Female control	Male - Female experimental
Structural chromosome aberrations	15	> 0.1	< 0.001
	40	< 0.1	> 0.1
	60	> 0.1	< 0.001
	90	> 0.1	< 0.001
Aneuploidy	5	> 0.1	< 0.001
	40	> 0.1	< 0.01
	60	> 0.1	< 0.001
	90	> 0.1	< 0.001

Chromosomal aberrations of c/c type (centromeric fusion) are known as Robertsonian translocations. They cause structure forms as meta- and submeta-centric chromosomes through centric exchanges of two acrocentrics. This phenomenon is due to single strand's breaks in the minor SAT DNA located in the centromeric regions of mice chromosomes (Kipling *et al.* 1994). On day 15 9.9% c/c fusion were recorded in experimental males, 3.4 times more than in experimental females. On day 40 c/c fusions in both males and females were about 50% lower compared to those on day 15 and on days 60 and 90 identical percentages of c/c fusions were found in males and females, higher compared to day 40.

Telomere-telomer fusions (t/t) were also detected in metaphases of the treated laboratory mice, albeit at lower frequency (Figure 2). In females they were rather less compared to males. The higher percentage was 3.7% in males on day 60. Telomeres are specialized nucleoprotein complexes localized at the physical ends of linear eukaryotic chromosomes that play an important role in protecting the ends of chromosomes and preventing chromosome fusion in normal cells. The DNA component of telomeres is characterized by being a G-rich double strand composed by short fragments with different tandem repeats depending on the species considered. Telomere loss results in sister chromatid fusion and prolonged breakage/fusion bridge cycles, leading to extensive DNA amplification and large deletions (Murnane and Sabatier 2004). It is interesting that in our experiment no t/t fusions were found in control mice (Figure 2).

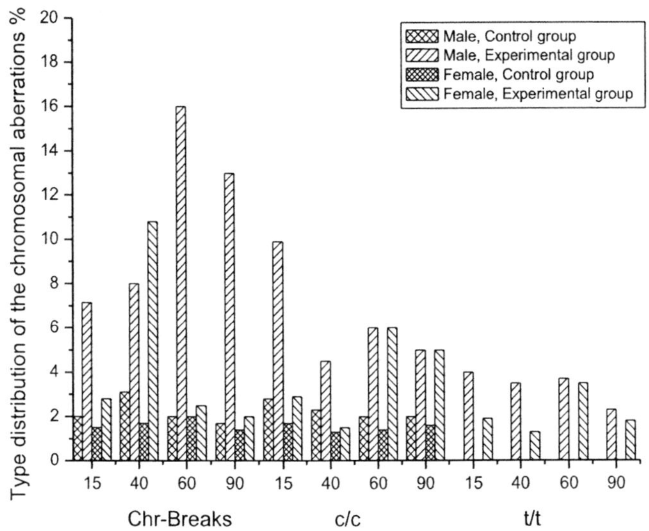

Figure 2. Distribution of the different types chromosomal aberrations found in laboratory mice BALB/c on days 15, 40, 60, and 90 of the exposure to Pb and Cd containing industrial dust.

Chromatide breaks were in highest percentage compared to other structural chromosome changes. In males they were 8%, 16%, and 13% on days 40, 60, and 90, respectively. In females they were bellow 3%, except on day 40, when they were 10.8%. It seems that females react rather tardily to metal toxicity compared to males, in which still on day 15 a strong reaction was observed (7.14%). However, in females the intoxication probably stimulates a stronger and effective repair.

Isochromatide breaks happen usually at a little time before the cell division, while chromatide breaks occur in the chromosomes, which replicate later in interphase. The weak presence of isochromatide breaks in the experiment may be due to the fact that Pb and Cd act on the chromosomes mainly in a later period of DNA replication. This assumption is confirmed also by the lack of DNA fragments because the fragments are parts from those chromosomes, which have been disturbed earlier in the S-period.

Polyploidy in experimental mice was almost not found. This fact suggests that Pb and Cd have not disturbed cell spindle significantly and consequently the normal distribution of the chromosome material has been not changed dramatically.

Significant high aneuploidy was found mainly in males (Figure 3). The aneuploidy in exposed females differed slightly when compared to controls on days 15, 40 and 60. When summarize the percentages of aneuploidy in experimental males and females for all time points of the observation, we obtain that the aneuploidy in males was exactly 3 times higher than that in females. Lead and cadmium compounds are known to affect one or more components of mitotic apparatus, which leads to an erroneous migration of chromosomes. Malsegregation occurs when a chromosome (or a chromatid) fails to migrate and remains at the metaphase plate. Nondisjunction implies the lack of dissociation between sister chromatids and a migration of both together to the same pole. Seoane et al. (2002) have adduced evidence that the aneugenic effect of some Cd and K salts in MRC-5 human cells is a consequence of malsegregation at anaphase but not of nondisjunction mechanisms. Aneugenic effect of Cd has been reported also in mollusca (Pacific oyster, *Crassosterea gigas*) (Bouilly et al. 2006). Aneuploidy as a component of Pb-induced genotoxicity has been observed in lymphocytes from peripheral blood samples of human (Shaik et al. 2006). Our results showed that the aneuploidy is a quite sensitive biomarker of Pb and Cd exposure: on day 15 a strong reaction was still observed; aneuploidy in males was 13.5%, the highest value compared to the percentages measured in other time points.

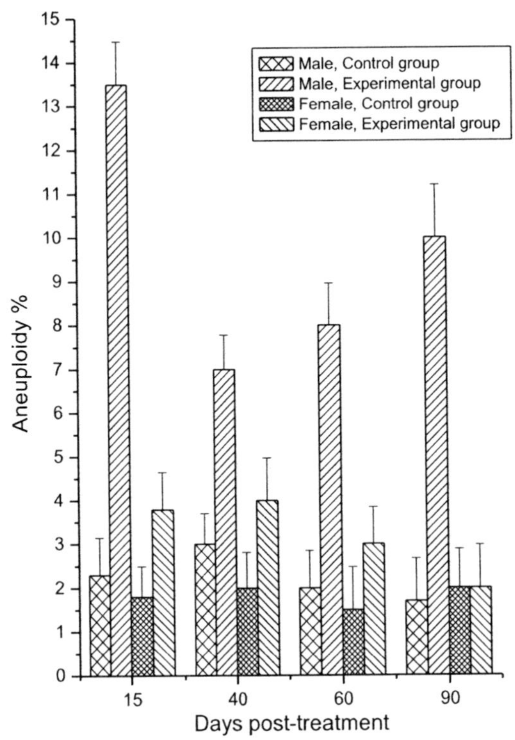

Figure 3. Aneuploidy found in laboratory mice BALB/c on days 15, 40, 60, and 90 of the exposure to Pb and Cd containing industrial dust.

Is the frequency of chromosomal aberrations due to the level of the Pb and Cd in bones? Lead is well known as a factor damaging the chromosomal structure in mammalian cells (Dhir *et al.* 1990). The Pb concentration in bone marrow is proportional to that found in bones (Demarque *et al.*, 1995) and the concentrations of Pb in the bones of the experimental animals was very high: over 1500 mg/kg after day 15 (Table 1). The sharp elevation of the bone Pb correlates with the increased percentage of aberrant cells in females: the CAF on day 40 were 2.14 times higher than on day 15 (Figure 1). It may be assumed that Pb cause predominantly chromatide breaks and centromeric fusions. However, Cd in high concentrations also induces DNA strand break and inhibit the biosynthesis of DNA, RNA and proteins (Snow 1992). In our experiment the concentration of Cd observed in the bone on day 90 was 20.5 $\pm$ 2.9 mg/kg, while the Cd level in the liver on the same time point was 479.1 $\pm$ 28.9 mg/kg (Topashka-Ancheva *et al.* 2003b). Besides, one could expect that Cd from the polymetal dust probably does not appear as a strong factor causing chromosome damage because of the specific distribution of Cd in cells. A small quantity of Cd is found in nucleus and mitochondria and the greeter part is in cytosol, where Cd is bound to metallothionein (Ershov and Pleteneva 1989). According to data obtained from different authors, induction of chromosome aberrations by cadmium compounds are ambiguous (Çelik *et al.* 2005).

Another problem is, whether lead and cadmium interact with genome directly or indirectly. Valverde et al. (2005) have not detected direct induction of DNA strand breaks by lead acetate, cadmium chloride, or a mixture of both metals, all at low non-cytotoxic concentrations.

Moreover basal DNA damage has been different in the different organs. The authors have found an induction of lipid peroxidation and increase in free radicals in of male mice and have concluded that an induction of genotoxicity and carcinogenicity is possible by indirect interactions, such as oxidative stress. The exploration of the influence of metal-induced oxidative stress on the genome will be of great concern to explain a more wide range of problems related to human cancerogenesis. Another form of indirect interaction is the disturbance of DNA repair. Beyersmann and Hechtenberg (1997) have noted an inhibition of DNA repair process by Cd. Hartman and Hartwig (1998) have indicated that Cd blocks DNA repair disturbing protein interactions involved in DNA damage recognition. An action of incision nuclease during DNA repair has been observed because of interaction of cadmium chloride with chromatin condensation (Privezentsev *et al.* 1996). Lead also is known to cause reduction in DNA repair capacity and thus, lead exposed workers will be more prone to DNA damage (Karakaya et al. 2005).

It is quite possible that Cd and Pb in high concentrations affect DNA directly. Snow (1992) has noted that direct genetic damages occur in high concentrations of Cd compounds. Interacting with chromatin, Cd causes DNA strand break, crosslinks or alteration in DNA. Also Privezentsev *et al.* (1996) have established that direct interaction between Cd and DNA leads to single strand breaks in DNA. Shaik et al. (2006) have supposed that Pb can induce single strand DNA breaks, possibly by competing with metal binding sites.

Our data suggest a possible direct metal-DNA interaction. The chromatide breaks were significant and they were the most numerous alterations in the genetic material of bone marrow cells (Figure 2). There are no reasons to suspect an essential disturbance in DNA repair because no significant increase of CAF was observed during the chronic exposure, i. e. no dose-dependence effect was recorded. The CAF in males on day 60 was high (25.7%) but then slightly decreased ($P < 0.02$). In females the CAF on day 40 was 16.3%, two times higher when compared to that of day 15 (7.6%). After day 40, however, the CAF level in females incessantly decreased (Figure 1).

An interesting problem is to understand why the reactions of males and females to exposure of toxic metals are different. It is seen that the CAF in males were higher than the CAF in females (Figure 1, Table 3). The CAF in males were on average 1.8 times higher compared with those in females. The CAF in males had their maximum on day 60, while the CAF in females showed a maximum on day 40. It seems that the chromosome apparatus in males is more sensitive to the metal intoxication: a rapid and strong elevation of CAF occurred after the beginning of the exposure. On days 90 CAF in females was 2 times lower compared with CAF in males. Probably in females a stronger defense reactions developed to protect the chromosomal structure against metal-caused damages. These results suggest a higher sensibility and lower repair in males compared to females.

The scientific literature lacks studies targeted to the elucidation of the differences between the genetic reaction (particularly chromosome aberrations) to toxic agents in male and female individuals. The most investigations of genotoxicity have been carried out on male specimens and thus there are not many results providing comparative analyses. Dubinin (1986) suggested, in the context of experiments studying the dose influence on the mutations rate in mouse oocytes, that the mutation rate in females is lower than that in males, because of more effective repair in females.

Recently Scheirs et al. (2006) investigating effects of environmental pollution on wood mice (*Apodemus sylvaticus*) found that the genetic damage was higher in male mice than in female mice at the most polluted site, but not at the other areas. This is supported to some extent by our data regarding the higher presence of chromosome aberration in males than in females.

Resting Metabolism and Body Temperature in the Context of Thermoregulation

The rate of resting metabolism is calculated by the rate of oxygen consumption per gramme per hour [O_2 g^{-1} h^{-1}]. In the ecological bioenergetics the oxygen consumption, a summary result from all energetic expenses in an animal, is an important parameter. It level could be considered as a factor representing the total physiological status of living organism, depending strongly on the external temperature, season, age and body weight of the animal.

By determination of oxygen consumption rate the zone of temperature comfort of mammalian organism (the thermo-neutral zone) can be established. In this zone the metabolism goes most economically, i.e with minimal O_2 consumption. With the decrease of the ambient temperature O_2 consumption increases. As much as wider is the thermo-neutral zone and as much as smaller is the **r**ate of **r**eduction of **r**esting **m**etabolism (RRRM), so much more adapted is a given species to environmental changes. The anthropogenic pollution generates uncertain conditions and it is of interest to study animals' reactions to the polluted environment. Our exploration showed that the heavy metal contamination reflects in disturbance of some physiological parameters, such as thermoregulation.

The resting metabolism is determined in conditions of a full repose in post-absorptive state. In the rodents the resting metabolism is difficult to measure because they feed often and their stomach is permanently full. Thus, in rodents the resting metabolism is measured as a relative entity: standardly animals are not feed for two hours.

In the vivarium a temperature of 20^0C was maintained. Besides, at each time point of measurements the experimental animals were kept per 20 minutes at ambient temperatures of 0^0, 5^0, 10^0, 15^0, 25^0, and 30^0C ($\pm$ 1^0C) via climatic chamber. The resting metabolic rate was measured using Kalabuchov-Skvorcov closed-circuit respirometer (Gorecki, 1968). The volume of the chamber used was 0.5 L. The oxygen consumption rate was determined in cm^3 O_2 g^{-1} h^{-1}. Body temperature was measured by a Harvard Rat Homeothermic Blanket System supplied with a rectal thermistor (Harvard Apparatus Ltd., Kent, UK) immediately after the measurement of the oxygen consumption at the different ambient temperatures. The thermistor was inserted 20-22 mm into mouse's rectum. The mean ± SD for each group was calculated. The results for control and exposed animals were compared using 1-way ANOVA-SPISS 7.5.

In Figure 4 the dependence of O_2 consumption *r*ate (OCR) by the ambient temperature is displayed. Further we shall sign the sets of values of OCR, measured at the different ambient temperatures, in control group and at each time point in experimental group as follows: $OCR_{Control}$, OCR_{15}, OCR_{40}, OCR_{60}, and OCR_{90}.

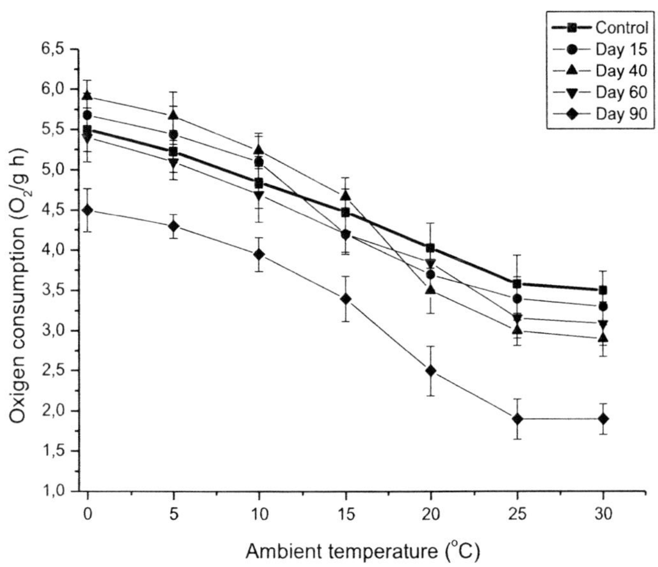

Figure 4. Oxygen consumption rate [O_2 g^{-1} h^{-1}], measured over the range of ambient temperatures 0–30°C in BALB/c mice on days 15, 40, 60, and 90 of the exposure to Pb and Cd containing industrial dust.

In the interval 25–30°C no statistically significant differences in the values of control group were observed ($P > 0.1$) and the average rate of resting metabolism was 3.54 O_2 $g^{-1}h^{-1}$. In wood mouse (*Apodemus sylvaticus*) a value of 3.66 O_2 g^{-1} h^{-1} has been determined (French et al. 1978). At 0°C OCR was 5.5 O_2 $g^{-1}h^{-1}$. This is an expected result; at low ambient temperatures the resting metabolism in mammals increases. OCR_{40} displayed rather higher values compared to $OCR_{Control}$ in the range 0–10 °C ($P < 0.01$, Table 4). In the same range OCR_{60} was lower than OCR_{15} and OCR_{40} ($P < 0.02$). At temperatures over 20°C OCR_{40} and OCR_{60} decreased rather bellow $OCR_{Control}$ (Table 4). The values of OCR_{90} at all ambient temperatures were essentially lower compared to those of $OCR_{Control}$ as well as to those of OCR_{15}, OCR_{40}, and OCR_{60} ($P < 0.001$). At 25 °C OCR_{90} was 2.45 times lower than $OCR_{Control}$.

RRRM in control group was determined over the range of ambient temperatures 0-25 °C to be 7.67% per °C. In the interval 0-20 °C Gorecki (1968) has estimated in bank vole RRRM about 3% per °C. As expected, wild mammals demonstrate a higher tolerance to extreme conditions. In the mice BALB/c exposed during 90 days we calculated RRRM of 10.4% per °C.

Table 4. P values of a comparison of the resting metabolism (Oxygen consumption rate = OCR) and mean body temperature (T) measured over the range of ambient temperatures 0–30°C in control (Control) and experimental (15, 40, 60, 90) groups

Resting metabolism		Mean body temperature	
Comparison	P	Comparison	P
$OCR_{Control}/OCR_{15}$	> 0.1	$T_{Control}/T_{15}$	< 0.05
$OCR_{Control}/OCR_{40}$	< 0.01	$T_{Control}/T_{40}$	< 0.1
$OCR_{Control}/OCR_{60}$	< 0.05	$T_{Control}/T_{60}$	< 0.02
$OCR_{Control}/OCR_{90}$	< 0.001	$T_{Control}/T_{90}$	< 0.001
OCR_{40}/OCR_{60}	< 0.02	T_{40}/T_{60}	< 0.1
OCR_{15}/OCR_{90}	< 0.001	T_{15}/T_{90}	< 0.001
OCR_{40}/OCR_{90}	< 0.001	T_{40}/T_{90}	< 0.001
OCR_{60}/OCR_{90}	< 0.001	T_{60}/T_{90}	< 0.001

At ambient temperatures between 20 and 30 °C the lower values of OCR_{40} and OCR_{60} compared to controls (Figure 4, Table 4) may be due to the reduced hemoglobin (Hb) content in erythrocytes (one of the main signs of lead-induced anemia). The decreased Hb leads to lower O_2 binding. The time course of Hb at 20 °C (Topashka-Ancheva et al. 2003b) is in agreement with the data of OCR. The fact that at lower ambient temperatures (0–10°C) OCR_{40} was over $OCR_{Control}$ may be due to a stimulated intake of oxygen. Toxicants when they are at not very high level stimulate respiration (Golubev et al. 1973). RRRM in 40 days-exposed mice was 11.64% per °C, i. e. 1.5 times higher than that in controls.

Severe and irreversible physiological disturbances were observed in mice exposed for 90 days. OCR_{60} was 12% lower than the $OCR_{Control}$ at 25°C; at the same temperature OCR_{90} was 47% lower than the $OCR_{Control}$.

On molecular level Pb alters membrane permeability and blocks the active centers-pumps (Venugopal and Lucky 1978). Lead ions (Pb^{2+}) readily bind to SH, PO_3 and COOH groups of membranes and thus decrease their resistance to osmotic shock (Lessler and Walters 1973). Lead bound to mitochondrial membrane represses its energy function (Simons 1986). Lead ions also lead to a decrease in concentration of reduced glutathione (Ershov and Pleteneva 1979), the substrate of the enzyme glutathione peroxidase that acts as an antioxidant. The attenuated function of this enzyme leads to an increase of free radicals (FR) in the organism. FRs damage cell membranes, particularly mitohondrial ones, and thus FRs can function as an additional pathway for reduction of the mitochondrial aerobic respiration. The effectiveness of these processes determines the resulting Pb toxicity. On the other hand, Cd is known as a factor uncoupling the electron flow and ATP synthesis (Ershov and Pleteneva 1979). Thus, the total ATP pool in organism decreases. The reduced of supply energy to the muscles combined with the significantly decreased Hb on day 90 makes the respiration difficult and this affects the oxygen consumption.

Resting metabolism is not among the most sensitive biomarkers indicating heavy metal burden. Oxygen consumption manifests strong decrease in the entire range of life-compatible ambient temperatures only in conditions of severe poisoning.

In Figure 5 the mean body temperature of the mice is displayed in the range of ambient temperatures 0–30°C. We shall sign the sets of body temperatures, measured at the different ambient temperatures, in control group and at each time point in experimental group as follows: $T_{Control}$, T_{15}, T_{40}, T_{60}, and T_{90}. $T_{Control}$ is almost constant in the interval 20–30 °C and it is about 38.4°C. The interval 25–30°C could be assumed as a thermo-neutral zone for the laboratory mice reared in vivarium. In decrease of ambient temperature to 15°C control animals maintained homeothermy. At 15°C the mean body temperature of controls was 38.0°C. The mean body temperature in control mice decreased significantly with the decrease of ambient temperature bellow 15°C ($P < 0.01$). At 0°C the mean body temperature in controls was 36.1°C. T_{15} and T_{60} were slightly lower compared to $T_{Control}$ in the range 0–15 °C (Table 4). The values of T_{90} were much lower compared to $T_{Control}$ as well as to T_{15}, T_{40}, and T_{60} in the entire range of 0–30°C ($P < 0.001$) showing that the thermoregulation in 3 months-exposed mice is disrupted. At 30 °C, 25 °C, 20 °C, and 0°C T_{90} was lower 5%, 6.5%, 7.3%, and 10%, respectively, than $T_{Control}$. In 90 days-exposed mice no homeothermy was observed in the thermo-neutral zone. The values of body temperature at 20°C, 25°C, and 30°C were 35.4°C, 35.9°C, and 36.4°C, respectively. Bellow 10 °C T_{90} decreased dangerously and at 0 °C it was 32.7 °C.

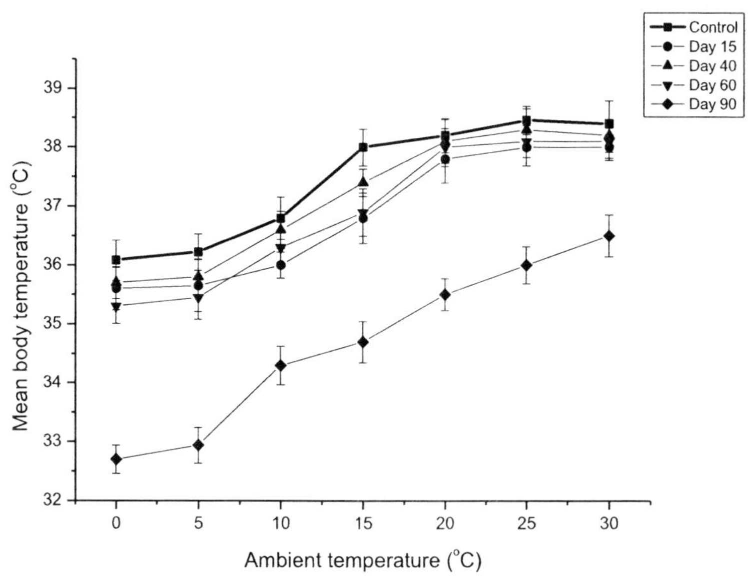

Figure 5. Mean body temperature, measured over the range of ambient temperatures 0–30°C in BALB/c mice on days 15, 40, 60, and 90 of the exposure to Pb and Cd containing industrial dust.

These data show clearly that Cd and Pb-loaded mammalian organism can not maintain homeothermy, when bioaccumulation of Pb and Cd corresponding to Cd concentrations in liver and kidney over 400 mg/kg and Pb concentrations in body and bones over 400 mg/kg and 1500 mg/kg, respectively.

Cooke and Johnson (1996) have reported a fall of body temperature in small mammals in conditions of Cd poisoning. However, we did not find in the literature data confirming or not an influence of toxicants, particularly heavy metals, on changes in body temperature of mammals, measured in a wide range of ambient temperatures.

The sharp decrease of body temperature in conditions of high dietary Pb and Cd may be explained by the reduced rate of cell metabolism and by probable impairment in central nervous system (CNS) thermoregulatory function.

An increase of lead exposure disturbs Cu metabolism and Cu excretion increases (Angle and Stelmak 1975). Cadmium interferes with Cu utilization by blocking enzymes that promote intestinal absorption of Cu (Peterling 1974). Our data are in accord with these findings: on day 90 of exposure a significant decrease of Cu concentrations was observed in body and organs of experimental mice (Table 1). Copper deficiency disturbs the normal functioning of cytochromoxidase, a key enzyme in mitochondrial respiratory chain, which leads to a reduction of the electron flow to oxygen. The total energy of the reduction-oxidative reactions in respiratory chain declines and thus, the heat production decreases.

Thermoregulation in mammals is a physiologic brain function governed by hypothalamic histamine. An increase in histamine level decreases body temperature (Sakata and Yoshimatsu 1997). Activation of histaminergic neurons in the hypothalamus plays an essential role under conditions of infection, inflammation and injury, in which hypotalamic histamine may attenuate excessive elevation of body temperature (Sakata et al. 1997). Lead is known as a neurotoxic agent (Goyer 1996). It is of concern to explore whether Pb can activate the histamine neuron systems in the brain. In such case an accelerated turnover rate of histamine could be cause thermoregulation disturbances under conditions of Pb poisoning, what we observed on day 90 of exposure.

As well as resting metabolism, also body temperature is not a very sensitive biomarker for detecting of heavy metal contamination in mammalian organism. A crucial decrease of body temperature in the entire range of 0-30°C occurred in 3 months-exposed animals.

Effect of Lead and Cadmium on the Growth of Laboratory Mice

Growth also can be considered as a biomarker of the physiological status of mammalian organism contaminated with heavy metals especially in the context of revealing of chronic effects in long-term exposure.

The data of the experiment are presented in Figure 6. On day 15 of the treatment the mean body weight (g) in experimental mice was 12% lower compared to that in control mice ($P < 0.001$). Up to day 60 the mean body weight in both control and experimental groups increased without statistically significant differences ($P > 0.1$). A sharp decrease of body weights of experimental animals was established on day 90, when the mean body weight of experimental mice was almost 30% lower than that of controls (Figure 6, Table 2). So, while the mean body weight in control group increased 1.34 folds, the mean body weight in experimental group increased only 1.08 folds during 3 months. Therefore a growth slow of 1.24 times has taken place.

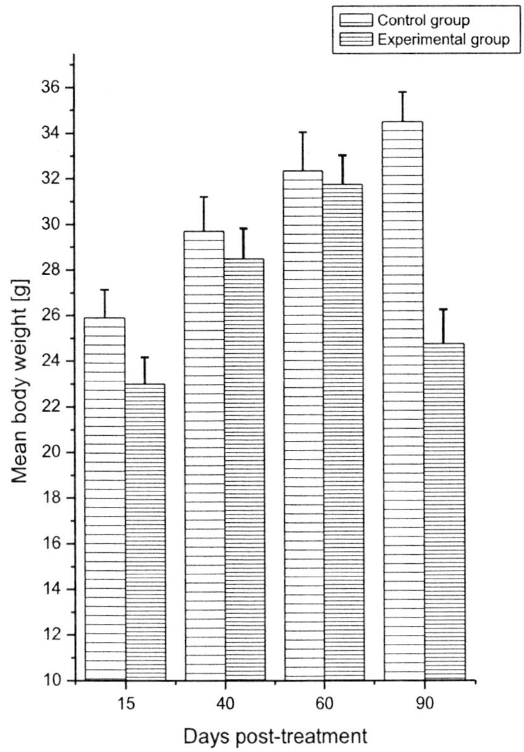

Figure 6. Mean body weight of BALB/c mice, determined on days 15, 40, 60, and 90 of the exposure to Pb and Cd containing industrial dust.

More dramatic reduction in the gain of body weight in BALB/c male mice 3 months old was observed by Eyden et al. (1978). They have treated the mice orally with Pb acetate in different concentrations and at level of 1.0% in animal food they have recorded a body weight of the treated mice 51% lower compared to controls.

An essential retardation in growth of the offspring of laboratory mice *Mus musculus* L., receiving food contaminated with Pb, was reported by Sawicka-Kapusta et al. (1987b). Lead has been administrated at concentrations of 20, 100, 500, and 1000 ppm to pregnant females. In the young mice on day 20 of their postnatal development a body weight gain of 4.88 folds has been measured, when the pregnant mice have fed with 1000 ppm diet. On day 20 the young mice with 20 ppm diet have shown a body weight gain of 7.74 folds, i. e. a growth slowing of 1.6 times.

Clear expressed decrease in gain of body weight was observed also in toxicological experiments with Cd. Der et al. (1977) have injected intramuscularly female rats with $CdCl_2$ in 50μg and 250 μg for 54 days. In 250 μg Cd group a significantly reduced body weight has been established. In the same time a splenomegaly and enlargement of the liver and kidney has been observed. The authors reported an increase of the final body weight 1.05 folds in 250 μg Cd group, while the respective increase in control group has been 1.13 folds. In contrast to our results the injected rats have shown a slower growth during the whole experiment, while the mice used by us demonstrated a severe weight loss at the end of experiment. However, because of the different experimental means of exposure and different chemical forms of administrated Cd these data are rather difficult to compare.

An association of body mass loss with significant increase in the relative spleen mass was reported also by Marques et al. (2006) but in Pb-exposed Algerian mice (*Mus spretus*). These authors have administrated Pb acetate via drinking water to the mice. Their result is in accord with our finding indicating a loss of body mass in exposed animals on day 15. However, after day 90 of exposure, body mass of Algerian mice has reversed to initial values and the authors consider a remarkable Pb poisoning resistance of mice. In our experiment a sharp decrease of body weight on day 90 was observed. We think this fact confirms the adverse role of Cd, when it is administrated in high concentrations, because the difference between liver levels of Pb in both experiments was not significant. The authors have measured a Pb concentration of 8.3 + 2.1 μg/g dry liver, while in our experiment we found Pb concentration of 9.8 + 0.7 mg/kg dry weight on day 90. The liver concentration of Cd was, however, 479.1 + 28.9 mg/kg in our BALB/c mice on day 90.

Cadmium in the form of $CdCl_2$ has been fed to male and female rats over a period of 3 months but in relatively low concentrations – up to 30 ppm (Loeser and Lorke 1977a). The authors have studied body weight, organ weights, food consumption, mortality, liver and renal functions, hematological indices etc. It is of concern that in such doses of dietary Cd no pathological changes have been recorded in none of these parameters. Similar results have been reported also regarding dogs (Loeser and Lorke 1977b). Hence the authors conclude that Cd concentrations up to 30 ppm over a period of 3 months are tolerated by mammals (especially rat and dog) without any side-effects.

On the contrary, our concept was to administrate enhanced doses of heavy metals in order to prove their strong toxic effects and so, to assess the maximum hazard for mammals. Up to day 90 of the treatment we established Cd concentration of 479.1 $\pm$ 28.9 mg/kg in liver and Pb concentration of 1625 $\pm$ 42 mg/kg in bone.

Usually, as most probable reason for weight-loss is considered the reduction in food and water intake. Ma (1996) notes that the loss of appetite is a clinically manifested sign of lead poisoning in mammals. O'Flaherty et al. (1986) indicate a reduced food intake in rats exposed to Pb. Lead attacks cell membranes (Ershov and Pleteneva 1989), influencing some enzyme systems, and in gastrointestinal tract this may lead to disturbance of the normal digestion. A reduction of food and water intake is pointed out also as one of the major effects of cadmium toxicity (Cooke and Johnson 1996). Cadmium diet inhibits the entering of Zn in the different tissues (Shroeder and Nason 1974) and thus disturbs the function of epithelial cells of the gastrointestinal tract. Cadmium inhibits also the biosynthesis of DNA, RNA and proteins (Snow 1992), which may result in a slow growth.

Food intake may be reduced also because of impairment of some functions of the CNS. Sakata and Yoshimatsu (1997) have established that hypotalamic neuronal histamine suppresses food intake through H_1-receptors in the ventromedial hypotalamic nucleus and the paraventricular nucleus. The turnover rate of hypotalamic histamine can be elevated by leptin, a protein secreted by adipose tissue. Thus, hypotalamic histamine has a role in the central regulation of feeding behavior controlled by leptin. Leptin deficiency leads to obesity (Yoshimatsu et al. 1999). In contrast, abnormally elevated brain histamine may lead to an abnormal increase of food intake suppression. Lead is known to cause modification of neuronal circuitry (Goyer 1996). We noted already that the disrupted homeothermy in the 3 months-exposed mice may be due to elevated histamine level. In this context it is of interest to explore, whether Pb could play some role as a factor raising the release of hypotalamic histamine.

Having in mind, however, the toxic effect of Cd, when it is in high concentration, and the consideration of Marques et al (2006) about a possible Pb poisoning resistance, we could assume that Cd might have a predominant role in body mass loss in cases of Pb-Cd poisoning.

The reduced water intake naturally leads to organism dehydratation. However, we suggest as a possible reason for body weight decrease in Cd- and Pb-exposed mammals also a dehydratation due to renal dysfunction. One of the effects of Cd and Pb poisoning is the renal toxicity. In the polyuric stage of nephropathy (Hegglin 1975) significant dehydratation takes place. The data about time course of hematocrit (Topashka-Ancheva et al. 2003) (Figure 7) support this suggestion. The sharp rise of Ht up to day 90 ($P > 0.001$) showed an elevated ratio RBC volume/total blood volume. The erythrocyte number changed not significantly after day 60 (Topashka-Ancheva et al. 2003b) and therefore, the Ht increase may be explained only as an effect of a developed dehydratation.

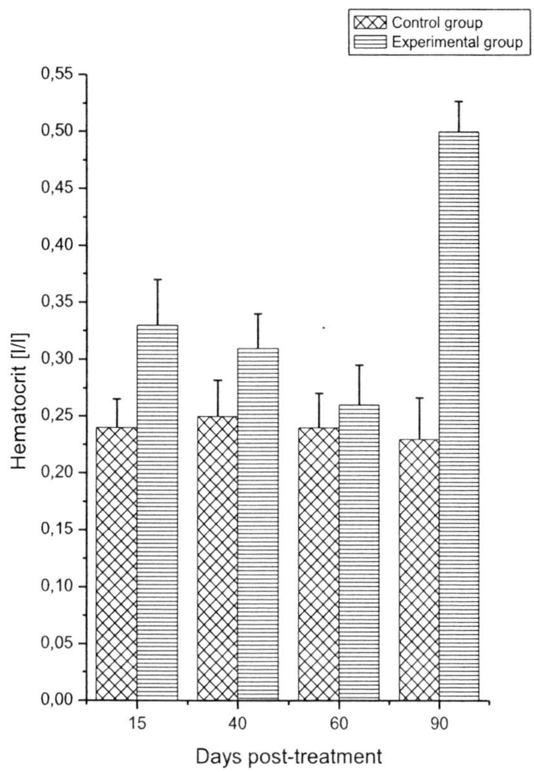

Figure 7. Hematocrit in BALB/c mice determined on days 15, 40, 60, and 90 of the exposure to Pb and Cd containing industrial dust.

Conclusion

Small mammals (laboratory mice BALB/c) were exposed to the industrial mixture in a chronic eco-toxicological experiment. Samples collected from the electrofilters of "KCM-Plovdiv" were used. The Lead-Zinc Plant "KCM-Plovdiv" is a significant source of environmental pollutants in Bulgaria. The Plant's polymetal industrial dust containing lead and cadmium is a harmful emission to the environment. Some aspects of the damaging action of heavy metals on the laboratory mice were studied to lay the groundwork for further assessment of the possible influence of the emission on populations inhabiting the polluted region.

The used heavy metal mixture contained Pb, Cd, Zn, and Cu. We concentrated our attention only on Pb and Cd toxicity because Zn and Cu in the body and organs decreased significantly on day 90, due to metal-metal interactions, discussed in a previous study (Teodorova et al. 2003). The most severe pathological changes were observed on day 90. In 3 months-exposed laboratory mice, the concentrations of Cd in the liver and the kidneys were 479.1 ± 28.9 mg/kg and 465.7 ± 11.1 mg/kg, respectively, and Pb concentrations in the bones and the body were 1625 ± 42 mg/kg and 475 ± 17.9 mg/kg, respectively.

The chromosome structure was influenced in several ways. We observed high chromosomal aberrations frequency (CAF): 25.7% in males on day 60 and 16.3% in females on day 40. The highest CAF, summarized on males and females, was recorded on day 60: 39.7%. We did not observe identical patterns of CAF change in males and females during the experiment. Females demonstrated significantly lower CAF compared to males and this may be due to a more effective DNA repair. The maximum values of CAF in males and females did not coincide in time. The most frequently encountered aberrations were chromatide breaks and centromeric fusions (c/c). The absence of fragments and the very small number of isochromatide breaks suggest that probably Pb and Cd influence the genome in a later period of DNA replication. The lack of polyploidy indicated that the cell spindle in conditions of Pb and Cd poisoning is not significantly influenced, thus the normal distribution of the chromosome material is not dramatically changed and, therefore, the cell division ability is retained.

Pathological changes were established in the resting metabolism and mean body temperature – parameters generally representing the physiological status in mammals. Dangerous decreases in both oxygen consumption and body temperature were observed on day 90 of exposure. The mean body temperatures of control mice and mice exposed up to day 60 did not differ significantly in the thermo-neutral zone 25–30°C ($P > 0.1$). This fact provides evidence that in 2-months-exposed animals the homeothermic thermoregulation is not yet disrupted. In thermo-neutral zone the changes in oxygen consumption almost mimicked the time course of hemoglobin determined in our previous paper (Topashka-Ancheva et al. 2003). The rate of reduction of the resting metabolism in 3-months-exposed mice was 10.4% per °C, while this rate in control mice was 7.7% per °C.

A sharp decrease in body weight was observed on day 90 of exposure, when the mean body weight in the exposed mice was 28% lower than the controls. Our result supports the findings of other authors, which have reported a significantly reduced gain in body weight in small mammals, exposed to lead and cadmium. Body mass loss in experimental mice may be due to a reduction in food and water intake, and to dehydration caused by developed renal dysfunction.

The results obtained showed that an industrial mixture containing high doses of lead and cadmium affects severely mammalian organism. CAF, and especially CAF in males, is a highly sensitive biomarker for heavy metal burden. The chromosomal apparatus responded quickly and at the same time it displayed a relative stability and repair potential. The stability of chromosome structure may be due to a relative balance between the processes of induction of aberrations by the mutagene and the elimination of the aberrant cells in the course of mitotic division. A lower level of chromosomal aberrations observed in females compared to males allows assume a more effective repair in females. Obviously, this is related to their reproductive function.

The resting metabolism, the body temperature, and body weight showed an essential dose-response effect after day 60 of exposure. The results showed that high doses of Pb and Cd (bellow 200 mg/kg Cd in liver and kidney and 400 mg/kg and 1500 mg/kg Pb in body and bones, respectively) were tolerated without significant effects on resting metabolism, body temperature, and growth rate in two-months-exposed mice. These parameters, however, were impaired dramatically in the third month of exposure. Thus, these parameters could be a reliable means for establishment of a range of physiological tolerance of mammals to heavy metal load.

These results could play a role to direct more precisely epidemiological and clinic investigation carried out on residents in the vicinity of such kind of industries, suggesting effects on human health endpoints. Also the findings could be useful for further explanations of the mechanisms damaging genetic structure and disturbing physiological status of mammal's organism by heavy metals, particularly lead and cadmium.

References

Andrews, S. M. and Cooke, J. A. (1984). Cadmium within a contaminated grassland ecosystem established on metalliferous mine waste. In: Osborn D., editor. *Metals in animals* (pp. 11-15). Cambrige, U.K: *Inst. Terrestial Ecol. Publ.* No. 12. Natural Environment Research Council.

Angle, C. R. and Stelmak, K. L. (1975) Lead and iron deficiency. *Trace Subst. Env. Health, 9*, 377-386.

Barry, PSI. (1975). A comparison of concentrations of lead in human tissues. *Br. J. Ind. Med., 32*, 119-139.

Beyersmann, D. and Hechtenberg, S. (1997). Cadmium, gene regulatory and cellular signaling in mammalian cells. *Toxicol Appl Pharmacol 144*, 247-261.

Bilban, M. and Jakopin, C. B. (2005). Incidence of cytogenetic damage in lead-zink mine workers exposed to radon. *Mutagenesis. 20* (3): 187-191.

Bouilly, K; Gagnaire, B; Bonnard, M; Thomas-Guyon, H; Renault, T; Miramand, P; Lapegue, S. (2006). Effects of cadmium on aneuploidy and hemocyte parameters in the Pacific oyster. *Crassostrea gigas. Aquatic Toxicol. 78* (2), 149-156.

Bremner, I. (1979). Mammalian absorption, transport and excretion of cadmium. In: Webb M., editor. *The chemistry, biochemistry and biology of cadmium* (pp. 175-193). Amsterdam: Elsevier/North-Holland Biomedical Press.

Çelik, A; Çömelekoglu, Ű; Yalin, S. (2005). A study on the investigation of cadmium chloride genotoxicity in rat bone marrow using micronucleus test and chromosome aberration analysis. *Toxicol. Industrial Health, 21*, 243-248.

Cooke, J. A. and Johnson, M. S. (1996) Cadmium in small mammals. In: W. N. Beyer, G. H. Heinz, A. W. Redmon-Norwood (Eds.) *Environmental contaminants in wildlife. Interpreting tissue concentrations* (pp. 377-388). Boca Raton New York London Tokyo: Lewis publishers.

Demarque, D; Jouanny, J; Poitevin, B; Saint Jean, Y. (1995). *Pharmacologie et matiere medicale homeopathique*. Boiron- C.E.D.H., France.

Der, R; Fahim, Z; Yousef, M; Fahim, M. (1977). Effects of cadmium on growth, sexual development, and metabolism in female rats. *Res. Comm. Chemical Pathol. Pharmacol., 16* (3), 485-505.

Dhir, H; Roy, AK; Sharma, A; Talukder, G. (1990). Modification of clastogenicity of lead and aluminum in mouse bone marrow cells by dietary ingestion of *Phyllanthus emblica* fruit extract. *Mutation Res. 241***,** 305-312.

Dodds-Smith ME; Johnson MS; Thompson DJ (1992) Trace metal accumulation by the shrew, *Sorex araneus*. II. Tissue distribution in kidney and liver. *Ecotoxicol.Environ. Saf. 24*, 118-13

Dubinin, NP (1986). *General genetics*. Moskow: Naouka Publish. Hause (In Russian).

Ershov, Y. A. and Pleteneva, T. V. (1989). *Mechanisms of toxic action of inorganic compounds*. Moskow: Medizina Publ. House (in Russian).

Eyden, BP; Maisin, JR; Mattelin, G. (1987). Long-term effects of dietary lead acetate on survival, body weight and seminal cytology in mice. *Bull. Environ. Contam. Toxicol., 78*, 266-271.

Fleming, DE; Chettle, DR; Webber, CE; O'Flaherty, EJ. (1999). The O'Flaherty model of lead kinetics: an evaluation using data from a lead smelter population. *Toxicol. Appl. Pharmacol. 161* (1), 100-109.

French, NR; Grant, WE; Grodzinski, W; Swift, DM (1976). Small mammal energetics in grassland ecosystem. *Ecol. Monogr., 46,* 201-220.

Golubev AA; Liublina, EI; Tolokontzev, NA; Filiv, VA. (1973). Quantitatively Toxicology. Leningrad: Medicina Publ. House (In Russian).

Gorecki, A (1968). Metabolic rate and energy budgets in bank vole. *Acta Theriol 13* (20), 341-365.

Goyer RA (1996) Toxic effects of metals. In: McGraw-Hill, editor. *Casarett and Doull's Toxicology. The basic Science of poisons*. Fifth edition (pp. 691-735). San Francisco Auckland Bogota Caracas Lisbon London Madrid Mexico City Milan Montreal New Delhi San Juan Singapore Sydney Tokyo Toronto: Health Professions Division, New York St. Louis.

Hartman, M. and Hartwig, A. (1998). Disturbance of DNA damage recognition after UV-radiation by nickel(II) and cadmium(II) in mammalian cells. *Carcinogenesis 19*, 617-621.

Hartwig, A; Schlepegrell, R.; Beyersmann, D. (1990) Indirect mechanism of lead-induced genotoxicity in cultured mammalian cells. *Mutation Res. 241***,** 75-82.

Hegglin, M (1975). Differentiale Diagnose der inneren Krankheiten. Stuttgart: Georg Thieme Verlag.

Hunter, BA; Johnson, MS; Thompson, DJ. (1987a). Ecotoxicology of copper and cadmium in a contaminated grassland ecosystem. II. Invertebrates. *J. Appl. Ecol., 24,* 587-599.

Hunter, BA; Johnson, MS; Thompson, DJ. (1987b). Ecotoxicology of copper and cadmium in a contaminated grassland ecosystem. III. Small mammals. *J. Appl. Ecol., 24,* 601-614.

Hunter, BA; Johnson, MS; Thompson, DJ. (1989). Ecotoxicology of copper and cadmium in a contaminated grassland ecosystem. IV. Tissue distribution and age accumulation in small mammals. *J. Appl. Ecol., 26*, 89-99.

International Commission for Protection Against Environmental Mutagens and Carcinogenic (1983).Report of Committee I. *Mutat Res 144*, 120-177.

Karakaya, AE; Ozcagli, E; Ertas, N; Sardas, S (2005). Assessment of abnormal DNA repair responses and genotoixic effects in lead exposed workers. *Am. J. Ind. Med. 47* (4), 358-363.

Kipling, D. (1994) Mouse centromeric mapping using oligonucleotide probe that detect variants of the minor satellite. *Chromosoma 103*, 46-54.

Lessler, MA; Walters, MI. (1973). Erythrocyte osmotic fragility in the presence of Pb or Hg. *Proc. Soc. Exp. Biol. Med., 142*, 548-553.

Loeser, E. and Lorke, D. (1977). Senichronic oral toxicity of cadmium. 1. Studies on rats. *Toxicology, 7*, 215-224.

Loeser, E. and Lorke, D. (1977). Senichronic oral toxicity of cadmium. 2. Studies on dogs. *Toxicology, 7*, 225-232.

Longhese, MP; Mantiero, D; Clerici, M. (2006). The molecular response to chromosome breakage. *Molecular Microbiol., 60* (5), 1099-1108.

Ma, WC; Denneman, W; Faber, J. (1991). Hazardous exposure of ground-living small mammals to cadmium and lead in contaminated terrestrial ecosystems. *Arch. Environ. Contam. Toxicol. 20*, 266-270.

Ma, WC (1996) Lead in mammals. In: W. N. Beyer, G. H. Heinz, A. W. Redmon-Norwood (Eds.) *Environmental contaminants in wildlife. Interpreting tissue concentrations* (pp. 281-296). Boca Raton New York London Tokyo: Lewis publishers.

Marcos AA; Bruzzone L; Apartin C; Rosenberg CE; Fink NE; Salibian A (2004). Biosensors of inorganic lead exposure and effect in an adult amphibian. *Arch Environ Contam Toxicol 46* (2), 224-230.

Marques, CC; Nunes, AC; Pinheiro, T; Lopes, PA; Santos, MC; Viegas-Crespo, AM; Ramalhinho, MG; Mathias, ML (2006). An assessment of time-dependent effects of lead exposure in algerian mice (*Mus spretus*) using different methodological approaches. *Biol. Trace Elem. Res. 109* (1): 75-90.

Martin, M. H. § Coughtry, P. J. (1982). Biological monitoring of heavy metal pollution. London: *Appl. Science* Publishers.

Metcheva, R.; Atanasov, N; Topashka-Ancheva, M. (1994) Bioaccumulation de metaux toxiques de l'usine de metallurgie non ferreuse pres de Plovdiv dans l'organisme du campagnol mediterraneen (*Microtus guentheri* Dan. et Als., 1880) (Microtinae, Rodentia). *Acta zool. bulg. 47*, 54-65.

Metcheva, R; Topalshka-Ancheva, M; Atanassov, N; Artinian, A; Minkova, V. (1995). Bioaccumulation and distribution of toxic elements, hematological indices and chromosomal changes in the organisms of indicator species of small rodents from Rila mountain. In: J. P. Carbonnel and J. Stamenov (Eds.) *Observatoire de Montagne de Moussala OM2, 3* (pp. 203-211). Sofia.

Metcheva, R; Artinian, A; Nikolova, E; Beltcheva, M; Laleva, R. (1997). Estimation of the environmental quality using monitor species small mammals from two regions with different antropogenic influences in Bulgaria. In: Peev D., Ammann K. Artinian A., editors *Ecomonitiring in Rozhen and Pirdop – Bulgaria* (pp. 113-119). Sofia: Ministry of Environment; SDC-Swiss Agency for Development and Cooperation.

Metcheva, R; Teodorova, S; Topashka-Ancheva, M. (2003). A comparative analysis of the heavy metal loading of small mammals in different regions of Bulgaria. I: monitoring points and bioaccumulation features. *Ecotoxicol Environ Safety 54*, 176-187.

Milton, A, Cooke, JA, Johnson, MS (2003). Accumulation of lead, zinc and cadmium in a wild population of *Clethrionomys glareolus* from an abandoned lead mine. *Arch Environ Contam Toxicol 44* (3), 405-411.

Murnane, J. P. and Sabatier, L. (2004) Chromosome rearrangements resulting from telomere dysfunction and their role. *BioEssays, 26*, (11), 1164-1174.

Műller, L; Abel, J; Ohnesorge, FK. (1986). Absorption and distribution of cadmium (Cd), copper and zinc following oral subchronic low level administration to rats of different forms of cadmium (Cd-acetate, Cd-metallothionein, Cd-glutathione). *Toxicology, 39*, 187-195.

Nayak, BN; Ray, M; Persaud, TVN (1989). Maternal and fetal chromosomal aberrations in mice following prenatal exposure to subembryotoxic doses of lead nitrate. *Acta. Anat. 135*, 185 -188.

O'Flaherty, EJ (1995a). Physiologically based models for bone-seeking elements. V. Lead absorption and disposition in childhood. *Toxicol. Appl. Pharmacol. 131*, 297-308.

O'Flaherty, EJ (1995b). PBK modeling for metals. Examples with lead, uranium, and chromium. *Toxicol. Lett., 82-83*, 367-372.

O'Flaherty, EJ (1998). Physiologically based models of metal kinetics. *Crit. Rev. Toxicol. 28* (3), 271-317.

O'Flaherty, EJ; Adams, WD; Hammond, PB; Taylor, E. (1986). Resistance of the rat to development of lead-induced renal functional deficits. *Toxicol Environ Health, 18* (1), 61-75.

O'Flaherty, EJ; Inskip MJ; Yagminas AP; Franklin CA (1996). Plasma and blood lead concentrations, lead absorption, and lead excretion in nonhuman primates. *Toxicol Appl Pharmacol, 138*, 121-130.

O'Flaherty, EJ (2000). Modeling normal aging bone loss, with consideration of bone loss in osteoporosis. *Toxicol. Sci., 55*, (1), 171-188.

Parker RE (1973). *Introductory statistics for biology*. London: Edward Arnold Publ. Ltd.

Pascoe, GA; Blanchet, RJ; Linder, G. (1996). Food chain analysis of exposures and risks to wildlife at a metals-contaminated wetland. *Arch. Environ .Contam. Toxicol .30*, 306-318.

Peterling HG (1974). The effect of cadmium and lead on copper and zinc metabolism. In: *Trace element metabolism in animals - 2* (pp. 311-325). Baltimore: University Park Press.

Preston, RY; Dean, LY; Galloway, Sh; Helden, H; Me Fee, A. (1987). Mammalian in vivo cytogenetic assays. Analysis of chromosome aberrations in bone marrow cells. *Mutation. Res. 189*, 157-165.

Privezentsev, KV; Sirota, NP; Gaziev AJ (1996). The genotoxic of cadmium studied in vivo. *Tsitologia i genetika 30* (3), 45-51.

Sakata, T. and Yoshimatsu, H. (1997). Hypothalamic neuronal histamine implications of homeostatic maintenance in its control of energy metabolism. *Nutrition, 13*, 403-411.

Sakata, T; Yoshimatsu, H; Kurokawa, M. (1997). Thermoregulation modulated by hypotalamic histamine in rats. *Inflamm. Res., 46, Supplement 1*, 35-36.

Sawicka-Kapusta, K; Gorecki, A; Lange, R. (1987a) Heavy metals in rodents from polluted forests in southern Poland. *Ecol. Pol. 35* (2), 345-354.

Sawicka-Kapusta, K; Czerwonka, M; Zakrzewska, M. (1987b) Effect of orally administrated lead on the postnatal development in laboratory mice. *Ecol. Pol. 35* (2), 385-397.

Scheirs, J; De Coen, A; Covaci, A; Beernaert, J; Kayawe, VM; Caturla, M; De Wolf, H; Baert, P; Van Oostveldt, P; Verhagen, R; Blust, R; De Coen, W. (2006). Genotixicity in wood mice (*Apodemus sylvaticus*) along a polluted gradient: exposure- age-, and gender-related effects. *Environ. Toxicol. Chem. 25* (8), 2154-2162.

Schroeder, H. A. and Nason, A. P. (1974). Interactions of trace metals in rat tissues. Cadmium and nickel with zinc, chromium, copper, manganese. *J. Nutr. 104,* 167-178.

Seoane, AI; Guerci, AM; Dulout, FM. (2002). Malsegregation as a possible mechanism of aneuploidy induction by metal salts in MRC-5 human cells. *Environ. Mol. Mutagen. 40* (3), 200-206.

Shaik, AP; Sankar, S; Ready, SC; Das, PG; Jamil, K (2006). Lead-indused genotoxicity in lymphocytes from peripheral blood samples of humans: In vitro studies. *Drug and Chemocal Toxicology 29* (1), 111-124.

Simons, TJB (1986). Cellular interactions between lead and calcium. *Br. Med. Bull. 42*: 431-434.

Snow, ET (1992). Metal carcinogenesis: mechanistic implications. *Pharmacology therapeutics, 53*, 31-65.

Sorsa, M; Wilbourn J; Vainio H. (1992). Human cytogenetic damage as a predictor of cancer risk. In: Vainio H,. Magee PH, McGregor DB, McMichael AJ, editors. *Mechanisms of carcinogenesis in risk identification* (pp 543-554). Lyons and International Agency for Reseach on Cancer.

Talmage, S. § Walton, B. T. (1991). Small mammals as monitors of environmental contaminants. *Rev. Environ. Contam. Toxicol. 119*, 47-145.

Teodorovà, S; Metcheva, R; Topashka-Ancheva, M (2003). Bioaccumulation and damaging action of polymetal industrial dust on laboratory mice *Mus musculus alba* I. Analysis of Zn, Cu, Pb, and Cd disposition and mathematical model of Zn and Cd bioaccumulations. *Environ Res 91,* 85-94.

Topashka-Ancheva, M; Avramova, F; Simeonova, V; Metcheva, R; Atanasov, N; Angelova, A; Stambolova, M (1995). Cytogenetical and biochemical studies on the effect of heavy metals on the organism of *Microtus guentheri* (Microtinae, Rodentia) in an ecologo-toxicological experiment. In: Annuaire de L'Universite de Sofia "St. Kliment Ohridski", Faculte de biologie", vol. 88 (pp. 69-80). Sofia: Presses Uuniversitaires "St. Kliment Ohridski".

Topashka-Ancheva, M; Metcheva, R; Atanasov, N (1998). Bioaccumulation and clastogenic effects of industrial dust on *Microtus guentheri* (Microtinae, Rodentia) in an ecologo-toxicological experiment. *Acta zool. bulg. 50*, 117-122.

Topashka-Ancheva, M; Metcheva, R; Teodorova, S (2003a). A comparative analysis of the heavy metal loading of small mammals in different regions in Bulgaria II: chromosomal aberrations and blood pathology. *Ecotoxicol Environ Safety 54*: 188-193.

Topashka-Ancheva, M; Metcheva, R; Teodorova, S (2003b). Bioaccumulation and damaging action of polymetal industrial dust on laboratory mice Mus musculus alba II. Genetic, cell, and metabolic disturbances. *Environ Res 54*: 152-160.

Tsuchia, K (1979) Lead. In: Friberg L., editor *Handbook on the toxicology of metals* (pp. 451-484). Amsterdam: Elsevier/North-Holland Biomedical Press.

Valverde, M; Trejo, C; Rojas, E (2005). Is the capacity of lead acetate and cadmium chloride to induce genotoxic damage due to direct DNA-metal interaction? *Toxicology and Industrial Health 21* (10): 243-248.

Venugopal B. and Luckey T. D. (1978) Metal toxicity in mammals, Vol 2. New York: Plenum Press.

Wern, CD (1986). Mammals as biological monitors of environmental metal levels. *Environ .Monit. Assess. 6*, 127-144.

World Health Organization (WHO) (1985). Environmental Health Criteria 51. Guide to Shoot-Term for Detecting utagenic and Cancerogenic Chemicals. Geneva: *WHO*, 1000-1114.

Yoshimatsu, H; Itateyama, E; Kondou, S; Tajima, D. Himeno, K; Hidaka, S; Kurokawa, M; Sakata, T (1999). Hypothalamic neuronal histamine as a target of leptin in feeding behavior. *Diabetes, 48,* 2286-2291.

In: Environmental Research Trends
Editor: Marvin A. Cato, pp. 231-270
ISBN: 978-1-60021-556-8

Chapter 7

PRIMARY CULTURE AND LEUKOCYTE MIGRATION AS NEW TOOLS TO EVALUATE THE EFFECTS OF PERSISTENT ORGANIC POLLUTANTS (POPS) IN FISH

***Ciro Alberto de Oliveira Ribeiro*[1]*, Helena Cristina Silva de Assis*[2]*, Francisco Filipak Neto*[1]*, Anna Lúcia Miranda*[1] *and Claudia Turra Pimpão*[2]**

[1]Departamento de Biologia Celular
[2]Departamento de Farmacologia, Caixa Postal 19031. Centro Politécnico, Universidade Federal do Paraná, CEP: 81.531-990, Curitiba – PR, Brasil

ABSTRACT

Organisms are continuously exposed to a variety of anthropogenic toxicants daily released to the environment and, at the present, the realistic effects on cells are still a challenge. In this chapter we discuss the use of new methods to evaluate the effects of pollutants in cells using *in vitro* and *in vivo* studies. The uses of primary cultured cells allow to evaluate the toxic mechanisms and cell responses in order to investigate the exposure to rational concentrations. The method consists in extracting from hepatic and/or muscle lipids the mixture of pollutants (organochlorine compounds (OCs) or polychlorinated biphenyls (PCBs)), which were chronically bioaccumulated in biota from field, and test their effects on cells *in vitro*. Cells such as hepatocytes are isolated through non-enzymatic perfusion protocol and different concentrations of OCs, PCBs or organo metals have been tested. Cell death, oxidative stress and ultrastructural parameters are evaluated. The following immunological parameters such as, production of nitric oxide, macrophage activity and cell attachment were considered to evaluate the *in vitro* exposure to leukocytes extracted from fish blood and macrophages from peritoneal cavity and head kidney of fish. The study of leukocyte migration into peritoneal cavity of fish is a new approach adapted from mammals to tropical fish specie. This method utilizes the peritoneal exudates after lipopolysaccharide (LPS, E.coli, 0111:B4)-induced leukocyte migration to evaluate the systemic response. These new methods developed have been used in recent investigations to study the effects of pollutants in tropical fish with a very satisfactory result, increasing the knowledge about the toxic mechanisms involved with the exposure to chemicals in aquatic organisms.

INTRODUCTION

Environmental stressors and their associated risks have always been an inherent element of society. The negative effects of ecological alterations are not always straight obvious, given the difficulties to understand the human-ecological interactions.

The classical approach to establish the hazards of toxic chemicals to wildlife is to determine the amount of a chemical present and then compare that value with those found to do harm in experimental animals. Unfortunately, there is no evidence that the same effects observed in the experimental condition to one chemical will be similarly found in individuals chronically exposed, even though the levels of given chemical are identical. The explanation is obvious, the effect of an isolated chemical can be completely different if associates with other from the same category of distinct groups in different concentrations.

This chapter describes some experiences using not classical methods to investigate the effects of persistent pollutants found in the environment. Although the use of primary culture and aspects of immunological endpoints have contributed to increase the information about toxicity mechanisms of pollutants, a limited number of studies have reported the effects of complex mixtures. In addition, the challenge here is to discuss the reasonable effects on potential target cells and introduce these new approaches, focusing also in the risk assessment for human populations. Since there is a large variety of pathway in which a substance can interact with molecules, tissues or organs to cause toxic responses, it is strongly difficult to define a set of chemical characteristics that render a chemical toxic to lifing organisms. A toxic exposure may be affected by the same substance due to differences in chemical and biochemical aspects, target cells or physiological conditions of the organism.

According to Manahan (1992) the toxicant can be placed into several main categories as listed below:

(1) Substances that exhibit extreme acidity, basicity, dehydrating ability or oxidizing power.
(2) Reactive substances that contain bonds of functional groups, which are particularly prone to react with biomolecules in a damaging way.
(3) Heavy metals, broadly defined, contain a number of members that are toxic by virtue of their interaction with enzymes, tendency to bond strongly with sulfhydryl (-SH) groups on proteins and other effects.
(4) Binding species are those that bond to biomolecules, altering their function in a detrimental way. The binding may be reversible or irreversible.
(5) Lipid-soluble compounds are frequently toxic because of their ability to cross cell membranes and similar barriers in the body. Lipid-soluble species frequently accumulate up to toxic levels through biouptake and biomagnification processes.
(6) Chemicals species that induce responses based largely on their chemical structures. Such toxicants often produce an allergic reaction as the body's immune system recognizes the foreign agent causing an immune system response.

POLLUTANT CHARACTERIZATION, BIOMARKERS AND EXPERIMENTAL MODELS

This chapter focuses the discussion on some pollutants, i.e. persistent organic pollutants (POPs), heavy metals and organometalloids, and polycyclic aromatic hydrocarbons studied in the last few years, and described our more recent experiences with organochlorine mixture extracted from fat tissues of fish and some pesticides exposed to hepatocytes and immune cells through *in vitro* studies.

Along with the technological advance, the use of a wide variety of new chemicals by industry intentionally or accidentally led to increased levels of these compounds in the environment. Among a myriad of chemicals, in addition to pesticides and industrial chemicals, many organic compounds such as PCDDs (polychlorinated dibenzo-*p*-dioxins) and PCDFs (polychlorinated dibenzofurans) were continuously emitted into the environment due to incomplete combustion of urban wastes, paper industries and during the manufacture of pesticides and chlorinated substances like PCBs (polychlorinated biphenyls).

Actually, thousands of persistent chemicals species are present in the environment and consequently a variety of them are bioaccumulated in natural biota and human populations. Many of those persistent chemicals are very stable in the environment, highly lipophilic and easily cross the biological barriers. Certain "families" of chemicals, e.g. PCBs congeners (209 in total) are widely dispersed through the planet, reaching regions theorically protected as the cold poles of Earth (Arctic and Antarctic) and high mountains such as Everest. These compounds are highly persistent with half-lives of years or decades in soil/sediments, accumulate in lipid rich tissues and biomagnify in food chain [Blus, 1996]. Most of their toxic effects to organisms are still unknown.

The Stockholm Convention in 2001 signed 12 POPs as highly dangerous to living being. In fact many of them are already banned or have a very restricted use in developed countries. Despite of that, some developing countries still apply them deliberately in agricultural and in the control of disease vectors, contributing to increase their levels in the global environmental.

The organochlorine compounds known as dichlorodiphenyltrichloroethane (DDTs) were first synthesized in 1874. In 1939, Paul Miller discovered its insecticide activity and subsequently wins the Nobel Prize [Carson, 1962]. The use of chemicals was intensified after the Second World War due to the development of chemical industry and the studies of new chemicals for both military and public use. The use of DDT to control of pests for agricultural crops and malaria-spreading mosquitoes are good examples. Thereafter, the environmental effects were dependent on the development of precise and accurate analytical techniques that could detect DDT and its main metabolites (DDDs and DDEs) in environmental samples. Many chlorinated chemical compounds were important for the agricultural development, e.g. HCH (Hexachlorocyclohexane), aldrin, dieldrin, heptachlor, chlordane, mirex and other.

Heavy metals include essential metals such as copper, iron, manganese and zinc but also nonessential metals like lead, cadmium and mercury. Cooper and zinc, despite of their participation in metabolic pathways, are potentially toxic when in excess. Heavy metals, essential or not, can be uptaken and accumulated by aquatic organisms, usually to concentrations higher than those in an equivalent amount of the surrounding medium [Rainbow, 1996].

Lead is a highly toxic heavy metal and all known effects on biological systems are deleterious. The physicochemical properties of lead have been exploited by humans for millennia [Pain, 1995]. The sources of lead in the environment are the fossil fuel combustion, industrial emissions, storage-battery plants, mines, older buildings and paints. Although biological magnification does not appear to occur with lead, many studies have recently reported its effects on biological systems [Rabitto et al, 2005; Gargioni et al, 2006; Alves Costa et al, 2006], and organisms living near punctual sources of lead emission are inevitably subject to greater exposure and enhanced risk of toxic effects.

Cadmium is a soft silver-white metal with a paint bluish tinge [Wren et al, 1995]. This element, which is found as impurity in ores of other metals, is used in stabilizers and batteries. The main sources of this metal are refinement and use of cadmium, copper and nickel smelting and fuel combustion. Cadmium accumulates in microorganisms, plant and animals tissues, and is scavengered by metallothionein complexes. The acute and chronic toxic effects of cadmium have been reported to different species of invertebrates, such as reduction of growth and survival of these organisms. In fish, the toxicity varies greatly among species and is influenced by life stage.

Mercury is a heavy metal used for centuries in the course of human history. The production of mercury has been substantial and it losses to the environment even more alarming. It is impossible to estimate the quantity of mercury used in gold mining activities throughout the world. The health concernment of methylmercury associated with environmental contamination began in 1956 in the region of Minamata – Japan [Wren et al, 1995]. Actually, although residual mercury from industrial activities and other sources still persist, now the general concern regarding mercury centers on more-widespread contamination form, i.e. by atmospheric transport (largely due to the combustion of fossil fuels). Also, increased levels of mercury in biota inhabiting lakes subject to acidification from acid rain have been reported. The increased levels of mercury in reservoirs of hydroelectrics are also significant, especially in Amazon region where mercury is still used to separate gold. The bioaccumulation and biomagnification of mercury in aquatic organisms are widely reported in different tissues [Oliveira Ribeiro et al, 1999; Mela et al, 2006] and have been associated with toxic effects.

Polycyclic aromatic hydrocarbons (PAHs) have two or more fused carbon rings, which can have substituted groups attached to the rings. According to Albers (1995) the crude oil contains 2-7% of PAHs. Most of them are formed by a process of thermal decomposition of organic molecules and subsequent recombination of the organic particles (pyrolysis). The primary source for atmospheric contamination by PAHs is the incomplete combustion of organic matter [Baek et al, 1991]. Accumulation of hydrocarbons is usually inversely related to the ability of the organism to metabolize them. The most important pathway to PAHs detoxification is associated with the mixed function oxygenase (MFO) system that is capable of metabolizing hydrocarbons. Petroleum can adversely affect organisms by physical and chemical actions, for example, by disturbing the cell membrane function and the enzymatic systems associated with the plasma membrane [Neff, 1985]. Although unmetabolized PAHs can have toxic effects, one of the major concerns in animals is the ability of reactive metabolites, such as epoxides and dihydrodiols to bind to cellular proteins and DNA [Pádros et al., 2003]. The induction of lesions and neoplasm in laboratory animals by metabolites of PAHs and the observations of lesions and neoplasm in fish from PAH-contaminated sites indicate serious potential heath problems for animals with a MFO system capable of metabolizing PAHs.

Biomarkers

There are only very few pollutants which have been confirmed to cause adverse effects. In most cases, causal relationships have not been established to a large group of persistent pollutants. According to Vasseur and Cossu-Leguille (2006), the simple reason to that is the complex chemical contamination of environmental compartments, which difficults to attribute harmful effects to any particular pollutant or category of pollutants.

Some pollutants like the POPs are very persistent in the environment and toxic to the exposed biota. These organic xenobiotics are able to travel long distances on air and water currents, becoming world-widely distributed. Their propensity to bioaccumulate in food chains, affinity and persistence in the fatty tissues of living organisms are responsible for serious health problems. Several POPs are also substrates for biochemical modifications during biotransformation process, resulting sometimes in the generation of very toxic metabolites.

Environmentally and chronically exposed organisms to POPs can offer at least, two important toxicological information. First, the acute and chronic effects of these pollutants to the organism can be accessed, for example, through pathological lesions to tissues and organs, functionality of important cells such as those from immune system and changes in enzymatic activities. Second, the level of exposure and the presence of some pollutants can be estimated through chemical analyses of biological samples, which represent an "end-result" of bioaccumulation and biotransformation processes. When the acquisition of basic information of "what xenobiotics are present", "where are they accumulated", "in which chemical form are they" and "how much are there of each chemical species", it is possible to design very interesting and innovative assays. One possibility is to reproduce the composition and the specific concentrations of each chemical species in a mixture of chemicals to which target cells can be exposed. Another possibility, which is the focus to be addressed in this chapter, is the application of the own extracted pollutants to expose selected cells.

The liver is an important target to and biotransformation site of POPs, since the POPs are bioaccumulated and biotransformated in the liver and this organ is sometimes victim of these processes. Then, primary hepatic cells culture (or hepatocytes culture) represent a very useful model to study the effect of these compounds. Through chromatographical protocols it is possible to identify and quantify some POPs in biological samples. There are several commercial standards to organochlorine compounds (OCs) and PCBs, in order that it is possible to identify whether some chemicals are present or not and in how amount. However, as there are not standards for all the possible organochlorine compounds, it is not achievable to know for sure all the chemicals present in the sample even after the use of extraction protocols. This is the main drawback for the application of extracted POPs on in vitro assays and need to be emphasized.

Two types of response can be expected when an organism is exposed to a pollutant: one resultant by the metabolism of the pollutant by the exposed organism and another resulting by the toxic mechanisms [Walker et al., 1996]. An example of the metabolism response is the induction of the protein expression responsible for the elimination of the xenobiotic.

Lipophilic xenobiotics, as organochlorines, are metabolized and/or conjugated with endogen molecules in order to render them more hydrosoluble and thus increasing their rate of excretion. The biotransformation enzymes from both phase I (e.g. monooxigenases of the endoplasmic reticulum) [Bucheli and Fent, 1995] and phase II (glutathione *S*-transferases and UDP glucuronyl transferases) are responsible for this process. These responses are related to the protection of the organism and the prevention of the xenobiotic toxic effects, by removing them before they interact with action sites. Another protective response is related to the repair of cellular damage. An important example is the stress proteins and the repair mechanisms of DNA [Walker et al., 1996].

The toxic effect of a xenobiotic begins at the molecule. Metabolism and behavior alterations, neurotoxicity, decrease in reproduction and growth, and immune suppression are frequently evaluated through physiological biomarkers, while damage in tissues and organs like tumors are better accessed by morphological biomarkers, e.g. histopathology. It is possible to have a biomarker for each level of biological organization. Therefore, it is in the basic levels, i.e. in the biochemical and molecular responses, that the early signals of the pollutants exposure are observed. These responses, if perceived early enough may prevent eventual damage to natural ecosystems [Hansen, 2003].

Biomarkers are predictive tools widely used to understand the effects of chemical stressors on cells. They have been largely used in the assessment of environmental quality and to give biological information in ecotoxicological investigation, i.e. the effects of pollutants on organisms. The effect of chemical contamination is related to both toxicity of each compound or mixture of compounds.

The most important potential use of biomarkers is to provide an early warning to imminent environmental problem. Changes in the genetic/molecular levels tend to occur first, followed by responses at cellular, tissue, organ and whole-body levels. Results of various tests to evaluate these changes should accurately reflect the differing status of the organism, thus providing a detailed picture of its health and also the status of the surrounding environment. Indeed, biomarker studies conducted at molecular as subcellular levels are often assumed to be more repeatable and predictable. On this way, a biomarker can be defined as a xenobiotic-induced variation in cellular or biochemical components or processes, structures or functions that is measurable or detectable in a biological system or sample [National Research Council, 1987].

Although some biomarkers are very specific to establish a cause-effect relationship to specific chemical (for example, acetylcholinesterase inhibition by organophosphorated pesticides), the real challenge today is, if possible, to find a biomarker able to measure specifically the effect of well-defined complex mixtures. For instance, the MFO system is induced by various organochlorines and PAHs and their induction is widely used as a biomarker despite of no correlated specificity. Biomarkers are often classified into two groups: biomarkers of exposure and biomarkers of effect [Shugart et al., 1992]. However, this division can be considered somewhat arbitrary. Both types represent a continuum of responses of organisms to chemicals, so that the classification might depend on how effects are defined. Biochemical responses in organisms exposed to toxic contaminants include the deployment of detoxifying enzymes and antioxidants systems. The suitability of various antioxidant parameters, such as the cellular content of different forms of glutathione (reduced and oxidized GSH), the production of lipid peroxides and the enzymatic activities of glutathione S-transferase (GST), superoxide dismutase (SOD), catalase (CAT), glutathione peroxidase (GPx) and glutathione reductase (GR), for use as biomarkers, has been widely examined in a variety of organisms. The toxicity of a wide multiplicity of xenobiotics appear to be related to their ability to induce oxidative stress, characterized by the increase in the production reactive species and/or stimulation of cellular oxidant stress responses [Kappus, 1987]. If the insult is either sustained or of sufficient intensity to acutely overcome cell defenses, damage to macromolecules can accumulate, leading to loss of cell function, membrane damage, and ultimately, to cell death [Sevanian and McLeod, 1997]. Antioxidant enzymes are present in the tissues of all organisms and function to remove oxygen reactive species (ROS). The enzymes include superoxide dismutase (SOD; EC 1.15.1.1), which converts superoxide anion ($O_2^{\cdot-}$) to hydrogen peroxide (H_2O_2) through the reaction $2O_2^{\cdot-} + 2H^+ \rightarrow H_2O_2 + O_2$, catalase (EC 1.11.1.6) which converts H_2O_2 to water ($2H_2O_2 \rightarrow 2H_2O + O_2$) and glutathione peroxidase (GPx; EC 1.11.1.9), which couples the reaction of H_2O_2 and reduced glutathione (GSH), resulting in the convertion of hydrogen peroxide to water and in the oxidation of glutathione to its oxidized form (GSSG) by means of the reaction $H_2O_2 + 2GSH \rightarrow GSSG + 2H_2O$.

Lipid peroxidation as an endpoint of oxidative damage in organisms exposed to various contaminants has been widely studied. Various experimental designs are used, including *in vitro* generation of lipid peroxides in organelle membranes (microsomes, mitochondria), lipid micelles or vesicles and other membrane models [DiGiulio et al. 1989]. Lipid peroxidation, a biomarker of cellular oxidative stress, has been recognized as a potential contributor to the oxidative damage caused by xenobiotic compounds. [Dargel, 1992]

The initiation of lipid peroxidation results in at least three changes in cells: (1) spreading oxidative damage in cells by chain reactions and oxidation of proteins; (2) forming adducts with DNA that might result in gene mutation and cancer and (3) changing membrane structure resulting in the modification of signaling pathways or cell death [Klaunig et al, 1997] Catalase is often induced concomitantly with SOD as a result of oxidative stress [Wennig and DiGiulio, 1988]. Thus, its detection in biological tissues is important if oxidative stress is suspected. Catalase can be measured in a number of ways including spectrophotometrically through the loss of H_2O_2 absorbance at 240 nm [Beers and Sizer, 1952] or the stoichiometric generation of O_2 from H_2O_2 [Del Rio 1977]. Glutathione peroxidase is recognized as one of the most important antioxidant defenses against oxygen toxicity in organisms. The enzyme is readily assayed in tissues by coupling glutathione peroxidase to excess glutathione redutase. The Table 1 resumes the oxidative responses in fish exposed to heavy metals and other pollutant organ persistent.

Biochemical endpoints associated with oxidative stress generally do represent primary responses to contaminants. Laboratory studies have established a strong causal link between exposure of fish to PAHs and coplanar PCBs and the expression of cytochrome P4501A1 and its associated ethoxyresorufin-*O*-deethylase (EROD) activity. The EROD induction is a classical biomarker and well established for MFOs and biotransformation in ecotoxicology [Livingstone, 1998].

Table 1. The response of some hepatic biomarkers of fish after exposure to chemical pollutants

Pollutant \ Biomarker	SOD	GPx	GR	CAT	GSH	GSSG	GST
Heavy metals	$\approx^{\mathrm{Lc}}$	$\uparrow^{\mathrm{Lc}}$	$\approx^{\mathrm{Lc}}$	$\downarrow^{\mathrm{Dl, Or}} / \uparrow^{\mathrm{Lc}}$			$\uparrow^{\mathrm{Lc}}$
Cd	$\approx^{\mathrm{Or}} / \uparrow^{\mathrm{Or}}$	$\downarrow^{\mathrm{Mu}} / \uparrow^{\mathrm{Or, On}}$	$\approx^{\mathrm{On}} / \downarrow^{\mathrm{On}}$	$\uparrow^{\mathrm{On, Or}} / \downarrow^{\mathrm{Fh, Dl, Or}}$	$\downarrow^{\mathrm{On}}$	$\approx^{\mathrm{Om}} / \uparrow_{\mathrm{Om, On}}$	$\approx^{\mathrm{Pl}} / \uparrow^{\mathrm{Or, On}}$
Hg	$\uparrow^{\mathrm{Ss, Gy}}$	$\approx^{\mathrm{Im}} / \uparrow^{\mathrm{Ss, Im}}$	$\approx^{\mathrm{Im}}$	$\approx^{\mathrm{Gy}} / \downarrow^{\mathrm{Or}}$			$\approx^{\mathrm{Im}} / \uparrow^{\mathrm{Ss, Im}}$
Cu				$\uparrow^{\mathrm{On, Cc, Br}} / \downarrow^{\mathrm{Dl, Or}}$	$\downarrow^{\mathrm{Im}}$		$\uparrow^{\mathrm{Cc, Li, Br}}$
Cr or Zn				$\downarrow^{\mathrm{Or}} / \uparrow^{\mathrm{On}}$			
Ag or Pb				$\downarrow^{\mathrm{On, Or}}$			
As	$\uparrow^{\mathrm{Cb}}$	$\uparrow^{\mathrm{Cb}}$	$\downarrow^{\mathrm{Cb}}$	$\uparrow^{\mathrm{Cb}}$	$\downarrow^{\mathrm{Cb}} / \uparrow^{\mathrm{Cb}}$	$\uparrow^{\mathrm{Cb}}$	
MeHg	$\approx^{\mathrm{Ss}}$	$\uparrow^{\mathrm{Ss}}$					
TBT	$\uparrow^{\mathrm{Se}}$	$\approx^{\mathrm{Se}}$		$\uparrow^{\mathrm{Se}}$	$\downarrow^{\mathrm{Sa}} / \approx^{\mathrm{Sa}}$		$\downarrow^{\mathrm{Sa}} / \approx^{\mathrm{Sa}} / \uparrow^{\mathrm{2xSe}}$
PAH (3MC)	$\approx^{\mathrm{Ll, Dl}} / \uparrow^{\mathrm{Dl}}$	$\downarrow^{\mathrm{Dl}} / \uparrow^{\mathrm{Dl}}$	$\approx^{\mathrm{Ll, Om}}$	$\approx^{\mathrm{Ll, Dl}} / \downarrow^{\mathrm{Ll}}$			$\approx^{\mathrm{Ll, Om}} / \downarrow^{\mathrm{Dl}}$
PAH (BaP)		$\uparrow^{\mathrm{Se}}$			$\downarrow^{\mathrm{Sa}} / \approx^{\mathrm{Sa}}$		$\approx^{\mathrm{Aa, Pv, Se, As, Pl}} / \downarrow^{\mathrm{Lm}} / \uparrow^{\mathrm{Sa}}$
PAH (BNF)	$\approx^{\mathrm{Om}}$	$\approx^{\mathrm{Om}}$	$\approx^{\mathrm{Zv}}$	$\approx^{\mathrm{Om}}$			$\approx^{\mathrm{Cc, Aa, 2x\ Om, Zv}} / \uparrow^{\mathrm{Fh, Om}}$
PAH (DNOC)				$\downarrow^{\mathrm{Aa}}$			
PAHs	$\uparrow^{\mathrm{Ll, Mb, Lx}}$	$\downarrow^{\mathrm{Mb}} / \approx^{\mathrm{My}}$	$\uparrow^{\mathrm{My}}$	$\uparrow^{\mathrm{Ip, Ll, Mb, My}}$	$\uparrow^{\mathrm{Ip}}$	$\uparrow^{\mathrm{Ip}}$	$\approx^{\mathrm{Aa, 2x\ Om, My, Ps}} / \uparrow^{\mathrm{Mb}}$
OCP (DDTs or DDE)			$\uparrow^{\mathrm{Om}}$				
OCP (hexachlorobenzene)	$\approx^{\mathrm{Cc}} / \uparrow^{\mathrm{Om}} / \downarrow^{\mathrm{Cc}}$	$\uparrow^{\mathrm{Om}}$	$\uparrow^{\mathrm{Om}}$		$\approx^{\mathrm{Cc}} / \downarrow^{\mathrm{Cc}} / \uparrow^{\mathrm{Om}}$		$\approx^{\mathrm{2x\ Om}} / \uparrow^{\mathrm{Om}}$
OCP (dieldrin)	$\approx^{\mathrm{Om}}$						
OCP (endosulfan)							$\approx^{\mathrm{Om}}$
OCP (2,4-DC and picloram)				$\approx^{\mathrm{Ip}}$			$\approx^{\mathrm{Ip}}$
OCPs	$\uparrow^{\mathrm{Mp}}$	$\uparrow^{\mathrm{Mp}}$	$\approx^{\mathrm{Mp}}$	$\uparrow^{\mathrm{Mp}}$			$\uparrow^{\mathrm{Mp}}$
PCB (3,3',4,4'-TCB)	$\approx^{\mathrm{Om}}$	$\uparrow^{\mathrm{Om}}$	$\uparrow^{\mathrm{Om}}$	$\downarrow^{\mathrm{Om}}$	$\approx^{\mathrm{Om}} / \uparrow^{\mathrm{Om}}$	$\uparrow^{\mathrm{Om}}$	$\approx^{\mathrm{Om}} / \uparrow^{\mathrm{Om}}$
PCB (Aroclor 1254)	$\approx^{\mathrm{Om, Mb, So}}$	$\uparrow^{\mathrm{Mb, So, Mu}}$	$\downarrow^{\mathrm{Mb}} / \uparrow^{\mathrm{So}}$	$\uparrow^{\mathrm{Mb, So}}$			$\approx^{\mathrm{Ip}}$
PCB (Clophen A40 or A50)			$\uparrow^{\mathrm{Om}}$				$\uparrow^{\mathrm{Pp, Om}}$
PCB (CB77, 126 or 156)							$\approx^{\mathrm{Pl}} / \uparrow^{\mathrm{Om}}$
PCBs	$\downarrow^{\mathrm{Gv}}$	$\downarrow^{\mathrm{Om}}$	$\approx^{\mathrm{Om}}$	$\uparrow^{\mathrm{Gv}}$	$\approx^{\mathrm{Om}}$	$\approx^{\mathrm{Om}}$	$\downarrow^{\mathrm{Om}} / \uparrow^{\mathrm{Ab}}$
PCDF (2,3,7,8-TCDF)	$\uparrow^{\mathrm{Af}}$	$\approx^{\mathrm{Af}}$		$\approx^{\mathrm{Af}}$	$\uparrow^{\mathrm{Ca}}$		$\uparrow^{\mathrm{Fh}} / \uparrow^{\mathrm{Ca}}$
PCDD (2,3,7,8-TCDD)					$\uparrow^{\mathrm{Ca}}$		$\approx^{\mathrm{Gm}} / \downarrow^{\mathrm{Om}}$
PAHs+Heavy metals	$\downarrow^{\mathrm{Se}}$	$\approx^{\mathrm{Se}}$	$\downarrow^{\mathrm{Aa}}$	$\downarrow^{\mathrm{Aa}}$	$\downarrow^{\mathrm{Aa, Sa}} / \approx^{\mathrm{Sa}}$		$\downarrow^{\mathrm{Om, Rr}} / \approx^{\mathrm{Sa, Pl}}$
PAHs+PCBs		$\approx^{\mathrm{Om, Bp, Lc, Cs}}$	$\approx^{\mathrm{Om, Bp, Lc, Cs}} / \uparrow^{\mathrm{Om}}$	$\downarrow^{\mathrm{Om}}$	$\approx^{\mathrm{Bp, 2x\ Cs}} / \uparrow^{\mathrm{Pv, Lb, Ps}}$		$\uparrow^{\mathrm{Bp, Om}} / \approx^{\mathrm{Aa, Cs, Ll, Pl, Om, Pl}}$
PCBs+OCPs							$\uparrow^{\mathrm{On, Cr}}$
PCBs+HCHs	$\uparrow^{\mathrm{On}}$	$\approx^{\mathrm{On}}$	$\downarrow^{\mathrm{On}}$	$\downarrow^{\mathrm{On}}$	$\approx^{\mathrm{On}}$		
PCBs+DDE	$\uparrow^{\mathrm{Am}}$	$\approx^{\mathrm{Am}}$		$\approx^{\mathrm{Am}}$	$\downarrow^{\mathrm{Am}}$		$\uparrow^{\mathrm{Am}}$
PAHs+PCBs+OCPs	$\approx^{\mathrm{Zo}} / \uparrow^{\mathrm{Sp}}$			$\uparrow^{\mathrm{Sp, Zo}}$	$\approx^{\mathrm{Rr}}$	$\approx^{\mathrm{Rr}}$	$\approx^{\mathrm{Pl}} / \downarrow^{\mathrm{Rr}} / \uparrow^{\mathrm{Gm}}$

Table 1. (Continued)

Pollutant \ Biomarker	SOD	GPx	GR	CAT	GSH	GSSG	GST
PAHs+PCBs+OCPs+PCDD/Fs							
PHAHs+PAHs	↑ Cc	≈ Cc	≈ Cc	≈ Cc	↓ Cc	↑ Cc	↑ Cc

Pollutant \ Biomarker	LPO	UDPGT	EROD
Heavy metals (Hg, Pb, Cd, Cu, As, Cr, Ag)	≈ Lc, Ss / ↑ Dl, Aa, Cp, Hf, Cb, Gy, Mu		≈ Pl / ↓ Dl / ↑ Dl
Organometals (TBT, TPT, MeHg)	≈ Ss		≈ Sa / ↓ Cg, Aa, Om, Sf, Dl / ↑ Sa
PAH (3MC)		≈ Ll, Om, Dl	↑ Ll, Dl
PAH (BaP)			↑ Ng, 2xSa, Cc, Ll, 2x Aa, 3x Pl, 2x Dc, Sf, Om, Sm, An, Ol, Dl
PAH (BNF)		≈ Cc, Fh / ↑ Zv, 2x Om	≈ Hr / ↑ Ss, 2x Zv, Hs, Am, 2x Cc, Ip, 2x Fh, St, Pw, El, 4x Aa, Ll, 8x Om, Rr, Sg, An, Br
PAH (DNOC)		↑ Aa	
PAHs			≈ Ot / ↑ Gm, Bs, 3x Om, 2x Ss, Nc, Ip, 2x Sc, Sh, Aa, Pb, My, Lx, Pa
OCP (hexachlorobenzene)		≈ Om	≈ 2x Om / ↑ Om
OCP (dieldrin)	↑ Om		
OCP (endosulfan or 2,4-DC)	↑ Ca		↑ Om, Ip
OCPs	↓ Mp		↑ Mp
PCB (3,3',4,4'-TCB)		≈ Om / ↑ Om, St	↑ 5x Om, Sf
PCB (Aroclor 1254)	↑ Om, Mb, So, Mu	≈ Ip / ↑ Pp	↑ Cc, Ip, Om, On
PCB (Aroclor 1260, CB118 or CB169)			↑ Bp, Om
PCB (Clophen A40 or A50)		↑ Om	≈ Pp / ↑ Pl, 2x Om, Ga
PCB (CB77 or CB153)		≈ Om / ↑ Om	↑ Ll, Pi, 3x Om
PCB (CB126)		↑ Om	↓ Pf / ↑ Ip, 2x Om, Re, Pf
PCBs	↑ Gv	≈ Om	≈ Lo, Om / ↑ Pl, Br, Hs, Tt, Fh, Ms, Cn, La, Rr, Pa
PCDF (2,3,7,8-TCDF)	↑ Af	≈ Fh, Af	↑ Af, Fh, Om, Sf, El
PCDD (2,3,7,8-TCDD)		↑ Br	↑ Sf, 2x Cc, Gm, 6x Om, 2x Br, Ip, El
PCDD or PCB+Heavy metals			↑ Ca, La
PAHs+Heavy metals	↑ Aa	≈ Om, Rr	↑ Sa, Dl, Pl
PAHs+PCBs		≈ Pl / ↑ Bp, Cc, Cs, 2x Om	≈ Hs, 2x Pt, Pp, Hr / ↑ 4x Ll, 2x Aa, Pl, Pf, 3x Om, Gm, 2x Bp, Cc, Ip, Lc, 2x Pv, 2x Pt, Gg, Cn, 2x Lb, 2x Ps, Pl
PCBs+OCPs		≈ Pp, Am / ↑ On, Cr	↑ Cr, Ll, Pp, On , 2x Pt, In
PAHs+PCBs+OCPs		≈ Pl	≈ Rr, Sp / ↑ Gm, Ll, Aa, 2x Pt, Mp, Fh, El, Pp, Rr, Zo
PAHs+PCBs+PCDD/Fs		↑ Aa	↑ Sc, Ll, Cl, Aa, Mb
PHAHs+PAHs		↑ Cc	↑ 2x Cc, Po, Pf

Symbols: ↓ decrease, ≈ no (significant) response, ↑ increase. Abbreviations: biomarkers - SOD (superoxide dismutase activity), GPx (glutathione peroxidase activity), GR (glutathione reductase activity), CAT (catalase activity), GST (glutathione *S*-transferase activity), GSH (concentration of reduced glutathione), GSSG (concentration of oxidized glutathione), LPO (lipid peroxidation), UDPGT (UDP glucuronyl transferase activity), EROD (ethoxyresorufin *O*-deethylase activity); pollutants – TPT (triphenyltin), TBT (tributyltin), MeHg (methyl mercury), PAHs (polycyclic aromatic hydrocarbons), 3MC (3-methylcholanthrene), BaP (benzo(a)pyrene), BNF (β-naphthoflavone, a PAH-like compound), DNOC (dissolved natural organic carbon), OCPs (organochlorine pesticides), 2,4-DC (2,4-dichlorophenol), PCBs (polychlorinated biphenyls), PCDFs (polychlorinated dibenzofurans), PCDDs (polychlorinated dibenzo-*p*-dioxins), PHAHs (polyhalogenated aromatic hydrocarbons). The double letters following the symbols (i.e. $\downarrow^{Xx}$, $\approx^{Xx}$, $\uparrow^{Xx}$) correspond to the species of fish in which the biomarker was evaluated. The multiplier (i.e. 2x, 3x, 4x etc) before the symbol of fish species indicates the number of different studies in which the species was utilized to evaluate the biomarker's response. Symbols and corresponding fish species: Aa - *Anguilla anguilla* (Eel), Ab - *Abramis brama* (Bream), Af - *Acipenser fulvescens* (Lake sturgeon), Am - *Ameiurus nebulosus* (Brown Bullhead), An - *Acipenser naccarii* (Sturgeon), Bp - *Barbus plebejus* (Barbel), Br - *Brachydanio rerio* (Zebrafish), Bs - *Boreogadus saida* (Polar cod), Ca - *Carassius auratus* (Crucian carp), Cb - *Clarias batrachus* (Indian catfish), Cc - *Cyprinus carpio* (Carp), Cg - *Cottus gobio* (Bullhead), Cl - *Callionymus lyra* (Dragonet), Cn - *Chondrostoma nasus* (Nase), Cp - *Channa punctatus*, Cr - *Clarias anguillaris* (Mudfish), Cs - *Chondrostoma soetta* (Nase), Dc - *Dorosoma cepedianum* (Gizzard shad), Dl - *Dicentrarchus labrax* (Sea bass), El - *Esox lucius* (Pike), Fh - *Fundulus heteroclitus* (Killifish), Ga - *Gasterosteus aculeatus* (Stickleback), Gg - *Gobio gobio* (Gudgeon), Gm - *Gadus morhua* (Atlantic cod), Gv - *Girardinichthys viviparus* (goodeid fish), Gy - *Gymnogeophagus gymnogenys*, Hf - *Heteropneustes fossilis* (Cat-fish), Hr - *Holocentrus rufus* (Squirrelfish), Hs - *Haemulon sciurus* (Blue-striped grunt), Im - *Ictalurus melas* (catfish), Ip - *Ictalurus punctatus* (Channel catfish), La - *Lepomis auritus* (Redbreast sunfish), Lb - *Lepidopsetta bilineata* (Rock sole), Lc - *Leuciscus cephalus* (Chub), Li - *Leuciscus alburnoides* complex (Iberian endemic minnows), Ll - *Limanda limanda* (Dab), Lm - *Lepomis macrochirus* (Sunfish), Lo - *Lota lota* (Burbot), Lx - *Leiostomus xanthurus* (Spot), Mb -*Mullus barbatus* (Red mullet), Mp - *Mugil* sp. (Grey mullet), Ms - *Micropterus salmoides* (Largemouth bass), Mu - *Micropogonias undulatus* (Atlantic croaker), My - *Myoxocephalus scorpius* (Shorthorn sculpin), Nc - *Notothenia coriiceps* (Antarctic rockcod), Ng - *Notothenia gibberifrons* (Antarctic fish), Ol - *Oryzias latipes* (Japanese medaka), Om - *Oncorhynchus mykiss* (Rainbow trout), On - *Oreochromis niloticus* (Nile tilapia), Or - *Oreochromis mossambicus* (Tilapia), Ot - *Opsanus tau* (Toadfish), Pa - *Pseudopleuronectes americanus* (Winter flounder), Pb - *Platycephalus bassensis* (Sand flathead), Pf - *Perca fluviatilis* (Perch), Pi - *Pimephales promelas* (Fathead minnow), Pl - *Platichthys flesus* (Flounder), Po - *Ptychocheilus oregonensis* (Northern squawfish), Pp - *Pleuronectes platessa* (Plaice), Ps - *Platichthys stellatus* (Starry flounder), Pv - *Parophrys vetulus* (English sole), Pw - *Prosopium williamsoni* (Mountain whitefish), Re - *Raja erinacea* (Skate), Rr - *Rutilus rutilus* (Roach), Sa - *Salvelinus alpinus* (Arctic charr), Sc - *Serranus cabrilla* (Comber), Se - *Sebastiscus marmoratus* (Cuvier), Sf - *Salvelinus fontinalis* (Brook trout), Sg - *Siganus canaliculatus* (safi fish), Sm - *Scophthalmus maximus* (Turbot), So - *Scorpoena porcus* (Scorpion fish), Sp - *Sardina pilchardus* (Sardine), Ss - *Salmo salar* (Atlantic/Baltic salmon), St - *Salmo trutta* (Lake trout), St - *Stenotomus chrysops* (Scup), Tt - *Thymallus thymallus* (Grayling), Zo - *Zosterisessor ophiocephalu* (Goby), Zv - *Zoarces viviparus* (Viviparous blenny).

In Vivo Experimental Studies

The experimental tests involving the studies of pollutants effects in aquatic organisms are based in the recognition that the response of organisms to the presence of toxic chemical is dependent upon the exposure level of the toxic agent. Thus, the principle is the description of the relationship between concentration and response. There are essentially two different approaches in toxicological research, the *in vivo* and *in vitro* systems. Independently of the method, ethical principles of rational selection and use of animals and their welfare may be considered. Usually, the selection of a particular method depends on the overall objectives of the study [Zhang et al., 2003] and it offers limited information about a specific issue. As a consequence, new methodologies need to be integrated with classical methods to test specific hypotheses [Beis and Stainier, 2006] and access complementary data.

Two different tests are usually designed: acute and chronic exposure, respectively evaluating the concentration-response relationship for survival or identification of target molecules, cells or tissues; and evaluating sublethal effects such as growth, reproduction, behavior or biochemical effects. As an alternative to the identification of xenobiotics by their physicochemical properties, such compounds can be detected by their biological action in bioassays. These kinds of tests are particularly interesting when dealing with a range of structurally different compounds that share a specific effect, each one with its own potency [Hoogenboom and Kuiper, 1997].

In experimental conditions a control group is required. A diversity of doses/ concentrations can be tested through intraperitoneal injection of chemicals, waterborne and dietary expôsures. Recently, the subchronic trophic exposure has been used to test a more realistic situation of effects of heavy metals in fish [Rabitto et al, 2005; Mela et al, 2006; Alves Costa et al, 2006]. Thereafter, there are temporal and spatial limitations in chronic exposure that must be considered before starting any experiment. Usually, experimental artifacts increase in a temporal scale and the number and size of test individuals also must be evaluated.

Another limitation in experimental tests is the use of more then one stressor chemical agent. Pádros et al. (2003) established an experimental design to evaluate the associated effects of TBT and BaP in *Salvelinus alpinus*. According to the results, TBT interferes with the toxicity of BaP decreasing its genotoxic effects due to the inhibition of the P450 enzymes activity. Mixtures of pollutants can be experimentally evaluated using sediment from contaminated sites, but as described to field studies, there is no precise information about the related effect with the chemicals present in the sediment.

The data generated from experimental tests must be interpreted and extrapolated from lower to higher levels of biological organization. This is a very important subject to validate all laboratory efforts in the development of new approaches to evaluate the effects of pollutants in tested individuals. There is no reasonable reason to study any endpoint using experimental tests if the data may not be extrapolated to populations in natural conditions.

Attempts to establish links between the different levels of biological organization are pertinent when it is considered the impact of chemical pollutants, because toxicity appears firstly in the individuals before populations are affected [Vasseur and Cossu-Legniele, 2006]. The extrapolation of data to populations is a challenge to all scientists working on environmental impact by chemicals. Animal behavior may be affected by interference with neurotransmission, and changes in behavior may impair the capacity of organisms to feed, to escape predators and communicate, which in turn can affect growth and reproduction. Also, tissue dysfunction and physiological changes may occur and cause population effects. Other effects such as endocrine disruption and immune depletion may lead to increased sensitivity to environmental stressors. The genotoxic findings may also lead to total compromised reproductive success or to dysfunction(s) in the offspring if the gametes remain viable, but display deleterious mutation(s).

Concluding, elucidating the mechanism(s) of toxicity would require experimental studies in controlled and more realistic conditions to establish links between molecular, cellular and population effects.

In Vitro Experimental Studies

Over the last years, several different tools have been developed to obtain reliable knowledge concerning toxicological profiles of xenobiotics, i.e. metabolic biotransformation, chemical interactions, exposure levels, adverse effects to biological systems and potential toxic risk to biota. The scope and sophistication of some of these new research tools allow scientists to better identify hazards and understand basic processes that influence the expression of toxic responses in biological systems [Baksi and Frazier, 1990].

A particularly fruitful approach is to identify and to develop *in vitro* models which retain the basic characteristics of a more complex *in vivo* condition and that can be experimentally manipulated for research purposes [Baksi and Frazier, 1990]. The use of non-animal test methods, including computer-based approaches and *in vitro* studies, provides important tools to enhance the understanding of hazardous effects caused by chemicals and for predicting these effects [Broadhead and Combes, 2001] on living organisms. Historically, toxicological research started from different tissues to achieve a single cell, to clarify and to identify basic toxicity mechanisms. During the first years, the attention was mainly focused on ''things'' occurring inside the cells after contact with a xenobiotic, instead of understanding ''why'' and ''how'' they happened. Later, the interest was concentrated on toxic mechanism[s] of action, due also to the strong improvement of tissue culture methodologies. Lately, one of the main aims and challenges of *in vitro* systems has been the reproduction of original tissue characteristics and its cell-cell interactions, to simulate, as far as possible, the *in vivo* environment [Zucco et al., 2004].

The probable foremost trait of *in vitro* systems is the accurate control and maintenance over experimental conditions, which are not possible *in vivo*. Physicochemical conditions of the cellular environment, concentration of physiological ions and non-physiological conditions can be rigorously controlled and manipulated. Biological factors that influence cellular responses, such as hormones and mediators, can be studied individually and in combinations which would not be possible *in vivo*. In addition, the concentration of toxic chemicals can be readily determined and controlled to more precisely define dose-response relationships. Also, *in vitro* systems eliminate interactive systemic effects that can complicate the experimental situation and thus, the interpretation of data is simplified. Often, *in vitro* systems exhibit reduced variability between experiments, since it is possible to control environmental conditions and to have internal controls, i.e., each experiment can act as its own control and it is possible to incorporate positive and negative control test chemicals into the experimental design to, in a sense, calibrate the system for each experiment [Baksi and Frazier, 1990].

The ability to obtain repeated samples during the course of an experiment adds an important dimension to *in vitro* studies. Time-course studies can provide valuable information concerning the sequence of cellular events during the toxicological response to xenobiotics. Other advantages of *in vitro* model systems are of a more practical nature. *In vitro* systems usually require significantly smaller quantities of test chemicals to conduct complete dose-response relationships and are often cheaper and provide data more rapidly than *in vivo* model systems. Since smaller quantities of test chemicals are used in these studies, there is less toxic waste to dispose at the end of a study. Finally, multiple types of experimental conditions can be initiated with populations of cells obtained from a single animal, thus reduced number of animals are usually necessary for some investigations [Baksi and Frazier, 1990; Guillouzo et al., 1997; Segner, 1998; Zhou et al., 2005]. One important aspect in research is the ethical use of animal models. Usually, *in vitro* systems require fewer specimens to conduct the studies and as it utilizes "parts of the animal" (cells, slices of tissues or organs) the suffering of animal is limited by its sacrifice. Thus, at the ethical point of view, *in vitro* models are indicated to investigate several specific issues instead of the *in vivo* studies.

Despite of the advantages of *in vitro* systems aforementioned, these models also have some limitations in comparison with *in vivo* systems. Interactions between different compartments, i.e., tissues and organs inside an organism are generally avoided; it is not an easy task to extrapolate the data to the organism situation; *in vivo* environment is frequently much more complex than the recreated situation *in vitro*; when cells are transferred from an *in vivo* environment to an *in vitro* system they usually have phenotypic alterations in long-term cultures; etc. Then, *in vitro* and *in vivo* studies must be integrated for a holistic acquisition of data about toxicity.

Typically, *in vitro* models include subcellular fractions (such as macromolecules, cell organelles, subcellular fractions), cellular systems (such as cell suspensions, primary cultured cells, genetically engineered cells, immortalized cells/ cell lines, cells in different stages of transformation or differentiation, stem cells, co-cultures of different cell types and barrier systems), and whole tissues (including organotypic systems, perfused organs, tissue slices and explants) [Eisenbrand et al., 2002]. Except for subcellular fractions, these systems are characterized by preserved integrated cell metabolism. None of them can be completely replaced by another [Červenková et al., 2001] and each one is important for obtaining specific data, which is fundamental for understanding toxicological aspects as a whole.

Scientifically, one of the major areas of potential value of *in vitro* approaches is to obtain mechanism-derived information [Eisenbrand et al., 2002]. These systems are considered to be of additional significance beyond "hazard identification" and hence it is important to consider their application to other elements of the risk assessment paradigm. However, a prerequisite for the successful application of *in vitro* approaches is the availability of appropriate validated test systems [Balls et al., 1990, 1995; OECD, 1996]. *In vitro* studies also provide information about the overall potential of some mixtures of xenobiotics (such as some POPs) to act as endocrine disruptors through interaction with specific signaling pathways, without requiring wide spectra of standards necessary for chemical analysis. The screening of complex mixtures from the environment enables prioritizing samples samples of interest that require further detailed chemical analysis [Janošek et al., 2006]

The chemical binding of xenobiotics within the *in vitro* system is an essential datum to be estimated since, sometimes, it is a potential source of error to consider the xenobiotic availability equal to the free chemical fraction in blood or plasma [Rane et al., 1977; Lin et al., 1978, 1980]. The identification of synergistic, additive and antagonistic interactions in xenobiotic combinations is another similarly important aspect. Actually, organisms are generally continuously exposed to a complex pool of xenobiotics and little is known about effects of these mixtures on biota. It is essential to note that some interactions such as synergistic response have their occurrence and magnitude determined by specific conditions of the system and by specific concentrations of each component of the pools of xenobiotics.

In vitro systems have been used extensively to characterize metabolic pathways in fish and could, in principle, be used to evaluate large numbers of chemicals and species. But, the relationships between *in vitro* and *in vivo* rates of metabolism are generally unclear. Toxicologists who study fish are faced, therefore, with the same fundamental challenge of pharmacologists and toxicologists who study mammals; namely, the need to extrapolate *in vitro* metabolism data to the *in vivo* condition and then predict the impact of this activity on chemical kinetics on exposed animals [Nichols et al. 2006]. When performing some comparisons, an effort should be made to work at environmentally relevant xenobiotic concentrations, including the relative proportions of different metabolites.

Areas for future research include determination of mechanism of action of xenobiotics, the role of metabolism in mediating organism toxicity, and the toxicity of complex mixtures and environmental samples [Baksi and Frazier, 1990]. The novel approaches of *in vitro* toxicology are focused on the development of molecular markers tha allow to detect effects at levels of exposure to potentially toxic chemicals lower than those that cause the onset of clinically observable pathological responses [Eisenbrand et al., 2002].

The rapid progress in genomics and proteomics, in combination with the power of bioinformatics, creates a unique opportunity to form the basis of better hazard identification, for increasing the understanding of underlying mechanisms and for a more relevant safety evaluation. Methodologically, a major advance would be the introduction of relevant biomarkers for the identification of potential hazards of chemicals and their metabolites formed in the body. Expression of stress response or other genes and ensuing biochemical alterations may be potential markers for compound-induced toxicity. In addition, the measurement of the transcription and translation products of gene expression can reveal valuable information about the potential toxicity profile of chemicals. Of equal importance is the critical utilization of biomarkers for genetic susceptibility and for protection, factors of major importance in determining individual response [Eisenbrand et al., 2002].

Immunotoxicological Models

Varieties of leukocyte types are involved in nonspecific cellular defenses of fish, and include monocytes/macrophages, granulocytes, and nonspecific cytotoxic cells (NCCs). Macrophages and granulocytes are mobile phagocytic cells found in the blood and in secondary lymphoid tissues. These cells are particularly important in inflammation, which is the cellular response to microbial invasion and/or tissue injury leading to the local accumulation of leukocytes and fluids [Secombes, 1996].

Table 2. The response of some immunological biomarkers of fish after exposure to chemical pollutants

Pollutants\biomarker	Lymphocyte proliferation	Leukocyte number	Respiratory burst	Phagocytic activity	Phagocytic index	Number of MMC	O_2^- production
Arsenic (As)		$\downarrow^{Cb}$		$\downarrow^{Cb}$			
Cadmium (Cd)			$\downarrow^{2x\ Om} / \uparrow^{Dl}$	$\downarrow^{2x\ Om,\ Dl} / \uparrow^{Om}$	$\approx^{Pl}$		
Chromium (Cr)					$\downarrow^{Sf}$	$\uparrow^{Pp\ (spleen0}$	
Copper (Cu)		$\downarrow^{Br}$	$\downarrow^{Om} / \uparrow^{Ca}$		$\approx^{Om} / \downarrow^{Br}$		
Zinc (Zn)		$\downarrow^{Br}$			$\downarrow^{Om} / \uparrow^{Br}$		
Nickel (Ni)			$\approx^{Om}$	$\approx^{Om}$			$\downarrow^{Ol}$
Manganese (Mn)			$\uparrow^{Cc}$	$\uparrow^{2x\ Cc}$			
Mercury (Hg)	$\downarrow^{Tt} / \uparrow^{Tt}$		$\downarrow^{Om}$	$\downarrow^{Dl}$			
Lead (Pb)						$\uparrow^{Hm\ (head\ kidney0}$	
Heavy metals			$\downarrow^{Om}$				
TBT - tributyltin		$\downarrow^{Mf,\ Ma,\ Bb,\ Mp}$	$\uparrow^{Ot} / \downarrow^{Ip,\ Ot}$	$\approx^{Bb} / \downarrow^{Mf,\ Mp,\ Ma,\ Ot,\ Tm,\ Mu}$	$\uparrow^{Ma} / \approx^{Mf,\ Bb} / \downarrow^{Mp}$	$\approx^{Hm\ (head\ kidney0}$	
DBT - dibutyltin		$\downarrow^{Mf,\ Ma,\ Bb,\ Mp}$		$\approx^{Bb} / \downarrow^{Mf,\ Mp,\ Ma}$	$\approx^{Mf,\ Bb,\ Ma} / \downarrow^{Mp}$		
Cd+BaP			$\uparrow^{Dl}$	$\downarrow^{Dl}$	$\downarrow^{Dl}$		
3MC	$\downarrow^{Cc}$		$\uparrow^{Cc}$				
BaP	$\downarrow^{3x\ Ol}$		$\downarrow^{On} / \uparrow^{Dl}$		$\approx^{Dl}$		$\downarrow^{Ol}$
DMBA			$\downarrow^{On}$	$\approx^{On} / \downarrow^{Ot}$			
PAHs	$\downarrow^{2x\ Lx,\ Om}$		$\downarrow^{Lx,\ Tm,\ 2x\ Ll,\ Om} / \uparrow^{Fh,\ Pv}$	$\downarrow^{Ot,\ Lx}$		$\downarrow^{Pa\ (liver0}$	$\approx^{Ll}$
PCB 126			$\downarrow^{Ip}$				
PCBs			$\approx^{Ll}$	$\downarrow^{Md}$			$\approx^{Ll} / \downarrow^{Hp}$
PAHs +PCBs			$\approx^{Ll} / \downarrow^{Ll}$				
PCB+Hg			$\downarrow^{La}$				
TCDD				$\approx^{Om}$		$\uparrow^{Cc\ (spleen,\ head\ kidney0}$	
Endosulfan (OCP)		$\approx^{Bb,\ Mf,\ Ma}$		$\approx^{Bb} / \downarrow^{Mf} / \uparrow^{Ma}$	$\approx^{Bb,\ Ma} / \downarrow^{Mf}$		
Lindane (OCP)			$\approx^{Cc} / \downarrow^{Om}$	$\approx^{Cc}$			

Symbols: ↓ decrease, ≈ no (significant) response, ↑ increase. Biomarkers: Lymphocyte proliferation, leukocyte number, respiratory burst in macrophages, phagocytic activity (number of engulfing cells) of macrophages, phagocytic index (mean number of particles engulfed per cell) of macrophages, number of MMC (melano-macrophage centers) in liver, head kidney and spleen, O_2^{-} (superoxide anion) production in kidney leucocytes. Abbreviations: pollutants – 3MC (3-methylcholanthrene), BaP (benzo(a)pyrene), DMBA (7,12-dimethylbenzanthracene), PAHs (polycyclic aromatic hydrocarbons), PCBs (polychlorinated biphenyls), TCDD (2,3,7,8-tetrachlorodibenzo-p-dioxin), OCP (organochlorine pesticide). The double letters following the symbols (i.e. $\downarrow^{Xx}$, $\approx^{Xx}$, $\uparrow^{Xx}$) correspond to the species of fish in which the biomarker was evaluated. The multiplier (i.e. 2x, 3x) before the symbol of fish species indicates the number of different studies in which the species was utilized to evaluate the biomarker's response. Symbols and corresponding fish species: Bb -

Bidyanus bidyanus (Silver perch), Br - *Brachydanio rerio* (Zebrafish), Ca - *Carassius auratus* (Crucian carp), Cb - *Clarias batrachus* (Catfish), Cc - *Cyprinus carpio* (Carp), Dl - *Dicentrarchus labrax* (Sea bass), Fh - *Fundulus heteroclitus* (Killifish), Hm – *Hoplias malabaricus* (Trahira), Hp - *Hipoglossoides platessoides* (American plaice), Ip - *Ictalurus punctatus* (Channel catfish), La - *Lepomis auritus* (Redbreast sunfish), Ll - *Limanda limanda* (Dab), Lx - *Leiostomus xanthurus* (Spot), Ma - *Macquaria ambigua* (Golden perch), *Micropterus dolomieu* (Smallmouth bass), Mf - *Melanotaenia fluviatilis* (Rainbowfish), Mp - *Maccullochella peelii* (Murray cod), Mu - *Micropogonias undulatus* (Atlantic croaker), Ol - *Oryzias latipes* (Japanese medaka), Om - *Oncorhynchus mykiss* (Rainbow trout), On - *Oreochromis niloticus* (Tilapia), Ot - *Opsanus tau* (Toadfish), Pa - *Pseudopleuronectes americanus* (Winter flounder), Pp - *Pleuronectes platessa* (Plaice), PV - *Pleuronectes vetulus* (English sole), Sf - *Saccobranchus fossilis* (catfish), Tm - *Trinectes maculatus* (hogchoker), Tt - *Trichogaster trichopterus* (Blue gourami)

References: Bowser et al., 1994; Carlson et al., 2002, 2004a, 2004b; Clemons et al., 1999; Cossarini-Dunier, 1987, 1988; Dethloff and Bailey 1998; Duchiron et al., 2002; Dunier et al., 1994; Elsasser et al., 1986; Faisal and Huggett, 1993; Faisal et al., 1991; Ghosh et al., 2006; Hart et al., 1998; Harford et al., 2005; Hoeger et al., 2004; Holladay et al., 1998; Hutchinson et al., 2003; Karrow et al., 1999; Kelly-Reay and Weeks-Perkins, 1994; Khangarot et al., 1999; Kranz and Gercken, 1987; Lemaire-Gony et al., 1995; Low and Sin, 1998; Muhvich et al., 1995; Payne and Fancey, 1989; Prophete et al., 2006; Rabitto et al., 2005; Reynaud and Deschaux, 2006; Reynaud et al., 2003; Reynaud et al., 2002; Rice and Schlenk, 1995; Rice and Weeks, 1989, 1990, 1991; Rice et al., 1995, 1996; Rougier et al., 1994; Sanchez-Dardon et al., 1999; Seeley and Weeks-Perkins, 1991, 1997; Spitsbergen et al., 1986; Tahir et al., 1993; Van der Weiden et al., 1994; Warinner et al., 1988; Weeks and Warinner, 1984; Weeks et al., 1986; Wishkovsk et al., 1989; Zelikoff, 1998; Zelikoff et al., 1995, 2000.

Acute inflammation in fish has been studied following induction by a large variety of natural and experimental stimuli. These include injection with phlogistic agents such as bacteria, exposure to metazoan parasites, subcutaneous inoculation of fungi, intrapulmonary stimulation with carbon or latex, and wounding [Roberts, 1989; Suzuki and Lida, 1992; Woo, 1992]. The cellular response is typically biphasic. It leads to the increase of neutrophils and their extravasations in blood preceding the appearance of monocytes and macrophages [Ellis, 1986]. The Table 2 presents a review of immunological studies related with effects of chemical pollutants.

Immunotoxicology can be most simply defined as the study of adverse effects on the immune system resulting from occupational, inadvertent or therapeutic exposure to drugs, environmental chemicals and, in some cases, biological materials. To understand the nature of these adverse effects is important to recognize that the function of the immune system is the preservation of organism integrity. The immune system is entrusted to identify what is *self* (i.e. all the cells and tissues that belongs to one's body) and what is *nonself* (i.e. stranger entities to one's body, which include opportunistic pathogens such as bacteria, viruses, fungi, protozoans, transformed cells, and cells and tissues from individuals belonging to different, as well as to the same species) [Holsapple, 2002].

Immunotoxic molecules may lead to two main types of effects: immunosuppression which may result in an increased susceptibility to tumors or infectious diseases, or deregulation of the immune response leading to hypersensitivity and autoimmunity [Pallardy et al.,1998].

Several studies have set out to characterize the immunotoxic potential of a variety of POPs using standard immunotoxicological parameters. Also, the oxidative stress by increased production of reactive oxygen species such as superoxide [Ruiz-Leal and George, 2004] and hydrogen peroxide species has been implicated in the toxicity of POPs. Neutrophils and other phagocytes manufacture $O_2^{\cdot-}$ by the one-electron reduction of oxygen at the expense of NADPH. Most of the $O_2^{\cdot-}$ reacts with itself to form H_2O_2. From these agents a large number of highly reactive microbicide oxidants are formed, including HOCl (hypochlorous acid), which is produced by the myeloperoxidase-catalyzed oxidation of Cl^- by H_2O_2; $OH^{\cdot}$ (hydroxyl radical), produced by the reduction of H_2O_2 by Fe^{++} or Cu^+ (Fenton's reaction) or by the reaction of $O_2^{\cdot-}$ and H_2O_2 (Haber-Weiss' reaction); $ONOO^-$(peroxynitrite), formed by the reaction between $O_2^{\cdot-}$ and $NO^{\cdot}$ (nitric oxide/nitrogen monoxide); and many others. These reactive oxidants are manufactured for the purpose of killing invading microorganisms, but they also inflict damage on nearby molecules and cells, and are thought to be of pathogenic significance in a large number of diseases [Babior, 2000].

NOS (nitric oxide synthase) activity has been detected in channel catfish head kidney leukocytes following intraperitoneal injection with live *Edwardsiella ictaluri* [Schoor and Plumb, 1994], although the cells responsible for NOS activity were not identified. In goldfish, Neumann et al. (1995) have shown that a long-term macrophage cell line and primary cultures of kidney macrophages secrete $NO^{\cdot}$, as detected by nitrite accumulation after incubation with LPS (bacterial lipopolysaccharide).

Macrophages are important cells in disease resistance of vertebrates. They are the main phagocytic cells, while granulocytes are non-phagocytic or weakly so in many fish species [Wang et al 1995; Ainsworth and Dexiang, 1990; Sarmento 2004b]. Modulation of fish macrophage's activity by xenobiotics was also reported, showing diverse effects that depend on dose and chemical type [Zelikoff et al, 1995; Sanchez-Dardon et al., 1999; Bols et al., 2001].

The majority of the studies on the biology of fish macrophages has been conducted using primary cultured cells, established after isolation of the cells from kidneys [MacArthur et al., 1985; Howell, 1987, Sarmento 2004a]. Despite there is a high variability in the response of macrophages in cultures, the evidences suggest remarkable morphological and physiological similarity between fish, mammalian and avian macrophages [Chung and Secombes, 1987, 1988; Wang et al, 1995].

In macrophages of mammalian, nitric oxide can be produced in response to cytokines, bacterial lipopolysaccharide (LPS) or parasites by inducible NO synthases (iNOS) [Natthan, 1991]. In fish, NO production has been demonstrated in macrophages after LPS induction in goldfish, catfish and turbot [Neuman et al, 1995; Yin et al, 1997; Tafalla and Novoa, 2000].

METHODOLOGY

Lipid and *Lipophilic* Compounds Extractions

A simple method for preparation of total pure lipid extracts from various tissues is described. The method consists of homogenizing animal tissue (liver, muscle etc) in chloroform-methanol (2:1) or dichloromethane/methanol (2:1) to a final mass(g)/volume (ml) relation of 1/20. The homogenate is filtrated and the lipophilic compounds are isolated from the lipids [Folch et al. 1957 with minor modifications]. The extraction and the purification for SPE (Solid Phase Extraction) are made in column of florisil (MgO_3Si) according to method of EPA 3620 (Bond Elut Florisil), first with hexane to get less polar compounds (HCB, *pp'*-DDE among others) and later with a solution of hexane/diethyl ether for compounds with higher polarity (OCs and triazines). These compounds can be quantified in chromatographer equipped with detector of electron capture [Oliveira Ribeiro et al. 2005].

Various procedures and solvent combinations have been employed to quantitatively extract lipids, such as the traditional or modified Folch procedure [Folch, 1957], which employs chloroform/methanol (2:1 v/v), or the Hara and Radin method [Hara and Radin, 1978; Radin, 1981], which employs n-hexane/2-propanol (3:2 v/v). The latter solvent mixture is less effective because (it does not quantitatively extract complex lipids such as gangliosides; otherwise it is less toxic. Another rapid extraction procedure, used for samples at 80% water, employs a chloroform/ methanol/water mixture to separate the lipids from all the non-lipids and it is commonly called the Bligh and Dyer method [Bligh and Dyer, 1959].

Supercritical fluid extraction (SFE), employing supercritical carbon dioxide, can reduce the use of toxic organic solvents but, in many cases, quantitative extraction of the most polar components requires the use of a traditional solvent (such as methanol or ethanol) as polar modifier (co-solvent) [Gallina Toshi et al., 2003].

The protocols of extraction of POPs can provide extracts of xenobiotics applicable to *in vitro* assays, for example, in those with lymphoid cells. In these assays, interactions among xenobiotics from the extracts and the combined effects of these compounds can be analyzed. It may assist researchers to explain the effects of these compounds in the modulation of immune components and in citotoxicity. Two different populations of lymphoid cells -

peripheral blood leucocytes (PBLs) and head kidney phagocytes can be used [Betoulle et al., 2000].

Primary Hepatocytes Culture of *Hoplias malabaricus*

In fish, as in other vertebrates, the liver is the site as well as the target of complementary processes that keep liver functions in tune with the metabolic requirements of the whole body [Guillouzo et al., 1990]. *In vitro* models are generally target-organ based, and the most frequently encountered target-organ toxicity is the liver [Ulrich et al. 1995]. It continuously readjusts hepatocellular structures and functions such as metabolism of nutrients; storage of energy molecules (glycogen, lipid); synthesis and secretion of proteins (e.g. albumin, vitellogenin, lipoproteins etc); maintenance of plasma glucose levels; elimination of nitrogen components after urea, uric acid or ammonia formation; metabolism of hormones; metabolism of xenobiotics; bile formation etc to fluctuating environmental conditions [Segner, 1998]. Isolated liver cells provide an excellent system for evaluating many aspects of hepatic metabolism, including the biochemical and cellular processes involved in the activation of toxic chemicals and environmental pollutants [Moon et al., 1985; Baksi and Frazier, 1990; Blaauboer et al., 1994; Rogiers et al., 1995; Guillouzo, 1998; Kelly et al., 1998; Ferraris et al., 2002], since hepatocytes are richly endowed cells with xenobiotic metabolizing enzymes [Guillouzo et al., 1990].

Four experimental *in vitro* models are commonly used for studying hepatic toxicity: perfused liver, tissue slices, fresh isolated cells and cells in culture. Each model has its advantages and disadvantages, and their selection should be based on the specific problem to be solved [Ulrich et al. 1995]. Compared to fresh isolates, cultures have the advantages that they allow the investigation of longer-term, time-dependent processes, they permit studies of structures and functions which require intercellular contact and organization, and the cells have time to recover from damage sustained during the isolation procedure [Segner, 1998]. Considering the models of cells in culture, either cell lines or primary cultured cells can be utilized to address some questions of *in vitro* toxicology. However, for several studies in toxicological research, primary cell cultures may be of much more utility than cell lines. Certainly cell lines have advantages in terms of convenience of supply and long-term cultures [Baksi and Frazier, 1990] because of their capability to be fast and easily cultured. But, the technical aspects of cell lines cannot be the only ground of choice. Immortalized cell lines are composed by abnormal cells, which keep somewhat distant from normal tissue condition while growing *in vitro* [Masters, 2000]. Thus, serious questions arise as to whether responses of cell lines to xenobiotics are in fact representative of the response of differentiated cells *in vivo* and at how extent cell lines keep the tissue identity [Masters, 2000]. Also, many immortalized cells may become dedifferentiated and lose their capacity for xenobiotic biotransformation [Ulrich et al. 1995], which implies that they do not necessarily have all the characteristics of the tissue of their origin [Sandbacka et al., 1999]. These limitations of cell lines may be, partially, a consequence of their maintenance in culture for several years or even decades, imposing a strong *in vitro* selective pressure on them. Furthermore, as they have been distributed in several laboratories, the same cell lines might also have undergone various selection steps due to different feeding techniques, growing conditions etc [Birgersdotter et al., 2005].

Primary cultures of isolated hepatocytes, although not entirely normal due to disruption of cell-cell interactions, potential membrane damage during preparation, and dedifferentiation in long-term cultures, represent cells which are central to systemic metabolic regulation in the intact organism; primarily involved in xenobiotic metabolism and often the specific target for chemical toxins. Consequently, they play an important role in aquatic toxicological research as they do in mammalian toxicology [Baksi and Frazier, 1990]. Furthermore, the universality of many basic biological processes among eukaryotic organisms means that lower animals, including fish, can serve also as useful research models for problems even in human toxicology [Baksi and Frazier, 1990]. It appears that for short-term culture (up to 5 or 8 days), monolayer systems are appropriate to conserve liver-specific gene expression. However, to prevent phenotypic changes of piscine liver cells during prolonged culture, techniques seem to be required which mimic more closely the in vivo environment of hepatocytes [Segner, 1998].

One imperative aspect to keep in mind when employing cultures of liver cells from fish in toxicological research is that xenobiotic metabolism may occur also in other tissues such as gills, blood and gastrointestinal tract and this extra-hepatic biotransformation may in some cases have a large impact on chemical kinetics [Van Veld et al., 1988; Barron et al., 1989]. In particular, the gastrointestinal tract and the bacteria therein may play an important role for xenobiotics uptake through feeding (bioaccumulated in the organism that is the food) due to, for example, hydrolytic cleavage of compounds, formation of mercapturic acids, reduction of nitro-group-containing compounds etc [Hoogenboom and Kuiper, 1997], while epithelial tissue of gills may influence mainly the water-borne of hydrophilic xenobiotics. In the case of blood, the activities of enzymes present in it may contribute to extra-hepatic metabolism during the transport of xenobiotics. In mammals, phase I metabolic activities are concentrated in the liver while phase II activities tend to be more highly distributed. Based on limited information, similar patterns have been suggested for fish [Clarke et al., 1992; Leaver et al., 1992].

Liver parenchymal cells or hepatocytes can be obtained by one or two steps of liver perfusion. In the former case, liver is perfused only with a calcium ion chelator (for example, EDTA, ethylenediamine tetraacetic acid) [Berry et al. 1983; Wang et al. 1985; Meredith 1988; Seddon and Prosser, 1999] to dissociate cells through weakening the Ca^{2+}/Mg^{2+} -dependent cell-cell and cell-extracellular matrix attachments. In the later case, after chelator perfusion, a bacterial collagenase-containing solution is employed to break up extracellular collagen. Remarkably, in both perfusion systems the cells lose their polar character and change shape after dissociation.

Collagenase disrupts intercellular contacts and communication systems. Proteolysis due to collagenase action also damages the enzyme and receptor apparatus of cells. It impairs their biophysical characteristics and transport capabilities as well. Nevertheless, the cells in culture are capable of repairing membrane defects and may preserve the majority of their functions [Červenková et al., 2001]. There are evidences that hepatocytes recover from the metabolic stress of isolation after few hours in primary cultures [Blaauboer et al., 1994]. Trout hepatocytes, for example, are able to re-establish the cell-cell contacts [Blair et al., 1990; Braunbeck and Storch, 1992; Segner, 1998; Segner and Cravedi, 2001] and only if the cells manage to re-aggregate under *in vitro* conditions they maintain their *in vivo*-like metabolic activities (i.e. biotransformation capacity) for 3–8 days [Bisell et al., 1973; Bonney et al., 1974; Pariza et al., 1975; Yamada et al., 1980; Segner et al., 1994; Braunbeck et al.,

1995]. This confirms that the preservation of normal intercellular contacts is critical for the preservation of some important features of differentiated cells [Ferrini, et al., 1997].

After disruption and isolation of hepatic cells, hepatocytes can be separated from other cell types by differential centrifugation whenever necessary. The quantitatively dominating cells are the hepatocytes which occupy about 80% of original fish liver volume. The non-hepatocytes of fish liver include biliary epithelial cells, sinusoidal endothelial cells, persinusoidal fat storing cells and melanomacrophages. Although the relative number of non-hepatocytes is small, they are suspected to participate in a number of important toxic or neoplastic processes [Segner, 1998]. Up to now, little is known about the role of these cells in liver toxicity.

Isolated hepatocytes can be utilized in suspension for very short-term assays (2-4 h) because of the limited survival of cells during they exhibit metabolic activity [Segner, 1998; Červenková et al., 2001], or they can be seeded onto plastic culture flasks, with or without pretreatment with extracellular matrix components [laminin, collagen, fibronectin etc]. The seeded hepatocytes are allowed to attach, stabilize and reaggregate to form monolayers of usually non-proliferating granular epithelial cells. Cultures are frequently stable for at least one week and hepatocytes begin to die and detach very early. The growth of contaminating cells increases rapidly after 1-2 weeks, with occasional colonies of fibroblasts appearing in the culture. This has, however, a negligible effect on the culture behavior. The lifetime of the culture may be extended for several weeks when appropriate media supplements, such as growth factors and/or cytokines, are added. However, it requires previous experiments to identify the appropriate supplementation and to determine their correct concentration [Červenková et al., 2001]. Generally primary monolayer hepatocytes cultures exhibit only basal levels of cytochrome P450 expression but it still possess the ability to respond to cytochrome P450 inducers [Ferrini et al., 1997]. The transference of mammals hepatocytes to their new environment may result in complete loss of some functions (e.g. induction of some cytochrome P-450 isozymes) [Guzelian et al., 1977; Sirica and Pitot, 1980; Steward et al., 1985] and marked attenuation of others, e.g. secretion of plasma proteins within few days. At the same time, the cells begin to express fetal-like functions (fetal isozymes) [Guguen et al., 1975; Guguen-Guillouzo et al., 1978; Sirica et al., 1979]. The rates of transcription of liver-specific genes drop between 1 and 10% of those found in the liver after 24 h of culture [Clayton and Darnell, 1983]. Addition of foetal calf serum to the culture medium favors the attachment of hepatocytes to plastic, but liver-specific functions are still more rapidly lost [Jefferson et al., 1984]. For teleostean primary cultures, the available data indicate a greater stability of biotransformation enzymes in liver cells than in mammalian hepatocytes [Stegeman et al, 1993; Pesonen and Andersson, 1997; Segner, 1998].

The obvious advantage of cell cultures is the simplification of the experimental system. However, the extrapolation of the metabolic data to *in vivo* systems is sometimes a hard task, in part due to discrepancies between the two systems. Reasons for some discrepancies may include the metabolism in tissues other than the liver, the incorrect assumption of rapid equilibrium of xenobiotics between blood and hepatocytes, the presence of active transport through the sinusoidal membrane, the interindividual variability [Iwatsubo et al., 1997], the substantial binding of xenobiotics to biological components which may alter their bioavailability to cells in *in vivo* and *in vitro* systems, the levels of expression of some detoxifying enzymes *in vitro* etc. Established methods for measuring *in vitro* binding include equilibrium dialysis [Obach, 1997] and ultracentrifugation [Tang et al., 2002]. Solid phase microextraction devices may provide a third method for measuring free chemical concentrations *in vitro* that is particularly useful for high K_{ow} (*N*-octanol : water partition coefficient) compounds [Heringa et al., 2004]. Finally, there is a need to collect supporting information required to perform *in vitro–in vivo* metabolism extrapolations, including estimates of microsomal protein recovery, CYP isozyme content, and liver hepatocellularity (content of hepatocytes per total of hepatic cells). Because these values may differ with species, age, and gender, initial efforts to collect this information should be performed in conjunction with in vitro activity measurements [Nichols et al. 2006].

Isolation of Peripheral Blood Leukocytes (PBLs) of *Ciprinius carpio*

Blood is collected from the caudal vein of fishes using a heparinized syringe and quickly diluted with 5 volumes of culture medium (Leibovitz L15) containing heparin (10 $IU.ml^{-1}$), antibiotics and layered onto an equal volume of Ficoll–Paque. After 40 min of centrifugation at 1000 x*g* and 4°C, the interface containing PBLs is collected and washed twice (800 x*g*; 10 min; 4°C) in culture medium. The cells are resuspended in culture medium containing 0.5 µM of 2-mercaptoethanol (2-ME). Cell viability is determined by trypan blue (0.4%) exclusion method and the cell suspension is adjusted to $6x10^7$ $cells.ml^{-1}$ in culture medium containing 2-ME [Duchiron et al., 2002].

Isolation of Head Kidney Phagocytes of *Ciprinius carpio*

To avoid stress, fishes are anaesthetized by immersion in water containing 2-phenoxyethanol (0.3% $v.v^{-1}$). Head kidney is aseptically removed from fishes, passed through a 100 µm nylon mesh and homogenized in L15 medium containing antibiotics and heparin (10 $IU.ml^{-1}$). The cell suspension is layered onto a 34:51% (v:v) Percoll gradient and centrifuged at 800 x*g* for 25 min at 4°C. The cells at the 34–51% interface, which corresponded to phagocytes (macrophages and neutrophils, 20 and 55%, respectively), are collected and washed twice in L15 medium at 800 x*g*, 10 min, 4°C. Viable cells are counted by trypan blue (0.4%) exclusion test using an hemocytometer and the cell suspension are adjusted to $2x10^6$ $cells.ml^{-1}$ in incubation buffer (140 mM NaCl; 5 mM KCl, 1mM $MgCl_2$, 20 mM HEPES, 10 mM D-glucose, 1 mM Na_2HPO_4, 1 mM $CaCl_2$, pH 7.4) [Betoulle et al., 2000].

Isolation of Macrophage from *Ancistrus multispinis* Kidney

The fishes are anaesthetized with MS-222, bled and killed. The kidneys were removed aseptically, washed four times in ice-cold L-15 medium supplemented with 2% fetal calf serum, 1% penicillin/streptomycin and 20 $U.ml^{-1}$ heparin and pressed through sterile nylon mesh [Wang, et al, 1995; Neuman et al., 1995; Sarmento et al 2004a].

For the establishment of *A. multispinis* kidney macrophage cultures, the cells were washed by centrifugation at 400x *g* for 10 min at 4° C and suspended in culture medium. For purification of macrophages, the cell suspension was layered on a 40%–60% Percoll discontinuous gradient and centrifuged at 400x *g* for 30 min at 4° C in L-15 medium supplemented with 0.1% fetal calf serum, 1% penicillin/streptomycin and 20 $U.ml^{-1}$ heparin [Wang et al, 1995; Neumann et al 2000; Sarmento et al., 2004a]. The cells at the 40%–60% Percoll interface were transferred into clean tubes and washed twice by centrifugation at 400 x *g* for 10 min at 4° C in the culture medium. The cell number was adjusted to $2x10^6$ viable cells per ml of L-15 medium containing 0.1% fetal calf serum and 1% penicillin/streptomycin, and 100 µL of this suspension was added to each well of a 96 wells microplate and incubated for 24 h at 22-25° C to cell attachment [Tafalla and Novoa, 2000; Neumann et al, 2000; Sarmento et al, 2004a]. After 24 h, unattached cells are removed by washing the wells with sterile PBS (three times) and depending on the experiment, cells are incubated for 48h with supernatants, 1 or 10 $\mu g.ml^{-1}$ LPS and a pollutant of interest.

Cell viability is evaluated by reduction of 3-(4,5-dimethylthiazol-2-yl)-2,5-diphenyl tetrazolium bromide (MTT assay) [Mosmann, 1983; Green et al., 1982; Sarmento, 2004b]. MTT (10 µL of a stock solution at 5 $mg.ml^{-1}$) is added to microplate wells containing adherent macrophages in 100 µl medium. Cells are incubated for 2h at 20°C and 100 µl of acid-isopropanol is added to each well and mixed thoroughly to dissolve the formazan. After 5 min, absorbances are measured in an ELISA reader at 570 nm wave length.

Nitric Oxide ($NO^{\cdot}$) Production

The $NO^{\cdot}$ production of fish macrophages is measured according to Green et al., 1990, Neumann et al, 1995; Tafalla and Novoa, 2000, with modifications. After the treatment with a pollutant and LPS, the cells are assayed for nitrite using the Griess reaction [Stuehr and Marletta, 1985]. Aliquots of 100 µl of the supernatant are incubated with 100 µl of 1% sulfanilamide and 0.1% n-1-naphthylethylenediamine dihydrochloride in 2.5% H_3PO_4 at room temperature for 10 min. Absorbance are measured at 570 nm nitrite concentrations were quantified by comparing them with $NaNO_2$ standards.

ASSAYS WITH EXTRACTED LIPOPHILIC POLLUTANTS

The Example with *Hoplias malabaricus* Hepatocytes

Once OCs or PCBs are obtained from biological samples (liver or muscle), they can be utilized as individual mixtures (i.e. compounds from a single tissue of a single specimen) or as a pool of chemicals from the same tissues of several specimens. The last situation is indicated to increase the amount of chemicals to be used in cultures.

OCs and PCBs are very lipophilic (have high hydrophobicity). Then, it is necessary to use a solvent (e.g. methanol, ethanol, acetone, dimethylsulfoxide and other) to render them soluble in culture medium. With this regard, it is important to mention that sometimes the own chosen solvent can cause toxic effects to cells or alter some endpoints evaluated, depending on their concentration, the cell type, the donor organism and the culture conditions. The use of fetal calf serum is another aspect to be considered. The choice of using or not mammal serum must be based on the cells requirement to attach, to survive and to express some hepatic-specific functions. Also, the serum may affect the bioavailability of some compounds due to its content of lipoproteins and other molecules. Here, a really important issue to be incited is if the conjugation of xenobiotics to components from serum do or do not represent a realistic situation, since these compounds may be transported through blood stream and some conjugation may naturally occur during their course up to the intake by cells.

As already mentioned, hepatocytes can be isolated through hepatic perfusion with or without an enzyme-containing solution. Non-enzymatic perfusion of the liver is a very economical procedure [Seddon and Prosser, 1999] to obtain viable cells. An example of a very simplified protocol to isolate fish hepatic cells is the following: Firstly, fish is anesthetized, sacrificed and cleaned with antiseptic solution(s). Then its liver is carefully exposed and removed to a sterile petri dish in a flow hood. Using syringe and needle a cold sterile phosphate buffered saline containing 2 mM of EDTA (without CO_2 / O_2 gasification) is injected through blood vases until the liver is clean of blood (blanched) and is softened. Finally, the liver is minced, gently pressed against a filter screen for mechanical disruption and cells are collected in buffer, centrifuged and washed in PBS and/or culture medium to remove debris [Filipak Neto et al, 2006]. Following the isolation, hepatocytes must keep in culture to attach and to recover from eventual damages. The culture medium conditions such as pH, osmolarity and supplementations need to be the best possible to the assays and depends on the biological model species. After attachment and recovery, the cells may be exposed to the xenobiotics by replacement of the culture medium by one containing either OCs or PCBs or both of them. After exposure during the desired period of time, cells can be analyzed through several different assays (for example, through parameters that address the ultrastructure of the cell and of its components, the cytoplasmic redox state of cells, the death of cells and other) to investigate the effects of these mixtures.

Ultrastructure can be used to evaluate qualitatively some parameters such as integrity of organelles, activeness of cell (level of chromatin's condensation, nucleolus appearance, RER and mitochondria abundance and other), incidence of apoptosis or necrosis etc. However, under TEM the cellular parameters can not be quantitative evaluated due to the small area and the cuts thickness (very thin) of samples analyzed.

Redox parameters of cells can provide important clues of how POPs affect the cellular functioning. Nitrogen cold samples of cells can serve to know if antioxidant enzymes such as catalases, glutathione *S*-transferases, glutathione peroxidases, superoxide dismutases, glutathione reductases etc are affected by the presence of POPs, as well as to quantify prooxidant (hydrogen peroxide, superoxide anion and other) and antioxidant (e.g. reduced glutathione) species. Also, the level of damage to lipids (lipid peroxidation etc), proteins (protein carbonylation etc) and to DNA (adducts formation, DNA cleavage etc) can be assayed. These data are useful to speculate possible pathways by which POPs are causing or not oxidative stress to cells.

If viability tests show that cells are dying due to the presence of POPs, it is likewise useful to know by which process cellular death is occurring. Annexin V and viability assays through flow cytometry, for example, provide information of how many cells are viable, in early apoptosis and in late apoptosis or necrosis. Other assays to investigate apoptosis include TUNEL and caspases assays. All these assays address the ultimate level of toxicity caused by xenobiotics, that is, the complete stop of functioning of cell.

One very important aspect concerning the utilization of extracted POPs to expose cells is the assumption that they may be remobilized from fat storage during a period of increase in energy demand or decrease in food supplies. When this mobilization occurs, these chemicals may become again available to cells and may be able to interact with them or with cellular components. Then, when it is tested the effects of these extracted mixtures on cells, we try to simulate this situation *in vitro* and acquire data about what happen to cells and at some extent, how this may happen. This way, after the use of the extraction protocol presented in this chapter even if some compounds are not identified in these mixtures, this simulation is somewhat much more close to the *in vivo* situation than the application a surely chemically defined mixture of xenobiotics, which would not include some unidentified xenobiotics such as some products derived from biotransformation of identified parental compounds. However, whenever an effect is identified it may be implicated with the mixture of extracted compounds which contains at least the compounds A, B, C and D for example and not with a compound in special, since in a mixture the interactions among xenobiotics make this system much more complex and unpredictable that the situation of one or two xenobiotics in contact with cells.

The fish species *Hoplias malabaricus* is a predator tropical teleost species found in South America, which is an excellent in vivo model for trophic exposure bioassays [Rabitto et al., 2005; Oliveira Ribeiro et al., 2006]. Utilizing an easy protocol [Filipak Neto et al, 2006], hepatocytes from *H. malabaricus* may be isolated, cultured and exposed to pollutants. The procedure is very simple. After 3 days in culture (to allow cells to attach and recover from the stress of isolation), flasks containing the liver cells are rinsed in PBS and the culture medium (supplemented with 5-10% calf serum, 0.1-0.2 $U.ml^{-1}$ bovine-porcine insulin, 6 mM $NaHCO_3$, 15 mM HEPES and antibiotics, pH 7.8) is replaced by one containing either organometals (such as methylmercury) at found concentrations found in liver or lipophilic extracted compounds (such as POPs). In the case of lipophilic compounds, they are first dissolved in DMSO or other solvent and then added to the culture medium. The cells are exposed for 2 up to 4 days and analyses are performed through different assays (submitted data) such as those cited in this chapter. Since hepatocytes are cultured during up to 7 days, these short-term assays utilize cells before they dedifferentiate, which is fundamental to acquire data that represents as well as possible the normal tissue condition.

H. malabaricus is an overspread fish species found in many lakes and rivers of Brazil and it inhabits all the geographical regions of this country. Then, some specimens can be captured in different impacted areas or areas suspect of impact. The accumulated lipophilic compounds can be extracted and tested in the hepatocytes of the same fish species for a general screening of endangered environments, since chemical analyses are not a cheap procedure and do not say much about biological effects itself, especially due to complex mixtures of chemicals, into which interactions dictate a lot of the action of xenobiotics and cellular responses. Also, some data can be acquired about by which processes this mixture acts and how cell systems and/or components behave after exposure.

The Example with *Cyprinius carpio* Macrophages

The OCs mixture extracted from liver and muscle can also be used to evaluate the effects on cell immunology of macrophages and leukocytes of *Cyprinius carpio*. Biomarkers such as oxidative burst, phagocitic activity and cellular viability are measured. The Nitro Blue Tetrazolium (NBT) test is usually used to measure the oxidative burst as described below. To exposure the OCs mixture it was first solubilized in DMSO (not exceeding 0.1%). After incubation with MAF-containing supernatants, macrophages monolayer were washed and 100 mL of NBT or NBT+PMA solutions were added per well and incubated for 15 min in the dark at 21°C. NBT was dissolved at 1 mg/mL in PBS and PMA added at 1 μg/mL. The medium was then removed and the cells were fixed in methanol. Cells were washed in 70% methanol and allowed to air dry. The formazan in each well was then dissolved in 50 mL 2 M KOH and 75 mL DMSO, with mixing. The turquoise-blue-colored solution was then read in a multiscan spectrophotometer at 630 nm using KOH/DMSO as a blank. The absorbance values obtained with NBT+ PMA-stimulated cells were divided by the absorbance values obtained with cells incubated with NBT alone. This gave the index of stimulation of macrophage.

The Example of *Ancistrus multispinis* Peritoneal Leukocytes Migration

Fish received an i.p injection of vehicle (1% of Tween 80 in saline) or two different doses of the pollutants. For analyses of peritoneal exudates, four days after injection fishes from exposed groups to pollutants were injected with lipopolysaccharide (LPS, E. coli 0111:B4, $0.1\mu g.kg^{-1}$) and the same number of fishes from control group with sterile saline. The dose of LPS was based on previous works [Afonso et al, 1997; Ulich et al., 1995].

The peritoneal cells were collected from undisturbed peritoneal cavities (control) or stimulated with LPS four hours after injections (vehicle or LPS). Briefly, the abdominal side of the fish was cleaned with ethanol. PBS with osmolarity adjusted to 320-355 $mOsm.l^{-1}$ and supplemented with 20 $U.mL^{-1}$ of heparin and 3% (w/v) of bovine albumin was injected into the peritoneal cavity (2-3 ml per fish – adjusted to the size of the fish), in the ventral midline, midway between the pelvic and pectoral fins. This procedure was followed by the injection of air (about the same volume as used to PBS) through the same needle (20x5.5mm). The abdominal area was then massaged for about 30s to disperse the peritoneal cells in the injected PBS. PBS containing the peritoneal cells was then collected and placed on ice until processed [Afonso et al 1997]. Hemorrhagic peritoneal washings were discarded.

The peritoneal cell suspension was collected and the total leukocytes were counts under bright field microscopy using a Neubauer haemocytometer [Tavares-Dias and Moraes, 2004]. For this procedure, peritoneal exudates samples were diluted in Natt-Herrick's solution [Natt and Herrick,1952] and the results were reported as the number of cell per $.\mu l^{-1}$ of peritoneal exudate.

CONCLUSION

As persistent pollutants are rarely found alone, a mixture of them can be bioaccumulated in special tissues presenting similar physicochemical properties. This complexicity involves for example the exposure to DDE, PCBs and dioxin as responsible for the reproductive impairment in fish.

In general, the majority of the persistent compounds are highly lipophilic, chemically stable and slowly biodegraded, making urgent further studies to understand the importance of its toxic mechanisms. The potential long-term effect of chemicals and bioaccumulation along food chains increases for biological persistent compounds. Thus, attention must also be given to studies the effects of complex mixtures, and if possible evaluating long-term effects.

Many of pollutants are compounds that exhibit the ability to mimic, antagonize or alter the bioavailability and action of the endogenous molecules. Elucidating mechanisms would require experimental studies in conditions of exposure to establish links between molecular, individual and population effects. Finally, the further studies in reasonable conditions of exposure, i.e. realistic pollutant mixtures and concentrations, are not only recommended but absolutely necessary for extrapolation of experimental results and field application.

REFERENCES

Afonso, A; Ellis, AE; Silva, MT. The leucocyte population of the unstimulated peritoneal cavity of rainbow trout (*Oncorhynchus mykiss*). *Fish and Shellfish Immunology*, 1997, 7, 335-348.

Albers, PH. Petroleum and individual polycyclic aromatic hydrcarbons. In: Hoffman, DJ; Rattner, BA; Allen Burton Jr, G; Cairn Jr, J. *Handbook of Ecotoxicology*. USA: Lewis Publishers; 1995; 30-56.

Alves Costa, JRM; Mela, M; Silva de Assis, HC; Pelletier, É; Randi, MAF; Oliveira Ribeiro, CA. Enzymatic inhibition and morphological changes in *Hoplias malabaricus* from dietary exposure to lead (II) or methylmercury. *Ecotoxicology and Environmental Safety*, 2006 (In Press).

Atli, G; Alptekin, Ö; Tükel, S; Canli, M. Response of catalase activity to Ag^{+}, Cd^{2+}, Cr^{6+}, Cu^{2+} and Zn^{2+} in five ts of freshwater fish *Oreochromis niloticus*. *Comparative Biochemistry and Physiology Part C: Toxicology and Pharmacology*, 2006, 143(2), 218-224.

Babior, BM. Phagocytes and Oxidative Stress. *American Journal Medicine*, 2000,109,33– 44.

Baek, SO; Field, RA; Goldstone, ME; Kirk, PW; Lester, JN; PEny, R. A review of atmospheric polycyclic aromatic hydrocarbons, sources, fate and behavior. *Water Air Soil Pollutant*, 1991, 60, 279.

Baksi, SM; Frazier, JM. Isolated fish hepatocytes-model systems for toxicology research. *Aquatic Toxicology*, 1990, 16, 229–256.

Balls, M; Blaauboer, BJ; Brusik, D; Frazier, J; Lamp, D; Pemberton, M; Reinhardt, C; Robertfroid, M; Rosenkranz, H; Schmid, B; Spielmann, H; Stammati, AL; Walum, E. Report and recommendations of the CAAT/ERGATT workshop on the validation of toxicity test procedures. *ATLA*, 1990, 18, 313–337.

Balls, M; Blaauboer, BJ; Fentem, J; Bruner, L; Combes, RD; Ekwal, B; Fielder, R;J; Guillouzo, A; Lewis, RW; Lovell, DP; Reinhardt, CA; Repetto, G; Sladowski, D; Spielmann, H; Zucco, F. Practical aspects of the validation of toxicity test procedures. The report and recommendations of ECVAM Workshop 5. *ATLA*, 1995, 23, 129–147.

Bano, Y; Hasan, M. Mercury induced time-dependent alterations in lipid profiles and lipid peroxidation in different body organs of cat-fish *Heteropneustes fossilis*. *Journal Environment Science Health B,* 1989, 24(2), 145-166.

Barron, MG; Schultz, IR; Hayton, WL. Presystemic branchial metabolism limits di-2-ethylhexyl phthalate accumulation in fish. *Toxicol Appl Pharmacol*, 1989, 98, 49–57.

Basha, PS; Rani, AU. Cadmium-induced antioxidant defense mechanism in freshwater teleost *Oreochromis mossambicus* (Tilapia). *Ecotoxicology and Environmental Safety*, 2003, 56, 218–221.

Beers, RF; Sizer, IW. A spectrophotometric method for measuring the breakdown of hydrogen peroxide by catalase. *Journal Biology Chemistry*, 1952, 195, 133-140.

Beis, D; Stainier, DYR. In vivo cell biology: following the zebrafish trend. *TRENDS in Cell Biology*, 2006, 16(2), 105-112.

Berntssen, MHG; Aatland, A; Handy, RD. Chronic dietary mercury exposure causes oxidative stress, brain lesions, and altered behaviour in Atlantic salmon (*Salmo salar*) parr. *Aquatic Toxicology*, 2003, 65(1), 55-72.

Berry, MN; Farrington, C; Gay, S; Grivell, AR; Wallace, PG. Preparation of hepatocytes in good yield without enzymatic digestion. In: Harris RA, Cornell NW. *Isolation, characterization, and use of hepatocytes*. New York: Elsevier Science; 1983; 7–10.

Betoulle, S; Duchiron, C; Deschaux, P. Lindane increases in vitro respiratory burst activity and intracellular calcium levels in rainbow trout (*Oncorhynchus mykiss*) head kidney phagocytes. *Aquatic Toxicology*, 2000, 48, 211–221.

Bhattacharya, A; Bhattacharya, S. Induction of oxidative stress by arsenic in *Clarias batrachus*: Involvement of peroxisomes. *Ecotoxicology and Environmental Safety*. 2006, (In Press).

Birgersdotter, A; Sandberg, R; Ernberg, I. Gene expression perturbation in vitro - A growing case for three-dimensional (3D) culture systems. *Seminars in Cancer Biology*, 2005, 15, 405–412.

Bisell, DM; Hammaker, LE; Meyer, UA. Parenchymal cells from adult rat liver in nonproliferating monolayer culture I Functional studies. *Journal of Cell Biology*, 1973, 59, 722–734.

Blaauboer, BJ; Boobis, AE; Castell, JV; Coecke, S; Groothuis, GMM; Guillouzo, A; Hall, TJ; Hawksworth, GM; Lorenzon, G; Miltenburger, HG; Rogiers, V; Skett, P; Villa, I; Wiebel, FJ. The practical applicability of hepatocyte cultures in routine testing. *ATLA*, 1994, 22, 231-241.

Blair, JB; Miller, MR; Pack, D; Barnes, R; Teh, SJ; Hinton, DE. Isolated trout liver cells: establishing short-term primary cultures exhibiting cell–cell interactions. *In Vitro Cell Develop Biology,* 1990, 26, 237–249.

Bligh, EG; Dyer, WJ. A rapid method of total lipid extraction and purification. *Canadian Journal of Biochemistry and Physiology*, 1959, 37, 911-917

Blus, LJ. DDT, DDD and DDE in Birds. In: Beyer, WN; Heinz, GH; Redmon-Norwood, A.W. *Environmental Contaminants in Wildlife*. USA; Lewis Publishers; 1996. 49-72.

Bols NC; Brubacher JL; Ganassin RC; Lee LEJ. Ecotoxicology and innate immunity in fish. *Develop Comparative Immunology*, 2001, 25, 853–73.

Bonney, RJ; Becker, JE; Walker, RR; Potter, VR. Primary monolayer cultures of adult rat liver parenchymal cells suitable for study of the regulation of enzyme synthesis. *In Vitro*, 1974, 9, 399-413.

Boselli, E; Velazco, V; Caboni, MF; Lercker, G. Pressurized liquid extraction of lipids for the determination of oxysterols in egg-containing food. *Journal of Chromatography A*, 2001, 917(1-2), 239-244.

Bowser, DH; Frenkel, K; Zelikoff, JT. Effects of in vitro nickel exposure on the macrophage-mediated immune functions of rainbow trout (*Oncorhynchus mykiss*). *Bulletin of Environment Contamination Toxicology*, 1994, 52, 367–373.

Braunbeck, T; Hauck, C; Scholz, S; Segner, H. Mixed function oxygenase in cultured fish cells: contributions of substrates and response to hormones. *In Vitro Cell Develop Biology*, 1995, 30A, 306–311.

Braunbeck, T; Storch, V. Senescence of hepatocytes isolated from rainbow trout (*Oncorhynchus mykiss*) in primary culture: an ultrastructural study. *Protoplasma*, 1992, 170, 138–159.

Broadhead, CL; Combes, RD. The current status for food additives toxicity testing and the potential for application of the Three Rs. *ATLA,* 2001, 29, 471–485.

Bucheli, TB; Fent, K. Induction of cytochrome P450 as a biomarker for environmental contamination in aquatic ecosystems. *Critical Reviews in Environmental Sciences and Technolology*, 1995, 25, 201-268.

Carlson, EA; Li, Y; Zelikoff, JT. Exposure of Japanese medaka (*Oryzias latipes*) to benzo[a]pyrene suppresses immune function and host resistance against bacterial challenge. *Aquatic Toxicology*, 2002, 56(4), 289-301.

Carlson, EA; Li, Y;Zelikoff, JT. Benzo[a]pyrene-induced immunotoxicity in Japanese medaka (*Oryzias latipes*): relationship between lymphoid CYP1A activity and humoral immunosuppression. *Toxicology and Applied Pharmacology*, 2004a, 201, 40–52.

Carlson, EA; Li, Y;Zelikoff, JT. Suppressive effects of benzo[a]pyrene upon fish immune function: evolutionarily conserved cellular mechanisms of immunotoxicity. *Marine Environment Research*, 2004b, 58, 731–734.

Carson , R. Silent Spring. Boston: Houghton Mifflin; 1962.

Červenková, K; Belejova, M; Veselý, J; Chmela, Z; Rypka, M; Ulrichová, J; Modrianský, M; Maurel, P. Cell suspensions, cell cultures, and tissue slices – important metabolic in vitro systems. *Biomed. Papers*, 2001, 145(2), 57–60.

Chung, S.; Secombes, C J. Analysis of events occurring within teleost macrophages during the respiratory burst. *Comparative Biochemistry Physiology*, 1988, 89B, 539–544.

Chung, S.; Secombes, C. J. Activation of rainbow trout macrophages. *Journal of Fish Biology,* 1987, 31 (A), 51–56.

Clarke, DJ; Burchell, B; George, SG. Differential expression and induction of UDP-glucuronosyltransferase isoforms in hepatic and extrahepatic tissues of a fish, *Pleuronectes platessa*: immunochemical and functional characterization. *Toxicology and Applied Pharmacology*, 1992, 115, 130–136.

Clayton, DF and Darnell, JE. Changes in liver-specific compared to common gene transcription during primary culture of mouse hepatocytes. *Molecular Cell Biology*, 1983, 3, 1552-1561.

Clemons, E; Arkoosh, MR; Casillas, E. Enhanced superoxide anion production in activated peritoneal macrophages from English sole (*Pleuronectes vetulus*) exposed to polycyclic aromatic hydrocarbons. *Marine Environment Research*, 1999, 47, 71–87.

Cossarini-Dunier, M. Effects of manganese ions on the immune response of carp, (*Cyprinus carpio*) against *Yersinia ruckeri*. *Develop Comparative Immunology*, 1988, 12, 573–579.

Cossarini-Dunier, M. Effects of the pesticides atrazine and lindane and of manganese ions on cellular immunity of carp, *Cyprinus carpio*. *Journal of Fish Biology*, 1987 31, 67–73.

Dargel, R. Lipid peroxidation A common pathogenic mechanism? *Experimental Toxicology Pathology*, 1992, 44:169-181.

Dautremepuits, C; Paris-Palacios, S; Betoulle, S; Vernet, G. Modulation in hepatic and head kidney parameters of carp (*Cyprinus carpio* L.) induced by copper and chitosan. *Comparative Biochemistry and Physiology Part C: Toxicology and Pharmacology*, 2004, 137(4), 325-333.

Del Rio, LA; Ortega, MG; Lopez, AL; Gorge, JL. A more sensitive modification of the catalase assay with the Clark oxygen electrode. *Analytical Biochemistry*, 1977, 80, 409-415.

Dethloff, GM.;Bailey, HC. Effects of copper on immune system parameters of rainbow trout (*Oncorhynchus mykiss*). *Environmental Toxicology and Chemistry*, 1998, 17, 1807–1814.

Di Giulio, RT; Washburn, PC; Wenning, RJ. Biochemical Responses in Aquatic Animals: A Review of Determinants of Oxidative Stress. *Environmental Toxicology and Chemistry*, 1989, 8, 1103-1123.

Duchiron, C; Betoulle, S; Reynaud, S; Deschaux, P. Lindane increases macrophage-activating factor production and intracellular calcium in rainbow trout (Oncorhynchus mykiss) leukocytes. *Ecotoxicology and Environmental Safety*, 2002, 53(3), 388-396.

Dunier, M; Siwicki, AK; Scholtens, J; Dal Molin, S; Vergnet, C; Studnicka, M. Effect of lindane exposure on rainbow trout (*Oncorhynchus mykiss*) immunity. III. Effect on nonspecific immunity and B lymphocyte functions. *Ecotoxicology and Environmental Safety,* 1994, 27, 324–334.

Eisenbrand, G; Pool-Zobel, B; Baker, V; Balls, M; Blaauboer, BJ; Boobis, A; Carere, A; Kevekordes, S; Lhuguenot, J-C; Pieters, R; Kleiner. Methods of in vitro toxicology. *Food and Chemical Toxicology,* 2002, 40, 193–236.

Elia, AC; Galarini, R; Taticchi, MI; Dörr, AJM; Mantilacci, L. Antioxidant responses and bioaccumulation in *Ictalurus melas* under mercury exposure. *Ecotoxicology and Environmental Safety*, 2003, 55, 162–167.

Ellis, A. E. The function of teleost lymphocytes in relation to inflammation. *Journal of Tissue Reaction,* 1986, 8, 263-70.

Elsasser, MS; Roberson, BS; Hetrick, FM. Effects of metals on the chemiluminescent response of rainbow trout (*Salmo gairdneri*) phagocytes. *Veterinary Immunology Immunopathology*, 1986, 12, 243–250.

Faisal, M; Huggett, RJ. Effects of aromatic hydrocarbons on the lymphocyte mitogenic responses in spot (*Leiostomus xanthurus*). *Marine Environmental Research*, 1993, 35, 121–124.

Faisal, M; Marzouk, MSM; Smith, CL; Huggett, RF. Proliferative responses of spot (*Leiostomus xanthurus*) leukocytes to mitogens from a polycyclic aromatic hydrocarbon contaminated environment. *Immunopharmacology and Immunotoxicology*, 1991, 13, 311–328.

Ferraris, M; Radice, S; Catalani, P; Francolini, M; Marabini, L; Chiesara, E. Early oxidative damage in primary cultured trout hepatocytes: a time course study. *Aquatic Toxicology*, 2002, 59, 283–296.

Ferrini, JB; Pichard, L; Domergue, J; Maurel, P. Longterm primary cultures of adult human hepatocytes. *Chemistry Biology Interaction*, 1997, 107, 31–45.

Filipak Neto, F; Zanata, SM; Randi, MAF; Pelletier, É; Oliveira Ribeiro, CA. Hepatocytes Primary Culture from the Neotropical Fish Trahira *Hoplias malabaricus* (Bloch, 1794). *Journal of Fish Biology*, 2006 (in press)

Folch J, Less M, Sloane-Stanley GH. A simple method for the isolation and purification of total lipids from animal tissues. *Journal of Biology Chemistry*, 1957, 226, 497–509.

Gallina, TT; Bendini, A; Ricci, A; Lerc, G. Pressurized solvent extraction of total lipids in poultry meat. *Food Chemistry*, 2003, 83(4), 551-555.

Gargioni, R; Filipak Neto, F; Buche, Df; Randi, MAF; Franco, CRC; Paludo, KS; Pelletier, E; Ferraro, MVM; Cestari, MM; Bussolaro, D; Oliveira Ribeiro, CA. Cell death and DNA damage in peritoneal macrophages of mice (*Mus musculus*) exponed to inorganic lead. *Cell Biology Internacional*, 2006, 30, 615-623.

Ghosh, D; Bhattacharya, S; Mazumder, S. Perturbations in the catfish immune responses by arsenic: Organ and cell specific effects. *Comparative Biochemistry and Physiology Part C: Toxicology and Pharmacology,* 2006, 143(4), 455-463.

Gravato, C; Teles, M; Oliveira, M; Santos, M.A. Oxidative stress, liver biotransformation and genotoxic effects induced by copper in *Anguilla anguilla* L. – the influence of pre-exposure to β-naphthoflavone. *Chemosphere*, 2006 (In Press).

Green LC, Wagner DA, Glogowski J, Skipper PL, Wishnok JS, Tannenbaum SR. Analysis of nitrate, nitrite and 15-nitrite in biological fluids. *Analytical Biochemistry*, 1982, 126, 131–8.

Guguen, C; Gregori, C; Schapira, F. Modification of pyruvate kinase isozymes in prolonged primary cultures of adult rat hepatocytes. *Biochimie*, 1975, 57, 1065-1071.

Guguen-Guillouzo, C; Tichonicky, L; Szajnert, M; Schapira, F; Kruh, J. Changes in some chromatin and cytoplasmic enzymes in primary cultures of adult rat hepatocytes. *Biology Cell,* 1978, 31, 225-234.

Guillouzo, A. Liver cells models in in vitro toxicology. *Environ Health Perspective*, 1998, 106(2), 511–532.

Guillouzo, A; Morel, F; Langouet, S; Maheo, K; Rissel, M. Use of hepatocyte cultures for the study of hepatotoxic compounds. *Journal of Hepatology*, 1997, 26(2), 73-80.

Guillouzo, A; Morel, F; Ratanasavanh, D; Chesne, C; Guguen-Guillouzo, C. Long-term culture of functional hepatocytes. *Toxicology In Vitro*, 1990, 4(5), 415–427.

Guzelian, PS; Bissell, DM; Meyer, UA. Drug metabolism in adult rat hepatocytes in primary monolayer culture. *Gastroenterology*, 1977, 72, 1232-1239.

Hansen, PD. Biomarkers In: Market, BA, Breure, AM, Zechmeister, HG. *Bioindicators and Biomonitors.* USA; Elsevier Science; 2003; 203-220.

Hara, A and Radin, NS. Lipid extraction of tissues with a low-toxicity. *Analytical Biochemistry*, 1978, 90(1), 420-426.

Harford, AJ; O'Halloran, K; Wright, PFA. The effects of in vitro pesticide exposures on the phagocytic function of four native *Australian freshwater fish. Aquatic Toxicology*, 2005, 75(4), 330-342.

Hart, LJ; Smith, SA; Smith, BJ; Roberston, J; Besteman, EG; Holladay, SD. Subacute immunotoxic effects of the polycyclic aromatic hydrocarbon 7,12-dimethylbenzanthracene (DMBA) on spleen and pronephros leukocytic cell counts and phagocytic cell activity in tilapia (*Oreochromis niloticus*). *Aquatic Toxicology*, 1998, 41, 17–29.

Heringa, MB; Schreurs, RHMM; Busser, F; Van Der Saag, PT; Van Der Burg, B; Hermens, JLM. Toward more useful in vitro toxicity data with measured free concentrations. *Environment Science Technology*, 2004,38, 6263–6270.

Hoeger, B; Van den Heuvel, MR; Hitzfeld, BC; Dietrich, DR. Effects of treated sewage effluent on immune function in rainbow trout (*Oncorhynchus mykiss*). *Aquatic Toxicology,* 2004, 70(4), 345-355.

Holladay, SD; Smith, SA; Besteman, EG; Deyab, A.S; Gogal, RM; Hrubec, T; Robertson, JL; Ahmed, SA. Benzo[a]pyrene-induced hypocellularity of the pronephros in tilapia (*Oreochromis niloticus*) is accompanied by alterations in stromal and parenchymal cells and by enhanced immune cell apoptosis. *Veterinary Immunology and Immunopathology*., 1998, 64, 69–82.

Holsapple, M.P. Autoimmunity by pesticides: a critical review of the state of the science. *Toxicology Letters*, 2002, 127,101–109.

Hoogenboom, LAP; Kuiper, HA. The use of in vitro models for assessing the presence and safety of residues of xenobiotics in food. *Trends in Food Science and Technology*, 1997, 81, 157-166.

Howell, CJG. A Chemokinetic factor in the carp, *Cyprinus carpio*. *Develop Comparative Immunology,*1987, 11, 139-146.

Hutchinson, TH; Field, MDR; Manning, MJ. Evaluation of non-specific immune functions in dab, *Limanda limanda* L., following short-term exposure to sediments contaminated with polyaromatic hydrocarbons and/or polychlorinated biphenyls. *Marine Environmental Research,* 2003, 55(3), 193-202.

Hylland, K; Sandvik, M; Utne Skåre, J; Beyer, J; Egaas, E; Goksrøyr, A. Biomarkers in Flounder *(Platichthysflesus)*: an Evaluation of Their Use in Pollution Monitoring. *Marine Environmental Research,* 1996, 42(1-4), 223-227.

Iwatsubo, T; Hirota, N; Ooie, T; Suzuki, H; Shimada, N; Chiba, K; Ishizaki, T; Green, CE; Tyson, CA; Sugiyama, Y. Prediction of in vivo drug metabolism in the human liver from in vitro metabolism data. *Pharmacology Therapy*, 1997, 73(2), 147-171.

Janošek, J; Hilscherová, K; Bláha, L; Holoubek, I. Environmental xenobiotics and nuclear receptors - Interactions, effects and in vitro assessment. *Toxicology in Vitro*, 2006, 20, 18–37.

Jefferson, DM; Clayton, DF; Darnell, JE; Reid, LM. Post-transcriptional modulation of gene expression in cultured rat hepatocytes. *Molecular Cell Biology*, 1984, 4, 1929-1934.

Kappus, H. Oxidative stress in chemical toxicity. *Archives of Toxicology*, 1987, 60,144-149.

Karrow, NA; Boermans, HJ; Dixon, DG; Hontella, A; Solomon, KR; Whyte, JJ; Bols, NC. Characterizing the immunotoxicity of creosote to rainbow trout (*Oncorhynchus mykiss*): a microcosm study. *Aquatic Toxicology*, 1999, 43, 179–194.

Kelly, SA; Havrilla, CM; Brady, TC; Abramo, KH; Levin, ED. Oxidative stress in toxicology: established mammals emerging piscine model systems. *Environ Health Perspectives,* 1998,106, 375–384.

Kelly-Reay, K; Weeks-Perkins, BA. Determination of the macrophage chemiluminescent response in *Fundulus heteroclitus* as a function of pollution stress. *Fish Shellfish Immunology*., 1994, 4, 95–105.

Khangarot, BS; Rathore, RS; Tripathi, DM. Effects of chromium on humoral and cell-mediated immune responses and host resistance to disease in a freshwater catfish, *Saccobranchus fossilis* (Bloch). *Ecotoxicology and Environmental Safety*, 1999, 43, 11–20.

Klaunig, JE; Xu, Y; Bachowski, S; Jiang, J. Free-Radical Oxygen-Induced Changes in Chemical Carcinogenesis. In: K.B.Wallce, editor. *Free Radical Toxicology*. Taylor and Francis; 1997; 375-400.

Kranz, H; Gercken, J. Effects of sublethal concentrations of potassium dichromate on the occurrence of splenic melano-macrophage centres in juvenile plaice, *Pleuronectes platessa*, L. *Journal of Fish Biology*, 1987, 31A, 75–80.

Lange, A; Ausseil, O; Segner, H. Alterations of tissue glutathione levels and metallothionein mRNA in rainbow trout during single and combined exposure to cadmium and zinc. *Comparative Biochemistry and Physiology Part C*, 2002, 131, 231–243.

Leaver, MJ; Scott, K; George, SG. Expression and tissue distribution of plaice glutathione *S*-transferase A. *Marine Environmental Research*, 1992, 34, 237–241.

Lemaire-Gony, S; Lemaire P; Pulsford, AL. Effects of cadmium and benzo(a)pyrene on the immune system, gill ATPase and EROD activity of European sea bass *Dicentrarchus labrax*. *Aquatic Toxicology*, 1995, 31(4), 297-313.

Lin, JH; Hayashi, M; Awazu, S; Hanano, M. Correlation between in vitro and in vivo drug metabolism rate: oxidation of ethoxybenzamide in rat. *J. Pharmacokinet Biopharm*, 1978, 6, 327–337.

Lin, JH; Sugiyama, Y; Awazu, S; Hanano, M. Kinetic studies on the deethylation of ethoxybenzamide. A comparative study with isolated hepatocytes and liver microsomes in rat. *Biochem Pharmacol*, 1980, 29, 2825–2830.

Livingstone, DR. The fate of organic xenobiotics in aquatic ecosystems: quantitative and qualitative differences in biotransformation by invertebrates and fish, *Comparative Biochemistry Physiology*, 1998, 120, 43–49.

Livingstone, DR; Lemaire, P; Matthews, A; Peters, LD; Porte, C; Fitzpatrick, PJ; Förlin, L; Nasci, C; Fossato, V; Wootton, N; Goldfarb, P. Assessment of the impact of organic pollutants on goby (*Zosterisessor ophiocephalus*) and mussel (*Mytilus galloprovincialis*) from the Venice Lagoon, Italy: Biochemical studies. *Marine Environmental Research*, 1995, 39(1-4), 235-240.

Lopes, PA; Pinheiro, T; Santos, MC; Mathias, ML; Collares-Pereira, MJ; Viegas-Crespo, AM. Response of antioxidant enzymes in freshwater fish populations (*Leuciscus alburnoides* complex) to inorganic pollutants exposure. *The Science of Total Environment,* 2001, 280(1-3), 153-163.

Low, KW; Sin, YM. Effects of mercuric chloride and sodium selenite on some immune responses of blue gourami, *Trichogaster trichopterus* (Pallus). *The Science of Total Environment,* 1998, 214, 153–164.

Luo, Y; Shi, HH; Wang, XR; Ji, LL. Free radical generation and lipid peroxidation induced by 2,4-dichlorophenol in liver of *Carassius auratus. Huan Jing Ke Xue Huanjing Kexue*, 2005, 26(3), 29-32.

MacArthur, JI; Fletcher, TC. Phagocytosis in fish. In: MJ Manning and MF Tatner, editor. *Fish Immunology*. London: Academic Press; 1985; 29-46.

Manahan, SE. *Toxicological Chemistry*. 2nd edition. Chelsea, Michigan. USA: Lewis Publishers; 1992.

Masters, JR. Human cancer cell lines: fact and fantasy. *National Review of Molecular Cell Biology*, 2000, 1(3), 233–236.

Mela, M.; Randi, M.A.F.; Ventura, D. F; Carvalho, C.E.V; Pelletier, E; Oliveira Ribeiro, CA. Effects of dietary methylmercury on liver and kidney histology in the neotropical fish *Hoplias malabaricus. Ecotoxicology and Environmental Safety*. 2006 (In Press).

Meredith, MJ. Rat hepatocytes prepared without collagenase: prolonged retention of differentiated characters in culture. *Cell Biology Toxicology*, 1988, 4(4), 405–425.

Moon, TW; Walsh, PJ; Mommsen, TP. Fish hepatocytes: A model metabolic system. *Canadian Journal of Fish Aquatic Science*, 1985, 42, 1772-1782.

Mosmann T. Rapid colorimetric assay for cellular growth and survival: application to proliferation and cytotoxicity assay. *Journal Immunology Methods*, 1983, 65,55–63.

Muhvich, AG; Jones, RT; Kane, AS; Anderson, RS; Reimscheussel, R. Effects of chronic copper exposure on the macrophage chemiluminescent response and gill histology in goldfish (*Carassius auratus* L.). *Fish Shellfish Immunology*, 1995, 5, 251–264.

National Research Council – NRC. Committee on biological markers. *Environmental Health Perspective,* 1987, 74, 3.

Natt, MP; Herrick, CC.A new blood diluent for counting erythrocytes and leucocytes of the chicken. *Poultry Science*, 1952, 31, 735-738.

Natthan, C.F., Hibbs, J.B.Jr. Role of nitric oxide synthesis in macrophage antimicrobial activity. *Current Opinion Immunology*, 1991, 3, 65-70.

Neff, JM. Polycyclic aromatic hydrocarbons. In: Rand, GM and Petrocilli, SR Editors. *Fundamentals of Aquatic Toxicology*. New York: Hemisphere; 1985; Chap 14.

Neumann,N.F.; Fagan, D.;Belosevic, M. Macrophage activating factor(s) secreted by mitogen stimulated goldfish kidney leucocytes synergize with bacterial lipopolysaccharide to induce nitric oxide production in teleost macrophages. *Develop Comparative Immunology,* 1995, 19, 473–82.

Nichols, JW; Schultz, IR; Fitzsimons, PN. In vitro–in vivo extrapolation of quantitative hepatic biotransformation data for fish I. A review of methods, and strategies for incorporating intrinsic clearance estimates into chemical kinetic models. *Aquatic Toxicology,* 2006, 78, 74–90.

Obach, RS. Non-specific binding to microsomes: impact on scale-up of in vitro intrinsic clearance to hepatic clearance as assessed through examination of warfarin, imipramine, and propranolol. *Drug Metabolism Disposit*, 1997, 25, 1359–1369.

OECD. *Final report of the OECD workshop on harmonization of validation and acceptance criteria for alternative toxicological tests methods*. Organization for Economic Cooperation and Development. Paris; 1996.

Oliveira Ribeiro, CA, Rouleau, C, Pelletier, E, Audet, C, Tjalve, H. Distribution kinetics of dietary methylmercury in the artic charr (*Salvelinus alpinus*). *Environmental Science Technology,* 1999, 33, 902-907.

Oliveira Ribeiro, CA; Vollaire, Y; Sanchez-Chardi, A; Roche, H. Bioaccumulation and the effects of organochlorine pesticides, PAH and heavy metals in the Eel (*Anguilla anguilla*) at the Camargue Nature Reserve, France. *Aquatic Toxicology*, 2005, 74 (1), 53-69.

Oliveira Ribeiro, CA; Filipack Neto, F; Mela, M; Silva, PH; Randi, MAF; Costa, JRA; Pelletier, E. Hematological findings in neotropical fish *Hoplias malabaricus* exposed to subchronic and dietary doses of methylmercury, inorganic lead and tributyltin chloride. *Environmental Research*, 2006, 101, 74-80.

Padrós J, Pelletier É, Oliveira Ribeiro CA. Metabolic interactions between low doses of benzo[*a*]pyrene and tributyltin in arctic charr (*Salvelinus alpinus*): a long-term in vivo study. *Toxicology and Applied Pharmacology*, 2003, 192, 45-55.

Pain, DJ. Lead in the Environment. In: Hoffman, DJ; Rattner, BA; Allen Burton Jr, G; Cairn Jr, J. *Handbook of Ecotoxicology*. USA: Lewis Publishers; 1995; 356-391.

Pallardy,M, Kerdine, S,Lebrec, H. Testing strategies in immunotoxicology. *Toxicology Letters*, 1998, 102-103, 257-260.

Paris-Palacios, S; Biagianti-Risbourg, S; Vernet, G. Biochemical and (ultra)structural hepatic perturbations of *Brachydanio rerio* (Teleostei, Cyprinidae) exposed to two sublethal concentrations of copper sulfate. *Aquatic Toxicology*, 2000, 50, 109–124.

Payne, JF; Fancey, LF. Effect of polycyclic aromatic hydrocarbons on immune responses in fish: Changes in melano-macrophage centers in flounder (*Pseudopleuronectes americanus*) exposed to hydrocarbon-contaminated sediments. *Marine Environomental Research,* 1989, 28, 431–435.

Pesonen, M; Andersson, TB. Fish primary hepatocyte cultures: an important model for xenobiotic metabolism and toxicity studies. *Aquatic Toxicology*, 1997, 37, 253–267.

Prophete, C; Carlson, EA; Li, Y; Duffy, J; Steinetz, B; Lasano, S.; Zelikoff, JT. Effects of elevated temperature and nickel pollution on the immune status of Japanese medaka. *Fish and Shellfish Immunology*, 2006, 21(3), 325-334.

Pruell , RJ; Engelhardt, FR. Liver cadmium uptake, catalase inhibition and cadmium thionein production in the killifish (*Fundulus Heteroclitus*) induced by Experimental cadmium exposure. *Marine Environmental Research,* 1980, 3(2), 101-111.

Rabitto, IS; Alves Costa, JRM; Silva de Assis, HC; Pelletier, E; Akaishi, FM; Anjos, A; Randi, MAF; Oliveira Ribeiro, CA. Effects of dietary Pb(II) and tributyltin on neotropical fish, *Hoplias malabaricus*: histopathological and biochemical findings. *Ecotoxicology and Environmental Safety*, 2005, 60, 147–156.

Radin, NS. Extraction of tissue lipids with a solvent of low toxicity. *Methods in Enzymology*, 1981, 72, 5-7.

Raimbou, PS. Heavy metals in aquatic organisms. In : Beyer, WN; Heinz, GH; Redmon-Norwood, A.W. *Environmental Contaminants in Wildlife*. Boca Raton, USA: Lewis Publishers; 1996; 405-426.

Rana, S; Singh, R; Verma, S. Mercury-induced lipid peroxidation in the liver, kidney, brain and gills of a fresh water fish, *Channa punctatus*. *Japan. Journal of Ichthyology*, 1995, 42, 255–259.

Rane, A; Wilkinson, GR; Shand, DG. Prediction of hepatic extraction ratio from in vitro measurement of intrinsic clearance. *Journal of Pharmacology Experimental Theory*, 1977, 200, 420–424.

Reynaud, S; Deschaux, P. The effects of polycyclic aromatic hydrocarbons on the immune system of fish: A review. *Aquatic Toxicology*, 2006, 77(2), 229-238.

Reynaud, S; Duchiron, C; Deschaux, P. 3-methylcholanthrene inhibits lymphocyte proliferation and increases intracellular calcium levels in common carp (*Cyprinus carpio* L.). *Aquatic Toxicology*, 2003, 63, 319–331.

Reynaud, S; Marionnet, D; Taysse, L; Duchiron, C; Deschaux, P. The effects of 3-methylcholanthrene on macrophage respiratory burst and biotransformation activities in the common carp (*Cyprinus carpio* L.). *Fish Shellfish Immunology*., 2002, 12, 17–34.

Rice, CD; Banes, MM; Ardelt, TC. Immunotoxicity in channel catfish, *Ictalurus punctatus*, following acute exposure to tributyltin. *Archives of Environment Contamination Toxicology*, 1995, 28, 464–470.

Rice, CD; Kergosien, DH; Adams, SM. Innate immune functions as a bioindicator of pollution stress in fish. *Ecotoxicology and Environmental Safety*, 1996, 33, 186–192.

Rice, CD; Schlenk, D. Immune function and cytochrome P4501A activity after acute exposure to 3,3′,4,4′,5-pentachlorobiphenyl (PCB 126) in channnel catfish. *Journal of Aquatic Animal Health,* 1995, 7, 195–204.

Rice, CD; Weeks, BA. Influence of tributyltin on in vitro activation of oyster toadfish macrophages. *Journal of Aquatic Animal Health*, 1989, 1, 62–68.

Rice, CD; Weeks, BA. The influence of in vivo exposure to tributyltin on reactive oxygen formation in oyster toadfish macrophages. *Archives of Environment Contamination Toxicology*, 1990, 19, 854–857.

Rice, CD; Weeks, BA. Tributyltin stimulates reactive oxygen formation in toadfish macrophages. *Develop Comparative Immunology*, 1991, 15, 431–436.

Roberts, RJ. *Fish Pathology* 2nd Edition. London: Bailliere Tindall; 1989.

Rogiers, V; Blaauboer, BJ; Maurel, P; Philipps, I; Shephard, E. Hepatocyte-based in vitro models and their application in pharmacotoxicology. *Toxicology In Vitro*, 1995, 9, 685–694.

Rougier, F; Troutaud, D;Ndoye, A;Deschaux, P. Non-specific immune response of Zebrafish, *Brachydanio rerio* (Hamilton-Buchanan) following copper and zinc exposure. *Fish and Shellfish Immunology,* 1994, 4(2), 115-127.

Ruiz-Leal, M.; George,S. An in vitro procedure for evaluation of early stage oxidative stress in an established fish cell line applied to investigation of PHAH and pesticide toxicity. *Marine Environmental Research*, 2004, 58, 631–635.

Sanchez-Dardon, J; Voccia, I; Hontela, A; Chilmonczyk, S; Dunier, M; Boermans, H; Blakely, B; Fournier, M. Immunomodulation by heavy metals tested individually or in mixtures in rainbow trout (*Oncorhynchus mykiss*) exposed in vivo. *Environment Toxicology and Chemistry*, 1999, 18, 1492–1497.

Sandbacka, M; Pärt, P; Isomaa, B. Gill epithelial cells as tools for toxicity screening-comparison between primary cultures, cells in suspension and epithelia on filters. *Aquatic Toxicology,* 1999, 46, 23–32.

Sarmento, A. ,Marquesa, F., Ellisc, A. E., Afonso, A. Modulation of the activity of sea bass (*Dicentrarchus labrax*) head-kidney macrophages by macrophage activating factor(s) and lipopolysaccharide. *Fish Shellfish Immunology*, 2004a, 16, 79–92.

Sarmento, A., Guilhermino, L., Afonso, A. Mercury chloride effects on the function and cellular integrity of sea bass (*Dicentrarchus labrax*)head kidney macrophages. *Fish Shellfish Immunology,* 2004b, 17, 489 – 498.

Schoor, WP; Plumb, JA. Induction of nittic oxide synthase in channel catfish *Ictalurus punctatus* by *Edwardsiella ictaluri*. *Disease Aquatic ORG*, 1994, 19,153-155.

Secombes, CJ. The nonspecific immune system: Cellular Defenses. In: Iwana, G. and Nakanishi, T, editor. *The Fish Immune System*. USA; Academic Press; 1996; 63-103.

Seddon, WL; Prosser, CL. Non-enzymatic isolation and culture of channel catfish hepatocytes. *Comparative Biochemistry and Physiology Part A*, 1999, 123, 9–15.

Seeley KR ; Weeks-Perkins, BA. Suppression of natural cytotoxic cell and macrophage phagocytic function in oyster toadfish exposed to 7,12-dimethylbenz[a]anthracene. *Fish and Shellfish Immunology,* 1997, 7, 115–121.

Seeley, KR; Week-Perkins, BA. Altered phagocytic activity of macrophages in oyster toadfish from a highly polluted estuary. *Journal of Aquatic Animal Health*, 1991, 3, 224–227.

Segner, H. Isolation and primary culture of teleost hepatocytes. *Comparative Biochemistry Physiology*, 1998, 120, 71–81.

Segner, H; Blair, JB; Wirtz, G; Miller, MR. Cultured trout liver cells: utilisation of substrates and response to hormones. *In Vitro Cell Develop Biology*, 1994, 30(A), 306–311.

Segner, H; Cravedi, JP. Metabolic activity in primary cultures of fish hepatocytes. *ATLA*, 2001, 29, 251–257.

Sevanian, A; McLeod, L. Formation and Biological Reactivity of Lipid Peroxidation Products. In: K.B. Wallace, editor. *Free Radical Toxicology*. Taylor and Francis; 1997; 47-70.

Shugart, L; Bickman, J; Jackim, E; McMahon, G; Ridley, W; Stein, L; Steinert, SA. DNA Alterations. In: R.J. Hugget, R.A. Kimerle, P.M. Mehrle, Jr., and H.L. Bergman, editors. *Biomarkers: Biochemical, Physiological, and Histological Markers of Anthropogenic Stress*. Chelsea, MI; Lewis Publishers; 1992; 125-153.

Singh SM; Sivalingam PM. *In vitro* study on the interactive effects of heavy metals on catalase activity of *Sarotherodon mossambicus*. *Journal of Fish Biology*, 1982, 20,:683-688.

Sirica AE; Pitot HC. Drug metabolism and effects of carcinogens in cultured hepatic cells. *Pharmacology Review*, 1980, 31, 205-228.

Sirica AE; Richards, W; Tsukada, Y; Sattler, CA; Pitot, HC. *Fetal phenotypic expression by adult rat hepatocytes on collagen gel/nylon meshes*. Proc Academic Science USA, 1979, 73, 283-287.

Song, SB; Xu, Y; Zhou, BS. Effects of hexachlorobenzene on antioxidant status of liver and brain of common carp (*Cyprinus carpio*). *Chemosphere*, 2006, 65(4), 699-706.

Spitsbergen, JM; Schat, KA; Kleeman JM; Peterson, RE. Interactions of 2,3,7,8-tetrachlorodibenzo-p-dioxin (TCDD) with immune responses of rainbow trout. *Vet. Immunology Immunopathology*, 1986, 12, 263–280.

Stegeman, JJ; Miller, MR; Woodin, BR; Blair, JB. Effect of □-naphthoflavone on multiple cytochrome P450 forms in primary cultures of rainbow trout hepatocytes. *Marine Environmental Research,* 1993, 35, 209.

Steward, AR; Dannan, GA; Guzelian, PS; Guengerich, FP. Changes in the concentration of seven forms of cytochrome P-450 in primary cultures of adult rat hepatocytes. *Molecular Pharmacolgy*, 1985, 27, 125-132.

Stuehr, D.J.; Marletta M.A. Mammalian nitrate biosynthesis: mouse macrophages produce nitrite and nitrate in response to *Escherichia coli* lipopolysaccharide. Proc Academic Science, 1985, 82, 7738-7742.

Suzuki, Y; Lida, T. Fish granulocytes in the process of inflammation. *Annual. Review of Fish Disase,* 1992, 2, 149-160.

Tafalla C, Novoa B. Requirements for nitric oxide production by turbot (*Scophthalmus maximus*) head kidney macrophages. *Develop Comparative Immunology*, 2000, 24, 623–31.

Tagliari, KC; Cecchini, R; Rocha, JAV; Vargas, VMF. Mutagenicity of sediment and biomarkers of oxidative stress in fish from aquatic environments under the influence of tanneries. *Mutation Research/Genetic Toxicology and Environmental Mutagenesis*, 2004, 561(1-2), 101-117.

Tahir, A; Fletcher, TC; Houlihan, DF; Secombes, CJ. Effect of short-term exposure to oil contaminated sediments on the immune response of dab, *Limanda limanda* (L). *Aquatic Toxicology,* 1993, 27, 71–82.

Tang, C; Lin, Y; Rodrigues, AD; Lin, JH. Effect of albumin on phenytoin and tolbutamide metabolism in human liver microsomes: an impact more than protein binding. *Drug Metab Disposit,* 2002, 30, 648–654.

Tavares-Dias, M., Moraes, FR. *Hematologia de Peixes Teleósteos*. 1rst edition. Ribeirão Preto, SP: Eletrônica e Arte Final; 2004.

Thomas, P; Wofford, HW. Effects of cadmium and Aroclor 1254 on lipid peroxidation, glutathione peroxidase activity, and selected antioxidants in Atlantic croaker ts. *Aquatic Toxicology*, 1993, 27(1-2), 159-177.

Ulich, TR; Watson, LR; Yin, S; Guo, K; Wang, TH; Castillo, J. The intratracheal administration of endotoxin and cytokines: Characterization of LPS-induced IL-1 and TNF mRNA expression and the LPS, IL-1, and TNF-induced inflammatory infiltrate. *American Journal of Pathology*, 1995, 138, 1485-1496.

Van der Oost, R; Beyer, J; Vermeulen, NPE. Fish bioaccumulation and biomarkers in environmental risk assessment: a review. *Environmental Toxicology and Pharmacology*, 2003, 13(2), 57-149.

Van der Weiden, MEJ; Bleumick, R; Seinen, W; Van den Berg, M. Concurrence of P450 1A induction and toxic effects in the mirror carp (*Cyprinus carpio*), after administration of a low dose of 2,3,7,8-tetrachlorodibenzo-p-dioxin. *Aquatic Toxicology*, 1994, 29, 147–162.

Van Veld, PA; Patton, JS; Lee, RF. Effect of pre-exposure to dietary benzo(a)pyrene (BP) on the first-pass metabolism of BP by the intestine of toadfish (*Opsanus tau*): in vivo studies using portal vein-catheterized fish. *Toxicology and Applied Pharmacology*, 1988, 92, 255–265.

Vasseur, P; Cossu-Leguille, C. Linking Molecular interactions to consequent effects of persistent organic pollutants (POPs) upon populations. *Chemosphere*, 2006, 62:1033-1042.

Vega-López, A; Galar-Martínez, M; Jiménez-Orozco, FA; García-Latorre, E; Domínguez-López, ML. Gender related differences in the oxidative stress response to PCB exposure in an endangered goodeid fish (*Girardinichthys viviparus*). *Comparative Biochemistry and Physiology - Part A: Molecular and Integrative Physiology*, 2006 (In Press).

Viarengo, A; Bettella, E; Fabbri, R; Burlando, B; Lafauri, M. Heavy metal inhibition of EROD activity in liver microsomes from the bass *Dicentrarchus labrax* exposed to organic xenobiotics: Role of GSH in the reduction of heavy metal effects. *Marine Environmental Research*, 1997, 44(1), 1-11.

Walker, CH; Hopkin, SP; Sibly, RM; Peakall, DB. *Principles of Ecotoxicology*. Bristol, PA: Taylor and Francis; 1996.

Wang R.; Neumann, N.F.; Shen, Q.; Belosevic, M. Establishment and characterization of a macrophage cell line from the goldfish. *Fish Shellfish Immunology*, 1985, 5, 329–346.

Wang, C; Zhao, Y; Zheng, R; Ding, X; Wei, W; Zuo, Z; Chen, Y. Effects of tributyltin, benzo[a]pyrene, and their mixture on antioxidant defense systems in *Sebastiscus marmoratus*. *Ecotoxicology and Environmental Safety*, 2005a, In Press, Corrected Proofs.

Wang, CG.; Chen, YX.; Li, Y; Wei, W; Yu, Q. Effects of low dose tributyltin on activities of hepatic antioxidant and phase II enzymes in *Sebastiscus marmoratus*. *Bulletin of the Environment. Contamination Toxicology*, 2005b, 74 (1), 114–119.

Wang, S; Renaud, G; Infante, J; Catala, D; Infante, R. Isolation of rat hepatocytes with EDTA and their metabolic functions in primary culture. *In Vitro Cell Develop Biology*, 1995, 21(9), 526–530.

Warinner, JE; Mathews ES; Weeks, BA. Preliminary investigations of the chemiluminescent response in normal and pollutant-exposed fish. *Marine Environmental Research*, 1988, 24, 281–284.

Weeks, BA; Warinner, JE. Functional evaluation of macrophages in fish from a polluted estuary. *Veterinary Immunology and Immunopathology*, 1984, 12(1–4), 313–320.

Weeks, BA; Warinner, JE; Mason, PL; McGinnis, DM. Influence of the toxic chemicals on the chemotactic response of fish macrophages. *Journal of Fish Biology*, 1986, 28, 653–658.

Wennig, RJ; DiGiulio, RT. Microssomal enzyme activities, superoxide production, and antioxidant defenses in ribbed mussels (*Geukensia demissa*) and wedge clams (*Rangia cuneata*). *Comparative Biochemistry Physiology*, 1988, 90, 21-28.

Wishkovský, A; Mathews, ES; Weeks, BA. Effect of tributyltin on the chemiluminescent response of phagocytes from three species of estuarine fish. *Archives of Environment Contamination Toxicology*, 1989, 18, 826–831.

Woo, P.T.K. Immunological responses of fish to parasitic organisms. *Annual Reviw of Fish. Disease*, 1992, 2, 339-366.

Wren, CD; Harris, S; Harttrup, NA. Ecotoxicology of Mercury and Cadmium. In: Hoffman, DJ; Rattner, BA; Allen Burton Jr, G; Cairn Jr. *Handbook of Ecotoxicology*. USA; Lewis Publishers; 1995; 392-423.

Yamada, S; Otto, PS; Kennedy, DL. The effects of dexamethasone on metabolic activity of hepatocytes in primary monolayer culture. *In Vitro*, 1980, 16, 557–559.

Yin, Z; Lam, T.J., Sin, Y.M. Cytokine-mediated antimicrobial immune response of catfish, *Clarias gariepinus*, as a defense against *Aeromonas hydrophila*. *Fish Shellfish Immunology*, 1997, 7,93-104.

Zelikoff JT, Bowser D, Squibb KS, Renkel K. Immunotoxicity of low level cadmium exposure in fish: an alternative animal model for immunotoxicological studies. Journal of *Toxicology and Environmental Health*, 1995, 45, 235–48.

Zelikoff, JT. Biomarkers of immunotoxicity in fish and other non-mammalian sentinel species: Predictive value for mammals. *Toxicology*, 1998, 129, 63–71.

Zelikoff, JT; Bowser, D; Squibb, KS; Frenkel, K. Immunotoxicity of low level cadmium exposure in fish: an alternative animal model for immunotoxicological studies. *Journal of Toxicology and Environmental Health*, 1995, 45, 235–248.

Zelikoff, JT; Raymond, A; Carlson, E; Li, Y; Beaman, JR; Anderson, M. Biomarkers of immunotoxicity in fish: from the lab to the ocean. Toxicology Letters, 2000, 112-113, 325-331.

Zhang, Y; Bachmeier, C; Miller, DW. In vitro and in vivo models for assessing drug efflux transporter activity. *Advanced Drug Delivery Reviews*, 2003, 55, 31–51.

Zhou, B; Liu, W; Wu, RSS; Lam, PKS. Cultured gill epithelial cells from tilapia (*Oreochromis niloticus*): a new in vitro assay for toxicants. *Aquatic Toxicology*, 2005, 71, 61–72.

Zirong, X; Shijun, B. Effects of waterborne Cd exposure on glutathione metabolism in Nile tilapia (*Oreochromis niloticus*) liver. *Ecotoxicology and Environmental Safety*. 2006 (In press).

Zucco, F; Angelis, I; Testai, E; Stammati, A. Toxicology investigations with cell culture systems: 20 years after. *Toxicology in Vitro*, 2004, 18, 153–163.

In: Environmental Research Trends
Editor: Marvin A. Cato, pp. 271-301

ISBN: 978-1-60021-556-8

Chapter 8

RADIONUCLIDES AND PARTICULATE MATTER IN BELGRADE AIR

D. Todorovic, D. Popovic, S. Rajsic** and M. Tasic***
Institute of Nuclear Sciences Vinca, P.O.Box 522, 11000 Belgrade, Serbia
*Department of Physics and Biophysics, Faculty of Veterinary Medicine, University of Belgrade, Bul.Oslobodjenja 18, 11000 Belgrade, Serbia
**Institute of Physics, Pregrevica 118, 11080 Zemun, Serbia

ABSTRACT

The paper presents the results of the long term study on the concentrations of natural (^{7}Be, ^{210}Pb) and fission (^{137}Cs) radionuclides in ground level air in an urban area (city of Belgrade) from 1985 to 2005. The periods before the nuclear plant accident at Chernobyl in 1986, during the accident and afterwards are investigated. The seasonal trends (average monthly concentrations, seasonal indices and quarterly seasonal indices, min/max ratios, radionuclides loading indices) are considered due to weather conditions and microclimate on different sampling sites within the city. Activity of the radionuclides was determined on a high purity Germanium (HPGe) detector by standard gamma spectrometry.

The results of the study on concentrations, distribution and transport of suspended PM_{10} and $PM_{2.5}$ particles in air in Belgrade city center are presented, too. Trace metal concentrations in suspended particles were determined by Graphite Furnace Atomic Absorption Spectrometry (GF-AAS).

INTRODUCTION

Radionuclides in Air

Data on concentrations of cosmogenic and atmospheric radionuclides on global scale are not only important for comparing environmental impact of radioactivity from anthropogenic sources to natural ones but also for studying global atmospheric processes [3,22,28]. Radionuclides most of which are naturally occurring elements as ^{7}Be and ^{210}Pb are air pollutants, but in general they are proportionally small contributors compared to air pollutant chemicals. As after being formed most of air borne radionuclides are quickly attached to aerosols, their concentration in ground level air is manly due to behavior of aerosols in the atmosphere [27] and the rate of their removal from the atmosphere and deposition on ground and vegetation depends on the particle sizes they are attached on. Aerosols play an important role in the chemical and physical processes in the atmosphere, not only as carriers of atmospheric materials on local and global scale, but also as available surfaces for different chemical processes in the atmosphere, one of the most important being a catalytic distraction of ozone by chlorine. Therefore radionuclides can be used to trace the transport of ozone and aerosols in the atmosphere [23,24,36].

Air pollution is particularly harmful to residents of major city areas, the reason why the radioactivity air monitoring program has become a part of urban areas monitoring network since the II World War. Recent studies [24,50,66,67,68] confirmed that some cardiopulmonary diseases are highly correlated to concentration of aerosols in lower atmosphere, and not only to aerosol mass concentrations, but also to their size, surface area, particle number, chemical composition, etc [39]. Thus, strict air quality standards for particulate matter in urban areas have been introduced into European environmental protection legislative in the last decade.

Among the naturally occurring radionuclides in air, radon (^{222}Rn) and its short lived progenies (^{210}Pb, ^{210}Bi, ^{210}Po) are most significant, while ^{137}Cs, ^{90}Sr and ^{90}Pu are of major interest among the fission products.

^{7}Be (half life 53.28 days) is produced by cosmic rays in spallation processes with light elements (nitrogen, oxygen, carbon) in the upper troposphere (about 33%) and lower stratosphere (about 66%) with the maximum production rate at the height of 15 km. Once produced it readily attaches to aerosols with the mean diameter of about 0.3 μm [22, 24]. Variations in ^{7}Be annual mean concentrations are good indicator of the changes in the atmospheric production rate due to cosmic ray intensity and its seasonal patterns are correlated to stratosphere-to-troposphere exchange processes. At midlatitudes, concentrations of ^{7}Be in ground level air exhibit maxima in the spring/summers periods that are attributed to seasonal thinning of tropopause which allow stratospheric masses rich with ^{7}Be to enter the troposphere [1,4,6,17,29,30,34,47]. Cyclic variations of ^{7}Be concentrations in ground level air are due to many factors: the air exchange between stratosphere and troposphere, vertical mixing within the troposphere and the washout effects of ^{7}Be in the atmosphere due to precipitation. As a tracer of intrusion processes of stratosphere into troposphere ^{7}Be is also an indicator of the injection of ozone and upper atmospheric aerosols into lower atmosphere. Due to its origin the pattern of seasonal variations in ^{7}Be concentrations should be relatively stable on global scale [26].

^{210}Pb is an effective tracer of continental surface air masses history and is used to identify soil aerosols sources. It originates from the decay of primordial ^{238}U from the Earth crust into noble gas ^{222}Rn (half life 3.8 days) that escapes from soils into the atmosphere where one of its short lived daughters (^{218}Po) attaches to aerosols and decays into long lived ^{210}Pb (half life 22.3 years) [22,24,27]. There are also anthropogenic sources of radon as coal combustion and nuclear explosions but they contribute less than 1% to total radon and ^{210}Pb in air [88]. ^{210}Pb variations in concentrations could be used to trace continental emission, to estimate mean atmospheric residence time of aerosols and to investigate aerosols transport and removal processes [49, 84]. Deposition of ^{210}Pb varies with season and geographic position, and therefore data on ^{210}Pb concentrations are used in sedimentological, biogeochemical and geochronological studies [20,49]. The global distribution of ^{210}Pb is dominated by tropospheric distribution with occasional ingestion from stratosphere. Higher concentrations of ^{210}Pb are found in surface air at midlatitudes but its concentrations are rapidly decreasing towards higher latitudes. Due to its origin ^{210}Pb concentrations exhibit maxima during autumn that may be attributed to the increased emanation of ^{222}Rn in seasons when soils is less saturated with water [30]. The emanation of radon and therefore ^{210}Pb concentrations in air are affected by many factors, such as atmospheric pressure, temperature inversions, covering vegetation, snow and ice ground coverage, etc. [30]. Although a world database of global atmospheric distribution of ^{210}Pb has been established, there is still not sufficient data for Eastern Europe and Central Asia [38,60,64,80,83].

As ^{7}Be and ^{210}Pb are of different origin, their abundance in air on different heights in the atmosphere differs, too. Air masses transported from upper troposphere and lower stratosphere have higher concentrations of ^{7}Be while those in closer contact with ground are enriched with ^{210}Pb. ^{7}Be/^{210}Pb activity ratio depend on the altitude from which the air is transported, on continental emission and on the removal processes and it is a useful tool in studying air convection processes. The ratio exhibits seasonal pattern with summer maxima and winter minimums as both radionuclides have a similar seasonal trends, with higher activities in warm seasons due to enhanced vertical convection [2]. As ^{7}Be/^{210}Pb activity ratio in aerosols is mainly determined by air masses transport history, their combined activity defined by radionuclide loading index is a function of wet removal, particularly of precipitation scavenging. The particle size functions of both radionuclides are similar, and therefore on global scale their mean radioactive loading index defining their relative abundance should be stable (close to zero) [2].

Due to its long half life (30 years) and abundance in the process of fission ^{137}Cs is the most significant among fission products and a reliable indicator of anthropogenic radioactive pollution in the environment (nuclear weapon atmospheric tests, nuclear power plant accidents) since the II World War. Seasonal variations of ^{137}Cs concentrations in ground level air indicate the migrations from stratosphere to troposphere and could be used to detect the changes in the global climate. After the moratorium on the atmospheric nuclear weapon tests in the 70ties and the decrease in concentrations of Chernobyl cesium during the late eighties and afterwards as a result of migration of the Chernobyl deposited cesium into deeper layers of soils and washout effects, concentrations of ^{137}Cs in ground level air in the 90ties were of the order of $\mu Bq/m^3$, with one or two maxima in summers and winters [6,38]. Some authors attributed cesium winter maxima to the inversion weather conditions [5], but most concluded that in the 90ties the main source of ^{137}Cs and the main cause of its winter maxima was the soil dust resuspension in air from the Chernobyl fallout in 1986, while stratospheric contribution was much less significant [79]. That was confirmed by the trend of its summer max/min ratio indicating the slow decrease of Chernobyl deposited ^{137}Cs [5,29,79].

Air radioactivity monitoring program in Serbia started some forthy years ago, and it was focused generally on the measurements of ^{137}Cs concentrations in ground level air, while continuous measurements of ^{7}Be and ^{210}Pb concentrations in air started only in the last decade of the 20th century [52,54,56, 75,76,77,78,79,80]. The radioactivity control in the Institute of Nuclear Sciences »Vinca« (referred to as Institute Vinca in the text) in the vicinity of Belgrade started in 1959 within the Radioactivity Protection Department [75]. There were 4 meteorological stations where the radioactivity monitoring was performed, two within the

Intitute and two in the intermediate vicinity, in the direction of the dominant wind from the Institute towards the city of Belgrade, on the distances of 2 km and 7 km. The program comprised mesurements of radionuclide concentrations in aerosols, continuous background measurements, total beta activity measurements in aerosols and wet and dry deposition mesurements. Activity of plutonium in air was determined in 1965-69, and once again in 1986, from May 1st to May 15th [75].

The average monthly concentrations of ^{7}Be measured in air from 1991 to 1996 at Institute Vinca were in the range of 2.0-7.0 mBq/m^{3} with pronounced one/two maxima in summer or early autumn and a minimum in winter (maximum average quaterly seasonal index was calculated to 1.27 in summers). ^{137}Cs concentrations in air measured in the same period at the same site were in the range of 0.5 - 8.5 x 10^{-5} Bq/m^{3} with a spread maximum in spring-summer and another pronounced maximum in winter that was contributed to local resuspension effects (the average maximum seasonal quaterly index for ^{137}Cs was 0.9 in summers and 1.40 in winters). There was no correlation of ^{137}Cs concentrations in air with precipitation, while ^{7}Be concentrations were more closely related to the washout effect of rainfalls. ^{210}Pb concentrations in air in the same period were in the range of 2.5 – 34.0x10^{-4} Bq/m^{3}, with maxima during early/late autumn seasons [79,80,81,82,83].

Air Monitoring Program in Urban Areas in Serbia

Urban areas are specific eco-systems concerning pollution, due to high population density, high volume of vehicular traffic and high concentration of industrial plants located in the outskirts, in so called industrial zones, but often even located in a broader city centre, and therefore investigations of air pollution in city centers provide significant data for immediate and long-term health effects.

In the last decade, the radioactivity monitoring of the ground level air in urban areas become a part of the monitoring pollution program in the cities all over Europe. In Serbia, despite systematic radioactivity control/monitoring of the ground level air for more than four decades, there is yet no relevant database on the content of radionuclides in ground level air within the residential areas in the very centre of the city of Belgrade, as well as in the other major cities in the country.

During the nuclear plant accident in Chernobyl in 1986, the exposure dose rate was measured in the central Belgrade city area (Faculty of Veterinary Medicine). The maximum value of 5.5 pC/kgs was measured at May 1st, 1986. Shortly afterwards, on May 16th, the rate decreased to 3.5 pC/kgs, and obtained steady level of under 2 pC/kgs at the end of the year [59,78].

Besides ^{137}Cs, there are more data on radon concentrations in air in Belgrade dwellings and kindergartens measured on the grab air samples by alpha scintillation techniques (Lucas scintillation cell ZnS/Ag) in the early nineties [12,90]. Average instantaneous radon concentrations in Belgrade air were 59 Bq/m^{3} and were only half of the value measured in Slovenia (Ljubljana and Maribor, 177 Bq/m^{3} and 107 Bq/m^{3} respectively), but twice as high as those measured in Sarajevo (27 Bq/m^{3}). The differences in radon indoor concentrations in different regions of former Yugoslavia were attributed to different building materials and construction of the buildings. The mean effective dose equivalent of radon and its progenies in closed space in Belgrade dwellings was estimated to 0.074 mSv for radon and 1.22 mSv for its short lived daughters, in total to 1.3 mSv [51,55,58]. The values were lower than those estimated for other major cities in former Yugoslavia and for some of the cities in Serbia, too [40,41]. After 2000, there was also a number of studies dealing with the effects of DU ammunition on the urban environment [13,53,57].

Suspended Particulate Matter in Urban Air

Suspended atmospheric particulate matter, PM_{10} and $PM_{2.5}$ (aerodynamic diameter less than 10 μm and 2.5μm, respectively) are today a key issue in contemporary air pollution research. Their effect on public health and their activity in heterogeneous chemistry have been well recognized. In recent years, a number of epidemiological studies [15,67,89,95] have demonstrated that acute and chronic health effects are related to the inhalable PM_{10} exposure in the urban environment. Many studies have confirmed the hypothesis that apart the size, the chemical composition of the particles and their capacity to carry potentially toxic substances (such as organic substances or metallic compounds) adsorbed on their surfaces have a crucial role [85]. Some epidemiological data also seem to indicate possible seasonal effects of the particulate matter on human health. Such data are essential for the understanding of particle formation, transport, transformation and deposition mechanisms as well as the impact of particles inhaled by a respiratory system. This is especially important for urban aerosols, whose variety of size and composition make complete characterization a difficult task [72,73].

Therefore, it is important to study the concentration, physico-chemical characteristics and sources of atmospheric aerosol particles, as well as their spatial and temporal variation, especially in cities where large populations are exposed to elevated concentrations. European Commission (EC) has included PM_{10} monitoring and limit values in the Air Quality Directive, in 1999 [18]. The Directive establishes an annual limit for average value of 20 $\mu g/m^3$, and a daily limit value of 50 $\mu g/m^3$ (not to be exceeded more than 7 times per calendar year) to be met by 2010. Within the European programme for monitoring and evaluation of the long-range transmission of air pollutants (EMEP), measurements of PM_{10} and trace metals, as highly toxic species have been introduced [93]. The current focus of health-related sampling of particulate matter is on particles with aerodynamic diameter less than 10 μm (PM_{10}) but recent research pointed out the great health effect of fine particles $PM_{2.5}$, and even PM_1.

Urban anthropogenic particles are mainly in the $PM_{2.5}$ range (particulate matter, with aerodynamic diameter less than 2.5 μm) and its sampling diminishes the interference of natural sources and reduces the loss of potentially volatile components such as ammonia and chloride. As pollution-derived elements (such as trace metals) are often concentrated on fine particles, they could remain in the air with relatively long residence time and could efficiently penetrate human lungs. Recently, EC also established average $PM_{2.5}$ annual limit of 20 $\mu g/m^3$ [21]. In our country, limit values are still related to total suspended particles and for urban areas daily limit amounts 120 $\mu g/m^3$ and mean annual value is 70 $\mu g/m^3$ [9].

Although particulate matter (PM), above all particles with aerodynamic diameters less than 10 μm (PM_{10}) or less than 2.5 μm ($PM_{2.5}$), is of great concern for public health, no systematic studies have been performed in Belgrade until recently. The studies on the quality of air in urban atmosphere related to suspended particles PM_{10} and $PM_{2.5}$, and first measurements of their mass concentrations have been initiated in our country in 2002, and are still in progress. The results of preliminary investigations revealed the need for the continuous and long-term systematically sampling, measurements and analysis of interaction of the specific pollutants – PM_{10} and $PM_{2.5}$ and trace metals in the ground level [61,62,63,69,70,71,72,73]. In this chapter the results of the long-term project of the pollution monitoring in urban areas concerning suspended particle mass and trace metal concentrations in ambient air of the urban areas in Belgrade will be presented.

MATERIALS AND METHODS

Sampling Sites

Belgrade (44°47'N, 20°32'E; 205 m a.s.l), in West-Central Serbia on the junction of two rivers - Sava and Danube, is the capital of the Republic of Serbia, with about 2 millions inhabitants (Figure 1).

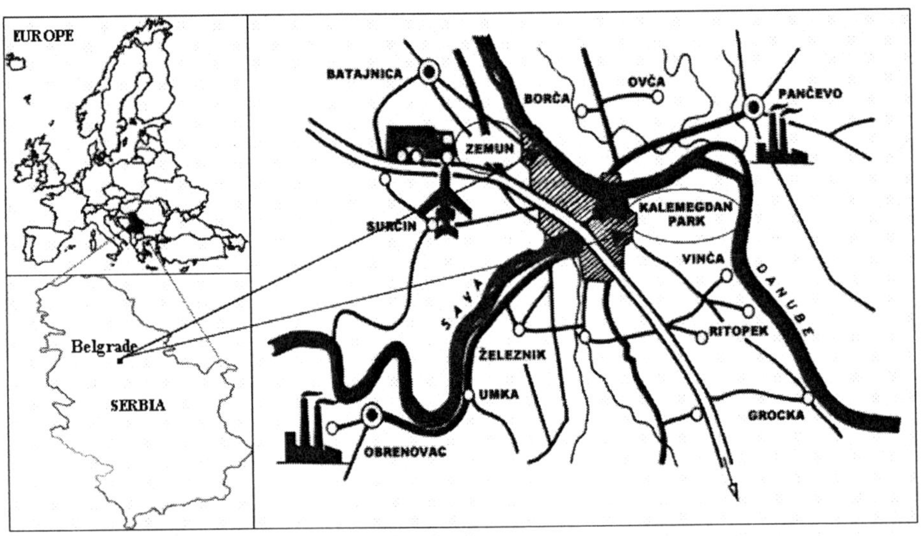

Figure 1. City of Belgrade and the Surrounding Area.

The climate of Belgrade is moderate continental with fairly cold winters and warm summers. The prevailing wind is N-NW, but a characteristic wind "Košava" (SE-ESE) blows with an annual frequency of 26% and an average speed of 4 m/s. In the winter, severe air pollution in the form of aerosol smog occurs frequently in the city center, particularly during meteorologically calm (wind speed < 2 m/s) and stable conditions. Passenger cars and city vehicles are often more than 10 years old, and leaded gasoline (0.4 g/l Pb) is still used in the country. This could be the significant source of ambient PM_{10} as diesel vehicles emit 10-20 times more particulate matter than gasoline vehicles [65]. There are 18 bigger heating plants within the city with a total capacity of 2018 MW, run with natural gas or crude oil and 59 smaller plants run only with crude oil (approximately 193 MW). Fuel used for domestic heating consists mainly of coal or crude oil.

The map of the central Belgrade city area with the three sampling sites (city parks) is presented in Figure 2. and the map of Institute Vinca is presented in Figure 3.

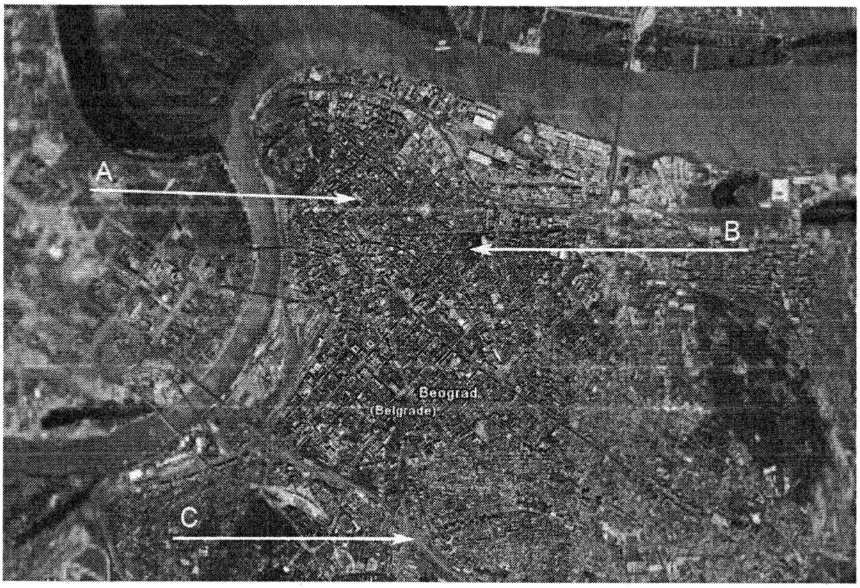

Figure 2. Sampling sites: A) Student Square, B) Botanic Garden, C) Karadjordje Park/Autokomanda (KP).

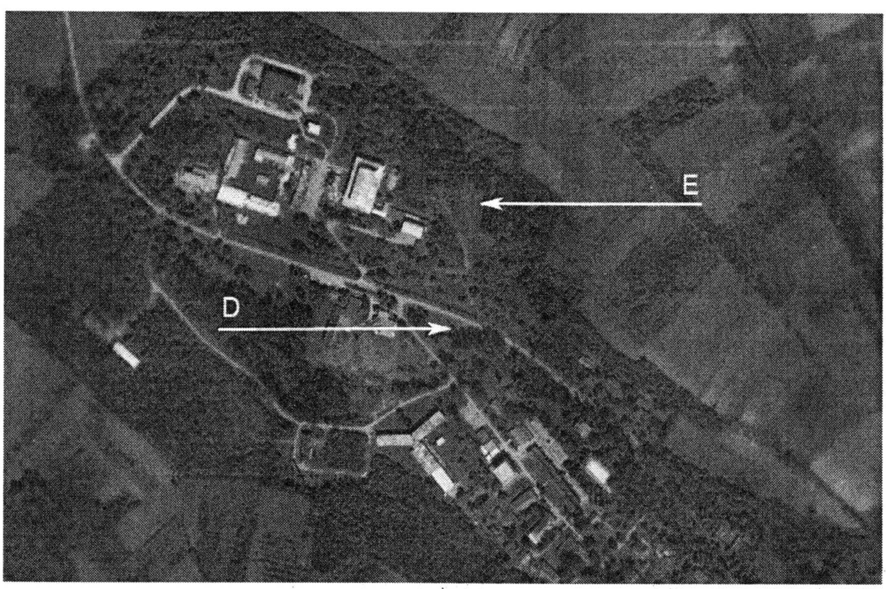

Figure 3. Sampling sites: E) Meteorological station M.Stub D) Institute.

Aerosol samples for radionuclides study were collected daily at two monitoring stations within the Institute of Nuclear Sciences Vinca: the M.Stub Meteorological Station and

Institute Meteorological Station and on the three representative "black spots" in the residential and business center in the Belgrade city area (Figure 2 and Figure 3).

Sampling of particulate matter PM_{10} and $PM_{2.5}$ started on three sites in the very urban area of Belgrade during the period of June-July, September-October and November-December 2002 and continued afterwards. The first sampling episode was carried out on the roof of the Rector's Office building of Belgrade University in Student Square (A), at a height of about 20 m, near a small city-park. The square has high traffic density and a bus terminal. As this sampling point is in the very city center, on the rooftop and the airflow is not blocked by any direction, it can be considered as representative for urban-background concentrations. The second sampling episode was run at about 6 m height in the Botanic Garden (B) about 50 m far from heavy-traffic streets. The third sampling site was the platform above the entrance stairs to the Faculty of Veterinary Medicine (C, referred in text as KP site) at a height of about 4 m from the ground, 5 m away from a street with heavy traffic and close to the big traffic junction (Autokomanda) with the main state highway (Figure 2). The traffic is controlled by street lights. This sampling point can be considered as traffic-exposed.

Radionuclides in Air Measurements

Aerosols samples were collected on filter papers (FILTRAK/Whatman 41/DDR, 15 cm diameter, relative efficiency for deposited dust 80%) by constant flow rate samplers (average air flow 20 m^3/h, average daily volume 600 m^3), ashed at temperatures below 400°C and a monthly composite sample containing 30-31 daily filters was formed (average volume $15x10^3$ m^3). The samples were measured in small metalic containers (inner diameter 25 mm, outer diameter 27 mm, height 4 mm; average air volume in a container $15x10^3$ m^3).

The activity of the radionuclides was determined on 3 HPGe detectors (Canberra and Ortec relative efficiency 23%, 20%, 18% respectively, resolution 1.89 keV at 1332 keV) by standard gamma spectrometry. The detectors (vertical, coaxial) were placed in a shielding cage (45x45x45 cm) of Pb bricks (width 10 cm) with layers of Cu (3 mm) and Fe (10 mm) on the inner side for first detector and 6 mm for the second detector. The third detector with the commercial shielding is a reverse-electrode Ge, with a thin Be window. Detectors are all placed in the middle of the shielding cage at the distance of about 17 cm from the cage walls [74].

Energy calibration was performed with a set of standard point sources (Coffret d'etalon gamma ECGS-2, Sacle, France, containing ^{133}Ba, $^{57,60}Co$ and ^{137}Cs: activities range 10^3-10^4 Bq, 25.11.1987). Geometric efficiency was determined with IAEA-083 (AIR4) stimulated air filter (spiked with solution of ^{60}Co: 2160 Bq/filter, ^{133}Ba: 846 Bq/filter, ^{137}Cs: 1182 Bq/filter and ^{210}Pb: 151 Bq/filter, uncertainty 5%, 1.1.86). A reference radioactive material within the SEV intercomparison project was used, too (aerosol powder matrix, 11 g in plastic box, with ^{137}Cs: 207 Bq, ^{54}Mn: 14 Bq, ^{65}Zn: 16 Bq, ^{57}Co: 5 Bq, ^{40}K: 7 Bq: 09.03.1988) [94].

Results are calculated for the middle of the sampling period and presented as "monthly means ± mean standard error" in Bq/m^3. Total standard error of the method was under 20% for ^7Be and ^{137}Cs, and up to 27% for ^{210}Pb.

Radionuclides were determined at gamma energies for ^{137}Cs: 661.6 keV, ^7Be: 477 keV and ^{210}Pb: 46 keV. Minimum detectable concentrations, were derived from the lower limit of detection as $LLD = k^2 \pm 2LC$, where k is the coefficient of normal distribution corresponding to the confidence level of 95%, while LC is the critical level depending on background photopeak counts. Minimum detectable concentrations (MDC) of the radionuclides in air were: 1.0 μBq/m^3 for ^{137}Cs, 10 μBq/m^3 for ^{7}Be and 20 μBq/m^3 for ^{210}Pb.

Counting time intervals were from 150,000 – 250,000 s. Background spectrum was recorded regularly after or before the sample counting and its integral mean count was 1.7 imp/s. Data were statistically analyzed on IBM/PS2 by SPECTRAN-AT and Genie 2000 program. Total standard error of the method (including relative errors in geometric efficiency estimation, photopeak counts estimation, sample volume determination, etc.) was estimated to be below 25%.

Precipitation data for the period were obtained from the Department of Meteorology, Institute of Nuclear Sciences Vinca [14].

Sampling and Analysis of Suspended Particles

Suspended particulate matter PM_{10} and $PM_{2.5}$ were collected on preconditioned (48 h at 20°C and constant relative humidity around 50%.) and pre-weighed Pure Teflon filters (Whatman, 47 mm diameter, 2 μm pore size) and Teflon-coated Quartz filters (Whatman, 47 mm diameter) using two MiniVol air samplers (Airmetrics Co. Inc., 5 l/min flow rate) provided with PM_{10} and $PM_{2.5}$ cut-off inlets. The sampling time was 24 h, yielding a sample volume of 7.2 m^3. Routine maintenance of the samplers and calibration of the flow meters were often conducted in order to ensure the sampling quality. After particle collection, the filters were sealed in plastic bags and kept in portable refrigerators, in a horizontal position during transportation back to the laboratory where they were reconditioned for another 48 h. The sampling methodology used in this study was described in detail by Rajsic et al. [61,62].

During the sampling, meteorological parameters including temperature, relative humidity, rainfall, wind direction and speed were provided at the Meteorological Station of the Hydro-Meteorological Institute of the Republic of Serbia (Hs = 132 m, φ = 44^0 48' N and λ= 20^0 28` E), located inside central urban area, very close (≈200 m) to the sampling site KP (Karadjordje park).

Analytical Methods

The samples were handled and processed in a Class 100 clean laboratory, at the Institute of Physics, Belgrade. All gravimetric determinations were made using a semi micro balance (Sartorius, R 160P), with a minimum resolution of 0.01 mg. Loaded and unloaded filters (stored in Petri dishes) were weighed after 48 hours conditioning in a desiccator, in the clean room at the relative humidity of 45-55% and the temperature of 20 ±2 °C. Quality assurance was provided by simultaneous measurements of a set of three ''weigh blank'' filters that were interspersed within the pre- and post-weighing sessions of each set of sample filters and the mean change in "weigh blank" filter mass between weighing sessions was used to correct the sample filter mass changes.

After completion of gravimetric analysis, PM samples were digested in 0.1 mol/dm^3 HNO_3 on an ultrasonic bath and analyzed for a set of trace metals by graphite furnace atomic absorption spectrometry (GFAAS) using the transversely-heated graphite atomizer (THGA; Perkin Elmer AA 600) with Zeeman-effect background correction. The THGA provided a uniform temperature distribution over the entire tube length, rapid heating and an integrated L'vov platform, which gave an improved signal/interference ratio and high analytical sensitivity. Analyte injection (20 μl) and the atomization were done in five steps controlled by the appropriate software and auto-sampler. For calibration, standard solutions containing all metals of interest were prepared using Merck certified atomic absorption stock standard solutions containing 1000 mg/dm^3 metal in HNO_3 (0.5 mol/dm^3) and Milli-Q quality deionized water, with no matrix modifier addition. Detection limits for the trace elements were found to be: 0.08 ng/m^3 for Cd, 0.2 ng/m^3 for Cr, 0.4 ng/m^3 for Cu, 1 ng/m^3 for Pb, 0.4 ng/m^3 for Zn, 0.8 ng/m^3 for Ni, 0.4 ng/m^3 for Mn, 10 ng/m^3 for V, 0.4 ng/m^3 for Fe and 4 ng/m^3 for Al. An extraction procedure with dilute acid was used for the evaluation of elements which can become labile depending on the acidity of the environment. This procedure gives valid information on the extractability of elements, since the soluble components in an aerosol are normally dissolved by contact with water or acidic solution in the actual environment [37].

RESULTS AND DISCUSSION

Concentration of ^{137}Cs, ^{7}Be and ^{210}Pb in Ground Level Air in Belgrade: Institute Vinca

Concentration of ^{137}Cs in Ground Level Air at Institute Vinca 1985-2005

During nuclear plant accident in Chernobyl in 1986, the activity of released ^{137}Cs was estimated to $3.7x10^{16}$ Bq, lower than in the nuclear weapon tests during the fifties and early sixties in the 20th century: $1.5x10^{18}$ Bq, but much higher than in other nuclear plant accidents ever: in the Windscale accident the released activity of ^{137}Cs was estimated to $4.4x10^{13}$ Bq, in the Three Miles Island accident it was under lower limit of detection [48,49]. However the scenario of the radioactivity release was rather different. In the nuclear plant accident in Chernobyl the radionuclides were injected into the troposphere from a single point, contrary to the nuclear weapon tests when the radionuclides were injected in the lower layers of stratosphere and transported into troposphere after a certain amount of time. In the Chernobyl accident it was estimated that 20-40% of ^{137}Cs from the reactor core was released during the accident itself, while the activity of ^{137}Cs released in 1986 was estimated to 74-85 PBq, and the leftover in 1996 was estimated to 68 PBq. Due to the large scale of the accident and the uneven weather conditions over former Soviet Union and Europe, the deposition of the fallout was highly uneven, too. Based on the average deposition of 70 kBq/m^2 the estimated effective dose in the North Hemisphere was 140 mSv and the collective equivalent dose was 620.000 man Sv [12,48,49].

In Serbia, the concentrations of ^{137}Cs in air before 1985 were of the order of 10^{-5} Bq/m^3 and rather constant since 1980. After the sharp increase during the Chernobyl nuclear plant accident, from May 1st to May 15th, the consequences of the accident in Serbia were apparent through 1986, 1987 and 1988, with the steady decrement of ^{137}Cs concentrations in the environment. Since 1989, the ^{137}Cs concentration in air was established to a level before the accident with seasonal variations due to migrations of air masses from stratosphere to troposphere and local meteorological and environmental conditions (precipitation rate, soil dust resuspension, etc) [10,11,81].

The highest ^{137}Cs concentrations in air and the integrated ^{137}Cs air activity measured in the Institute Vinca in 1986, 1987 and 1988, and the data for other parts of former Yugoslavia and Europe are presented in Tables 1. and 2. [54,59,77]. Values of highest concentrations of ^{137}Cs in air measured in former Yugoslavia in Table 1. are presented as the country averaged value.

Table 1. Highest concentrations of ^{137}Cs measured in air (Bq/m^3) in Belgrade (Institute Vinca) and former Yugoslavia after the Chernobyl accident

Site/Year	1986	1987	1988
Belgrade	$3.9x10^{-1}$	$15x10^{-4}$	$9.4x10^{-5}$
Former Yugoslavia	$8.8x10^{-1}$	$5.5x10^{-4}$	$8.0x10^{-5}$

Table 2. Integrated ^{137}Cs activity measured in air (Bq/m^3) in 1986 in Belgrade (Institute Vinca), former Yugoslavia and Europe

Belgrade	Former Yugoslavia	Europe
0.4	6.65	0.04 – 14.0

As can be seen from the data presented in Table 1. the highest concentrations of ^{137}Cs measured in air in the city of Belgrade in 1986 and 1987 were less than half of the values measured in other parts of former Yugoslavia. Higher concentrations measured in Belgrade air in 1988 were attributed to local conditions – soil dust resuspension from the fields in the immediate vicinity of the sampling site (Institute Vinca).) An integrated ^{137}Cs activity in Belgrade air was significantly lower than in other parts of former Yugoslavia, and within the lower part of values measured in Europe in 1986, mainly due to local weather conditions during the accident (the weather in Belgrade in May 1986 was mainly dry with heavy showers only at May the 1st and May 2nd).

Average monthly ^{137}Cs activities in air measured in Belgrade (Institute Vinca) from 1985 to 2005 are presented in Figure 4. The average monthly concentrations of ^{137}Cs in air in 1986 are presented in Figure 5.

After the moratorium on nuclear weapon tests in the atmosphere in 1964 and China accepting of the moratorium in the late 1970ties, the concentrations of ^{137}Cs injected into the atmosphere by nuclear weapons tests were low (10^{-6} Bq/m^3) [76]. In 1986 there was a sudden increase in concentration due to nuclear plant accident in Chernobyl on April 26th . Following the accident during May 1986, the concentrations of ^{137}Cs measured in air at Institute Vinca were increased up to 10^{-1} Bq/m^3, which is 10^5 times higher compared to values measured in 1985. During 1987 and afterwards, the concentrations of ^{137}Cs in air were decreasing slowly but constantly. Seasonal increases of ^{137}Cs concentrations in air in January 1987 and afterwards were attributed to local weather conditions and high resuspension rate of radioactive materials from soils in the neighboring fields [81]. The trend of decrease continued and in 1998 the concentration of ^{137}Cs in ground level air fell to the level before the accident. In the period from 1991 to 1996 the ^{137}Cs monthly concentrations were in the range of 0.5-8.5x10^{-5} Bq/m^3 with a spread maximum in late spring/early summer and another pronounced maximum in winter. Quarterly seasonal indices were in the range 0.8-1.4 with a maximum in winter [79]. Periodical increase events after 1998 were probably due to burning or radioactive waste (Spain, 1998) or periodical leakage from the nuclear plant installations in the region (Kozloduj, Bulgaria).

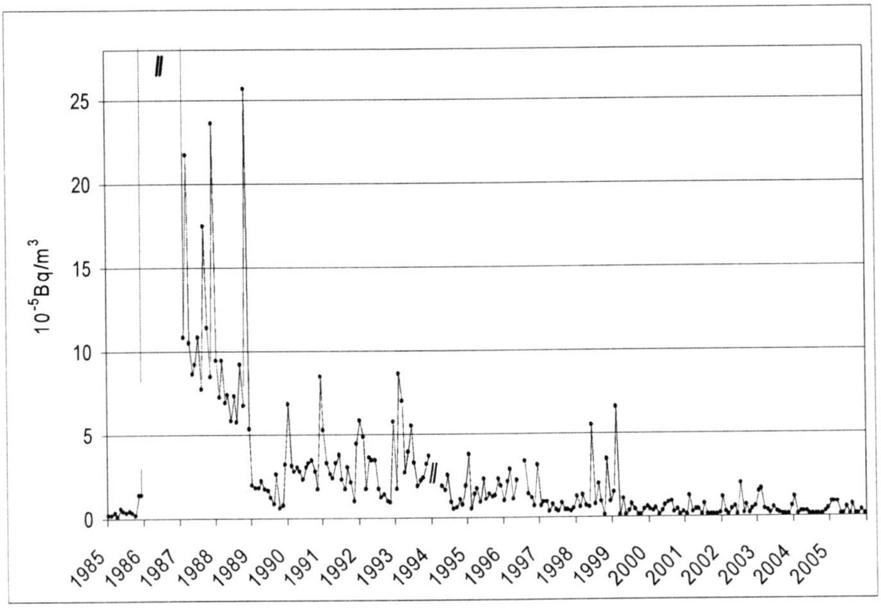

Figure 4. Average monthly concentrations of ^{137}Cs in air (Institute Vinca, 1985 -2005).

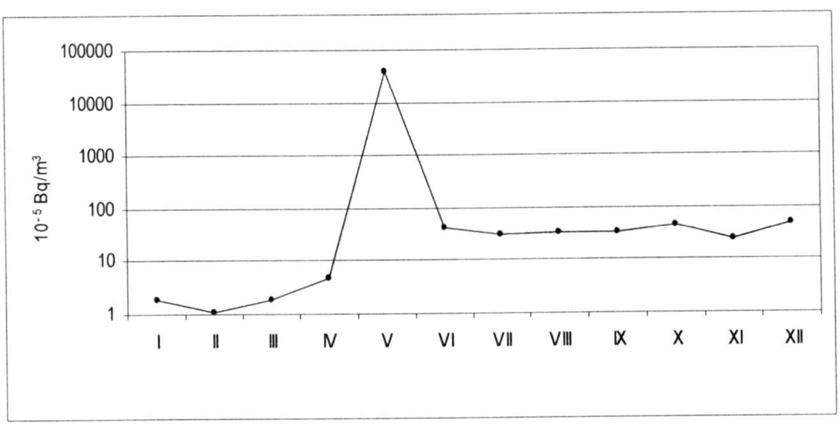

Figure 5. Average monthly concentrations of ^{137}Cs in air in Institute Vinca in 1986 (months I-XII).

Concentration of ^{210}Pb in Ground Level Air at Institute Vinca 1985-2005

Activity of ^{210}Pb in ground level air at Institute Vinca has been measured continuously since 1985 [80,81]. Average monthly concentrations of ^{210}Pb in air in the Institute Vinca from 1985-2005 are presented in Figure 6.

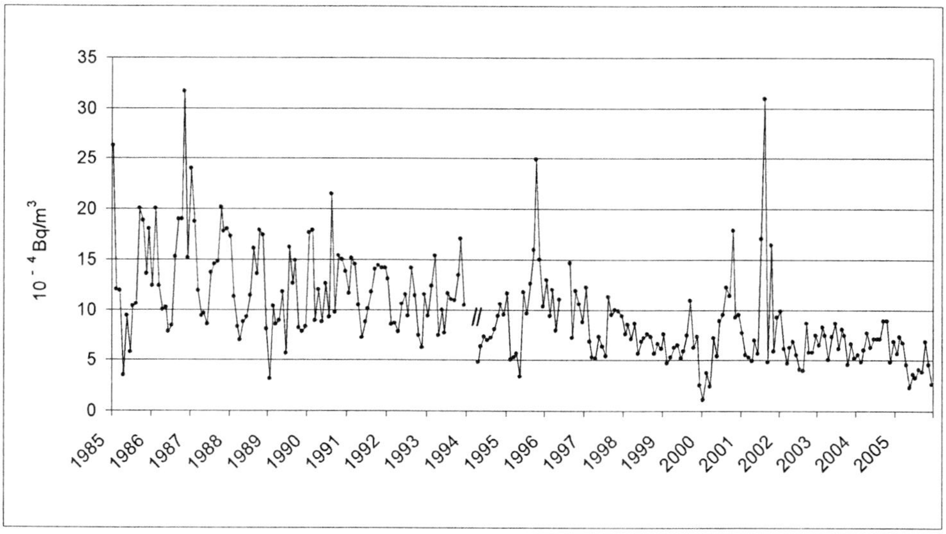

Figure 6. Average monthly concentrations of ^{210}Pb in air (Institute Vinca, 1985-2005).

Concentrations of natural ^{210}Pb measured in ground level air in the city of Belgrade (Institute Vinca) since 1985 were of the order of 10^{-4} Bq/m^3. This is within the range of the values measured at midlatitudes and reported by other authors elsewhere [2,16,23,24,27,31,34]. Average yearly concentrations of ^{210}Pb in ground level air in Europe are in the range of 2–7x10^{-4} Bq/m^3, close to the values measured on global level [22]. Higher values of ^{210}Pb measured in Belgrade air in some periods are probably due to the anthropogenic sources (heavy traffic run by leaded gasoline and city heating plants run by crude oil and/or coal). Seasonal variations of ^{210}Pb in air exhibited maxima in autumns and minimums in springs. The maxima are due to increased radon emanation from soils and inversely correlated with the amount of precipitation [83].

Concentration of ^{7}Be in Ground Level Air at Institute Vinca 1985-2005

Continuous monitoring of ^{7}Be activity in ground level air in Institute Vinca started in 1991. Average monthly concentrations of ^{7}Be in air from 1991 to 2005 are presented in Figure 7.

The values of ^{7}Be concentrations in ground level air measured at Institute Vinca from 1991 to 2005 were within the range of the values measured in Europe and elsewhere [17,23,27,31,34,45]. The seasonal pattern was stable with maxima measured in summers during stable pressure conditions and the minimums observed in winters. The maxima were correlated with the increment of temperature, while the minimums were linearly correlated with precipitation [31,35,79,81,83]. Two to three fold increase of ^{7}Be monthly concentrations in air in the non typical years of 2001 and 2003 are probably due to increased stratosphere-to-troposphere exchange.

Concentrations of ^{137}Cs, ^{7}Be and ^{210}Pb in Ground Level Air in Belgrade: the city parks

The pilot study on concentrations of radionuclides in ground level air in Belgrade central city area (three city parks in central city area) was performed in 2002-2003 [82,83]. The differences in concentrations of radionuclides in air were found to be 25 - 45% for ^{7}Be and

^{210}Pb due to different sites, and up to 80% for ^{137}Cs, but the overall seasonal variations pattern was similar to one established in the long term study of radionuclide concentrations in air at Institute Vinca, located outside the city area. The average values of radionuclides concentrations on three sites – Karadjordje Park (KP) Students' Square (SS) and Botanic Garden (BG) are presented in Table 3. The KP sampling site was actually located on the roof of the building of the Faculty of Veterinary Medicine, in the close vicinity of Autokomanda traffic junction.

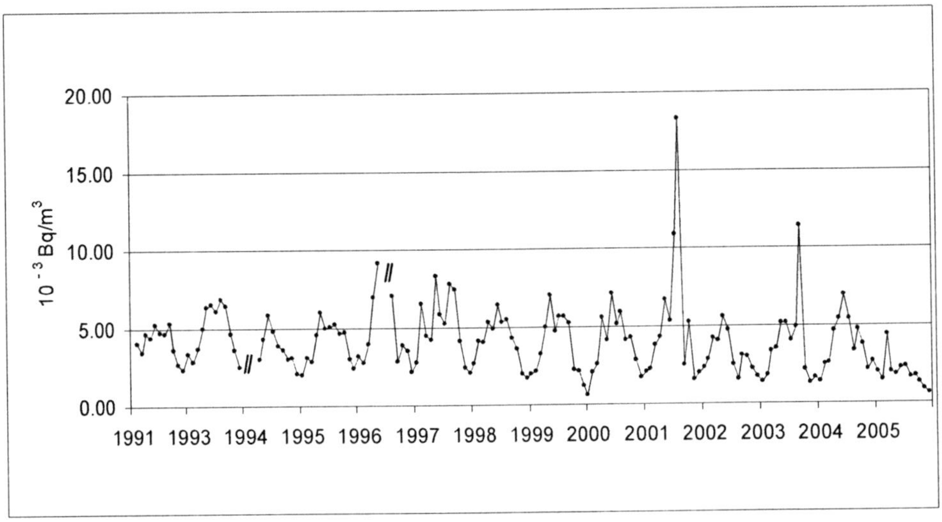

Figure 7. Average monthly concentrations of ^{7}Be in air (Institute Vinca, 1991-2005).

Table 3. Average concentrations of radionuclides in air (Bq/m^3) in Belgrade city parks, 2002-2003

Site/Radionuclide	^{7}Be	^{137}Cs	^{210}Pb
KP	$2.7x10^{-3}$	$0.5x10^{-5}$	$5.0x10^{-4}$
BG	$3.7x10^{-3}$	$3.6x10^{-5}$	$14x10^{-4}$
SS	$3.7x10^{-3}$	$4.2x10^{-5}$	$8.0x10^{-4}$

The average concentrations of radionuclides in air measured in city parks in the frame of the pilot study in 2002-2003 were in the range of values measured in Institute Vinca in the same period, but the differences between the sites were high, especially for ^{137}Cs and ^{210}Pb. The lowest values for both radionuclides were measured at KP site, as the site (the roof of the Veterinary Faculty building) was exposed to good ventilation, while higher values were obtained at sites with dense vegetation (BG) and poor ventilation (SS).

Since 2004, monitoring of air radioactivity continued on the KP site, as it was a location with heavy traffic and one of the most polluted areas within the center of the city. The sampling site itself was moved from the roof of the Faculty building some 50 m away, closer to a highway and was installed about 1 m from the ground.

The average monthly concentrations of ^{7}Be and ^{210}Pb in air measured at the KP site in 2004-2006 are presented in Figure 8. and Figure 9.

The results for ^{137}Cs concentrations in air are not presented, as more than 80% of the values were below the lower limit of detection ($1x10^{-6}$ Bq/m^3).

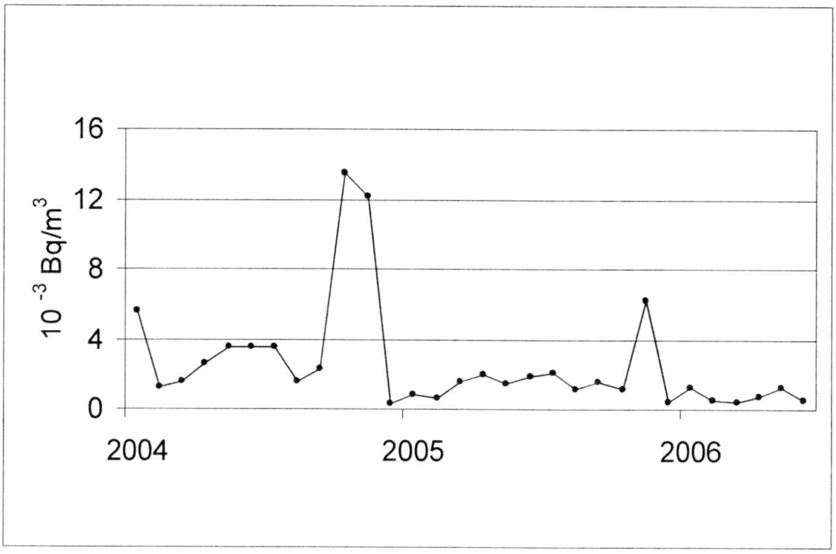

Figure 8. Average monthly concentrations of ^7Be in air (KP, 2004-2006).

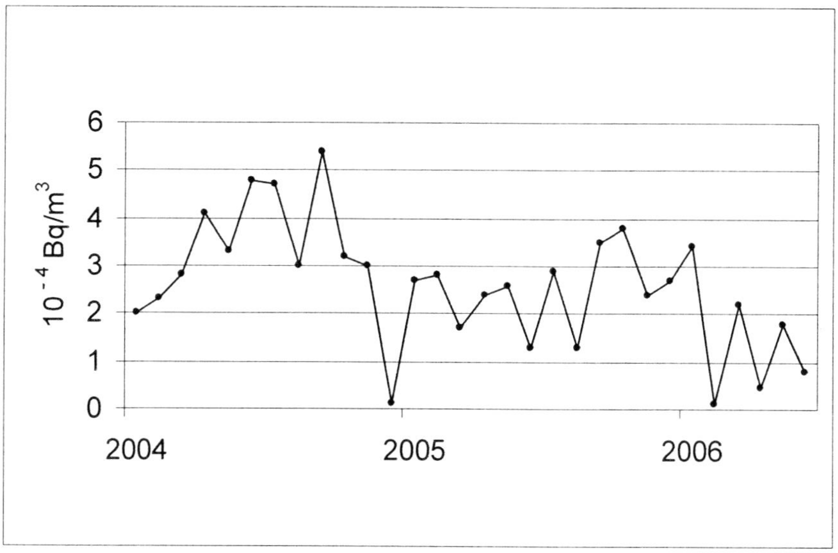

Figure 9. Average monthly concentrations of ^{210}Pb in air (KP, 2004-2006).

The concentrations of ^7Be and ^{210}Pb measured in ground level air in 2004 and 2005 in the center of the city (KP site) were in the range of values measured in the same period at Institute Vinca, but the pattern of seasonal variations was not the same. That could be due to local climate and short period of sampling (untypical years). As presented in Figure 8 and Figure 9 both ^7Be and ^{210}Pb exhibited seasonal variations with maximal values in (late) autumns. The maxima in seasonal pattern of ^7Be are more pronounced, while the variations of

^{210}Pb are unstable, as could be expected taking into the account the different origin of the radionuclides. The monthly concentrations of ^{7}Be are more correlated with the rate of precipitation : in 2004 the precipitation maxima were observed in June/July (above 100 mm), November (130 mm) and January (93 mm), in 2005 also in June/July (90-95 mm) and again in December (about 80 mm) and February (84 mm). Both radionuclides exhibited very low values in winters, the minimum for ^{210}Pb being slightly shifted in winter 2005/2006 compared to one in winter 2004/2005. To compare the data with the ones obtained at Institute Vinca in the same period the seasonal analysis was performed.

Seasonal Analysis of ^{7}Be and ^{210}Pb Concentrations in Ground Level Air in Belgrade in 2004-2006: Sites Institute Vinca and KP City Park

KP City Park

Average monthly concentrations of ^{7}Be and ^{210}Pb in 2004 and 2005 (site: city park KP) are presented in Figure 10 and Figure 11, respectively. (Overall average concentrations for the period are presented as a horizontal line). Monthly seasonal indices and quarterly seasonal indices are presented in Table 4. and Table 5.

Monthly seasonal indices were calculated as mean monthly concentrations divided by the overall average concentration for the monitoring period. Quarterly seasonal indices were calculated as seasonal means of the monthly seasonal indices: for winter (December, January and February), spring (March, April and May), summer (June, July and August) and autumn (September, October and November). Overall average monthly concentration for ^{7}Be in 2004-2005 (KP) was calculated as $3.04\times 10^{-3} Bq/m^3$. The seasonal pattern exhibit a maximum in autumn and a mild increase in summer, as presented in Figure 10. An extremely low value of average monthly concentration in December is probably due to local weather conditions (rate of precipitation and snow coverage).

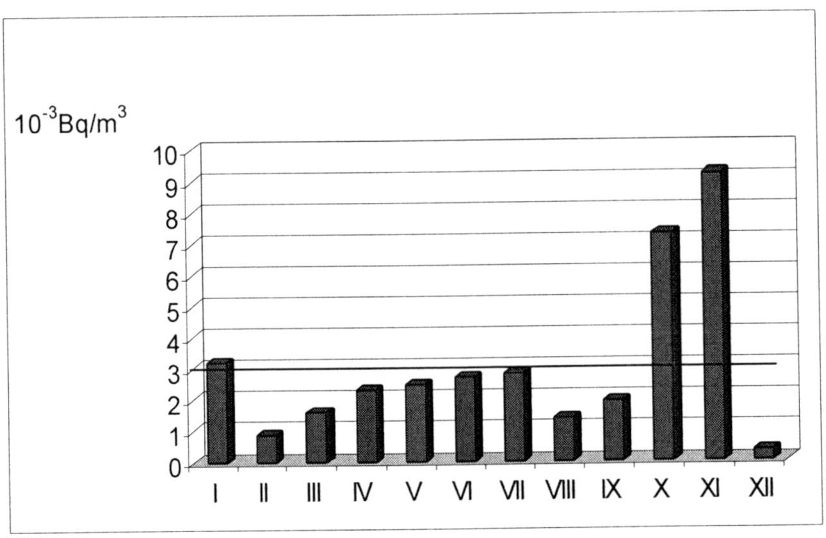

Figure 10. Average monthly concentrations of ^{7}Be in air (KP, 2004-2005) (months I – XII).

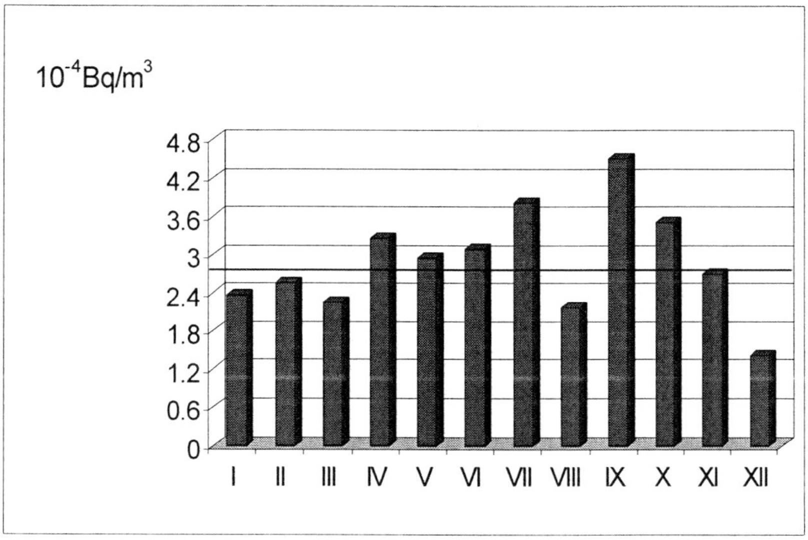

Figure 11. Average monthly concentrations of ^{210}Pb in air (KP, 2004-2005) (months I-XII).

The values of the quarterly seasonal indices for ^{7}Be presented in Table 4. are higher than those reported in the previous studies, especially in autumns [79]. Overall average concentration of ^{210}Pb in ground level air in 2004 and 2005 was 2.87x 10^{-4} Bq/m^3. A mild increase in ^{210}Pb concentrations during autumns could be seen and it is reflected in the values of the quarterly seasonal indices, too. Generally, the pattern of ^{210}Pb seasonal variations is unstable, probably due to local climate and vicinity of the highway. The values of the seasonal and quarterly seasonal indices are smaller than those reported in the previous studies [80].

Table 4. ^{7}Be monthly seasonal indices and quarterly seasonal indices (KP, 2004-2005)

Month	I	II	III	IV	V	VI	VII	VIII	IX	X	XI	XII
Monthly seasonal index	1.05	0.30	0.53	0.76	0.82	0.89	0.92	0.46	0.64	2.4	3.03	0.11
Quarter	Winter			Spring			Summer			Autumn		
Quarterly seasonal indices	0.49			0.70			0.76			2.02		

Table 5. ^{210}Pb monthly seasonal indices and quarterly seasonal indices (KP, 2004-2005)

Month	I	II	III	IV	V	VI	VII	VIII	IX	X	XI	XII
Monthly seasonal index	0.82	0.89	0.78	1.13	1.03	1.07	1.3 2	0.75	1.57	1.22	0.94	0.49
Quarter	Winter			Spring			Summer			Autumn		
Quarterly seasonal indices	0.73			0.98			1.05			1.24		

Max/min ratios of ^{7}Be and ^{210}Pb average monthly concentrations in air for 2004 and 2005 at KP site are presented in Table 6.

Table 6. Max/min ratios for ^{7}Be and ^{210}Pb (KP, 2004-2005)

Radionuclide/Year	2004	2005
^{7}Be	45	16
^{210}Pb	0.02	2.9

The ratios in the two years differ significantly, mainly due to extremely low values of both ^{7}Be and ^{210}Pb concentrations in air during winter 2004 and winter 2005.

Institute Vinca

Average monthly concentrations of ^{7}Be and ^{210}Pb in ground level air at Institute Vinca site in 2004 and 2005 are presented in Figure 12 and Figure 13, respectively. Monthly seasonal indices and quarterly seasonal indices are presented in Table 7. for ^{7}Be and in Table 8. for ^{210}Pb.

Overall average for ^{7}Be for the period was 2.81x 10^{-3} Bq/m^3. The trend of increase of concentrations in summer is more pronounced compared to the values measured at KP site in the same period, but the average monthly concentrations on both sites are of the same order of values. The seasonal pattern differs, too: at KP sites maxima are observed in autumns, while at Institute Vinca the maxima are exhibited in summers, as reported in the previous studies [79,83]. Generally, the values of seasonal indices at the Institute Vinca site are higher than those obtained for KP site and in good agreement with the values obtained before [79].

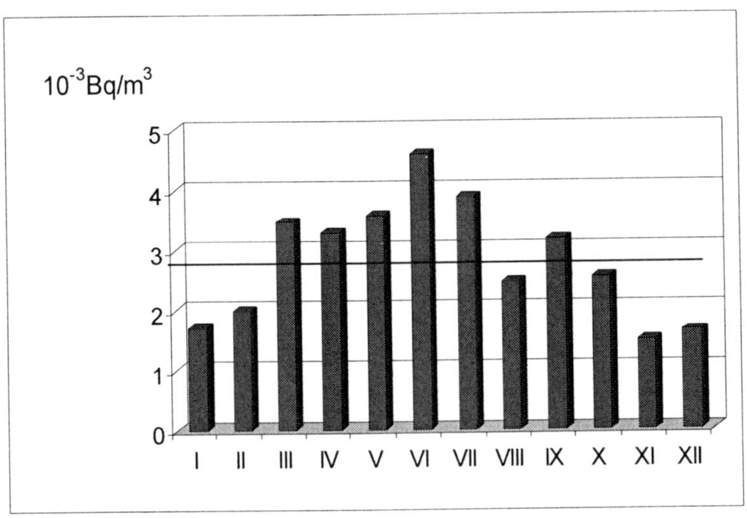

Figure 12. Average monthly concentrations of ^{7}Be in air (Institute Vinca, 2004-2005) (months I-XII).

Overall average concentration of ^{210}Pb in ground level air (horizontal line on Figure 13) for the period is 5.66 x 10^{-4}Bq/m^3, which is significantly higher than the value measured at KP site. Quarterly seasonal indices exhibit the maxima during autumns, but the differences between the indices for different quarters are not significant.

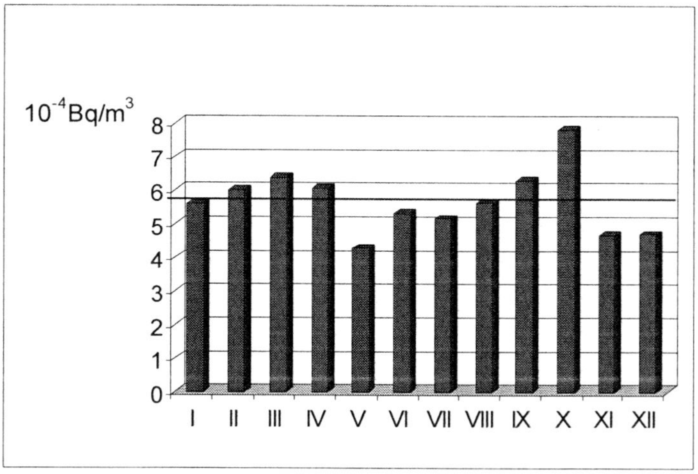

Figure 13. Average monthly concentrations of ^{210}Pb in air (Vinca Institute, 2004-2005) (months I-XII).

Table 7. ^{7}Be monthly seasonal indices and quarterly seasonal indices (Institute Vinca, 2004-2005)

Month	I	II	III	IV	V	VI	VII	VIII	IX	X	XI	XII
Monthly seasonal index	0.60	0.70	1.23	1.17	1.27	1.64	1.37	0.88	1.13	0.91	0.53	0.58
Quarter	Winter			Spring			Summer			Autumn		
Quarterly seasonal indices	0.63			1.22			1.30			0.86		

Table 8. ^{210}Pb monthly seasonal indices and quarterly seasonal indices (Institute Vinca, 2004-2005)

Month	I	II	III	IV	V	VI	VII	VIII	IX	X	XI	XII
Monthly seasonal index	0.98	1.06	1.13	1.07	0.75	0.94	0.91	0.99	1.11	1.38	0.83	0.83
Quarter	Winter			spring			summer			Autumn		
Quarterly seasonal indices	0.96			0.98			0.95			1.11		

Calculated max/min ratios for ^{7}Be and ^{210}Pb concentrations in ground level air measured at Institute Vinca in 2004-2005 are presented in Table 9.

Table 9. Max/min ratios for ^{7}Be and ^{210}Pb (InstituteVinca, 2004-2005)

Radionuclide/Year	2004	2005
^{7}Be	4.97	6.92
^{210}Pb	1.82	3.2

The max/min ratio for ^{7}Be measured at Institute Vinca is more stable than the one at KP site and is closer to the values reported before [79]; the max /min ratio for ^{210}Pb at Institute Vinca is in the same range as the one at KP site for 2005, but the value in 2004 at KP site is out of range.

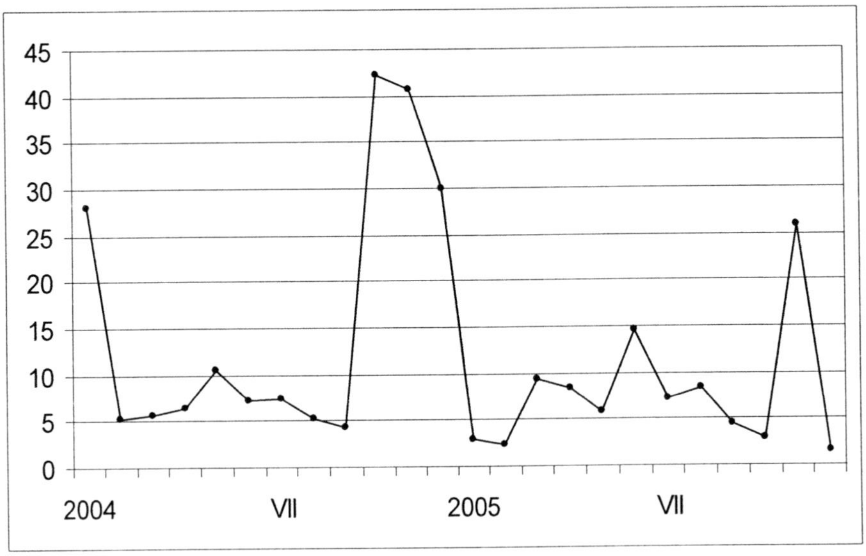

Figure 14. $^{7}Be/^{210}Pb$ monthly concentrations ratio (KP, 2004-2005).

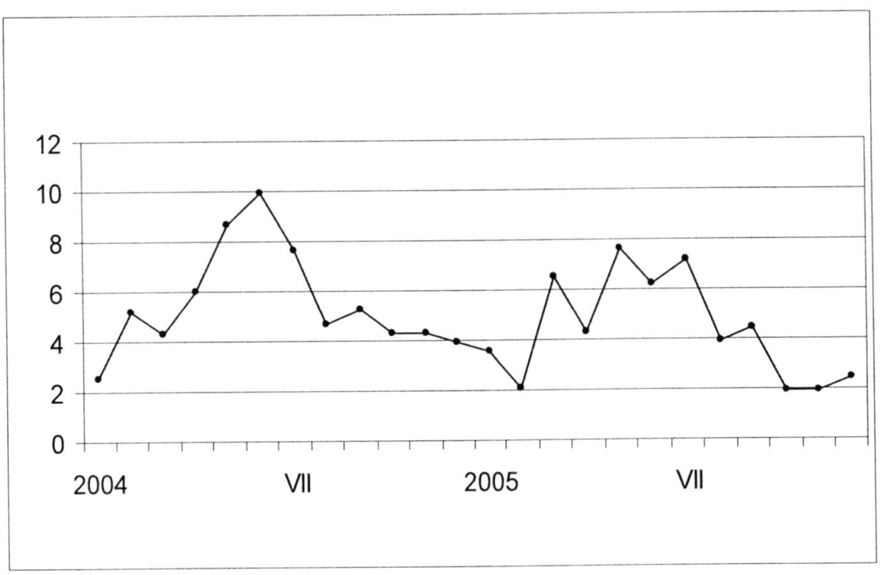

Figure 15. $^{7}Be/^{210}Pb$ monthly concentrations ratio (Institute Vinca, 2004-2005).

$^{7}Be/^{210}Pb$ Concentration Ratio in Ground Level Air in Belgrade and Radionuclide Loading Indices (RLI): KP and Institute Vinca, 2004-2005

The ratios of $^{7}Be/^{210}Pb$ concentrations in ground level air in 2004 and 2005 at two sites: KP and Institute Vinca are presented in Figure 14 and Figure 15, respectively. The values of the radionuclide loading indices (RLI) are presented on Figure 16 and Figure 17. The RLI values used to quantify the combined activity of the two radionuclides were calculated as the arithmetic means of the sum of normalized values of ^{7}Be and ^{210}Pb activities, as defined by Arimoto et al., 1999 [2]. Normalized value was calculated by subtracting the arithmetic mean activity for the period (2004-2005) from the individual activities of the radionuclide and dividing it by standard deviation.

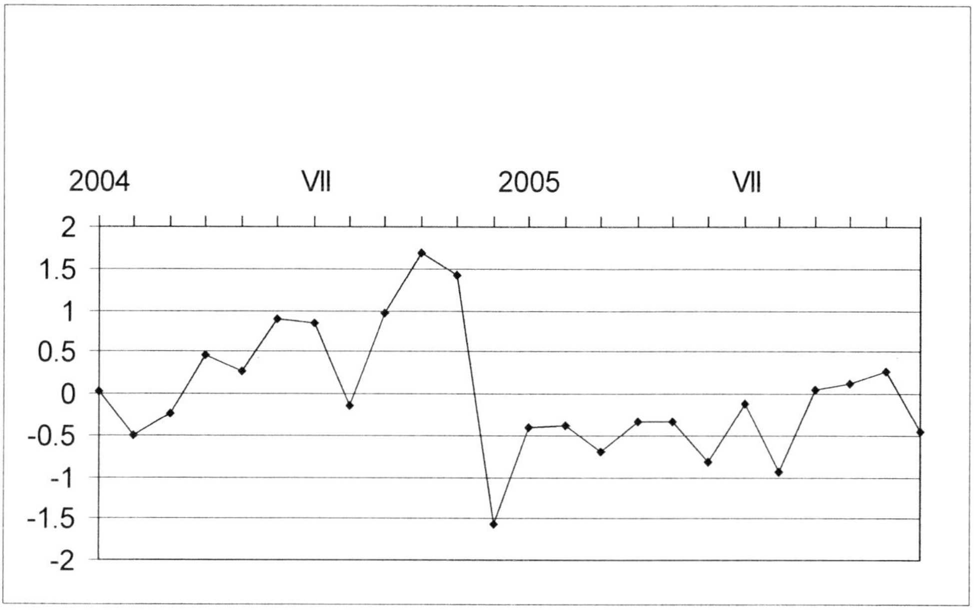

Figure 16. Radionuclides loading indices (KP, 2004-2005).

The variations of the $^{7}Be/^{210}Pb$ monthly concentrations ratio obtained in 2004 and 2005 at KP and Institute Vinca site differ both in values and pattern, mainly due to high concentration of ^{7}Be measured at KP in October and November 2004 and followed by the extremely small concentration measured in December 2004. Although the data cover just 2 year period the influence of the local climate on the radionuclides activity ratio pattern could be clearly seen.

As both radionuclides have similar particle size function [2] their mean RLI values although sensitive to removal processes particularly to precipitation scavenging, should approach zero, as presented in Figure 16 and Figure 17. For further analysis on RLI more data are required.

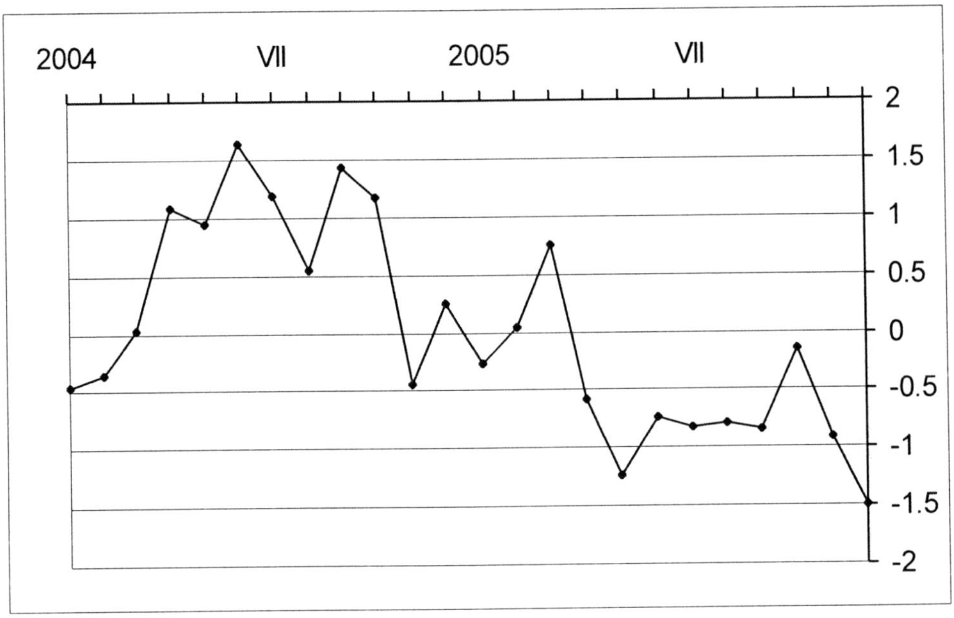

Figure 17. Radionuclides loading indices (Institute Vinca, 2004-2005).

Particulate Matter Measurements in Belgrade Ground Level Air

A first assessment of PM_{10} and $PM_{2.5}$ Particulate Level in the Ambient Air of Belgrade

Daily mass concentrations of 96 PM samples (PM_{10} and $PM_{2.5}$) were determined by gravimetric analysis of filters that were exposed to urban air in Belgrade during the year 2002. Statistical parameters of daily PM mass concentrations (μg/m^3) in Belgrade from June to December 2002 over the total sampling period and particularly over two main seasons: the season without heating - summer period and the winter period - heating season from 15 October to 15 April are presented on Table 10.

Table 10. Statistical parameters of daily PM mass concentrations (μg/m^3) in Belgrade from June to December 2002

	Period	N	X_A	S.D.	Max	Min	Median	98thPercentile
	Summer	22	56	37	164	12	54	158
PM_{10}	Winter	25	96	73	362	21	83	284
	Whole period	47	77	62	362	12	57	214
	Summer	17	35	16	78	3	36	68
$PM_{2.5}$	Winter	32	75	61	333	12	56	227
	Whole period	49	61	53	333	3	44	169

N – Number of samples; X_A – Arithmetic mean; S.D. – Standard arithmetic deviation.

The PM_{10} mean 24-hours mass concentration value, over whole measuring period was 77 $\mu g/m^3$, almost twice as much as the annual limit in European Union (40 $\mu g/m^3$) and more than 60% of days had mean daily concentrations above limit value of 50 $\mu g/m^3$. Average $PM_{2.5}$ mass concentration exceeded the EC annual limit of 20 $\mu g/m^3$ [21] by a factor of 3.

The suspended particles concentration in Belgrade, during the summer, was not as high as we expected taking into account that leaded gasoline was still in wide use in the country, as well as the presence of a number of old buses and trucks on the streets. We presumed, that during the summer, when the traffic density is reduced due to summer vacations, urban background PM level (56 $\mu g/m^3$) for Belgrade was slightly above daily limit value according to EU Directive [18].

On average, a seasonal variation was found, as higher concentrations occurred during the winter period. Mean 24 h mass concentrations were 96 $\mu g/m^3$ and 75 $\mu g/m^3$ for PM_{10} and $PM_{2.5}$, respectively in winter period (officially from 15 October to 15 April), being almost twice as much as average mass concentration for the summer period. This could possibly be attributed to the higher traffic density and combustion of fossil fuels for heating during winter, as well as to the prevailed meteorological conditions e.g. inversed layers, low temperature and stagnation of air masses. Higher winter PM_{10} concentrations were also recently presented in some other studies of European cities [7,8,25,42].

Seasonal variation is in agreement with the results for other air pollutants relevant for traffic valuation. The concentrations of CO, NO_2, Pb, HC and HCHO have been measured at several main cross-streets in Belgrade by the Institute of Public Health of Serbia "Dr Milan Jovanović-Batut". The concentrations of all pollutants were minimum during July and August (reduced traffic volume due to summer vacation) and maximum during the coldest winter months (November, December and January). For all pollutants, the annual mean values were above limit values.

The winter episode with simultaneous sampling of PM_{10} and $PM_{2.5}$ run at the Faculty of Veterinary Medicine (KP site), showed that PM_{10} and $PM_{2.5}$ concentrations data are very well correlated (Pearson's coefficient R=0.99), indicating the same source for fine and coarse particles. The relative contribution of fine, $PM_{2.5}$ to inhalable particles, PM_{10}, was determined for the period with simultaneously measurements of both fractions. On average, the mass percentage of $PM_{2.5}$ was 78% of PM_{10} i.e. 62% in summer and 85% in winter period. During the episode with intense photochemical processes, in summer, this ratio reached the value of 96% and during the high pollution period with the highest mass concentrations for PM_{10} and $PM_{2.5}$ particles (November, December), the value was up to 88%. Similar data have been obtained in some other cities as well.

Statistic analysis of daily mass concentrations and relevant meteorological parameters such as daily mean values of temperature, relative humidity, precipitation, atmospheric pressure, wind speed and direction, were performed; negative correlation with wind speed was observed giving the prominence to the efficiency of the atmospheric horizontal mixing as a dilution mechanism. For both seasons, the PM concentrations were higher when prevailing wind directions was south-westerly. One of the possible explanations might be that the most polluted location, Faculty of Veterinary Medicine, is opened from S-SW direction where the big crossroad with a highway is. Another reason could be the influence of the great source of air pollution, the complex of coaled fired power plants of Nikola Tesla A and Nikola Tesla B

- Obrenovac, 20 km south-west from Belgrade, which is significant source of secondary sulphate aerosols [91].

PM_{10} and $PM_{2.5}$ Mass and Trace Metal Concentrations

Descriptive statistics for daily mass (μg/m^3) and trace metal (ng/m^3) concentrations in PM_{10} and $PM_{2.5}$ sampled, between June 2003 and July 2005, in urban Belgrade, at two locations Student Square and Autokomanda (KP site), are presented in Table 11. The first point emerging from the statistics is the high mean and maximum levels of PM_{10} and $PM_{2.5}$. Thus, the PM_{10} mean mass concentration during the 2-year period (68.4 μg/m^3) exceeded the proposed EC annual limit of 40 μg/m^3 [18]. Of more concern was the average $PM_{2.5}$ concentration of 61.4 μg/m^3 for the 2-year period, which was three times higher than the EC annual limit of 20 μg/m^3 [19,21].

Table 11. Statistical parameters of daily mass (μg/m^3) and trace metal concentrations (ng/m^3) in Belgrade urban area from June 2003 to July 2005 in PM_{10} and $PM_{2.5}$

	PM_{10}	Pb	Cu	Zn	Mn	Fe	Cd	Ni	V	Al	Cr
N	209	210	210	210	210	210	210	210	210	210	167
Mean	68.4	46.5	71.3	1389.2	20.8	1462.9	1.4	17.7	36.6	873.8	10.2
S.D.	46.4	128.5	118.7	2313.4	15.9	1911.9	2.2	17.7	48.8	914.1	11.4
Maximum	333.8	1525.0	947.5	17239.6	72.7	20283.0	17.7	107.7	282.4	6104.2	58.0
Minimum	2.8	0.5	0.2	2.1	0.2	0.5	0.0	0.4	0.6	2.1	0.1
Median	57.6	19.1	42.4	400.2	14.6	746.7	0.5	12.0	10.8	598.5	7.5
98th percentile	193.0	322.4	556.6	7018.3	62.4	5749.1	7.9	71.6	166.2	4126.9	46.3

N-number of samples; S.D. - Standard arithmetic deviation.

	$PM_{2.5}$	Pb	Cu	Zn	Mn	Fe	Cd	Ni	V	Al	Cr
N	64	64	64	64	64	64	64	64	64	64	64
Mean	61.4	21.0	20.8	1998.0	15.2	1081.2	0.9	28.4	59.8	1180.3	6.2
S.D.	52.5	27.0	19.2	1846.4	13.7	1360.3	1.2	43.1	56.3	1657.4	3.8
Maximum	286.6	193.9	99.6	6642.4	50.7	5996.5	4.8	315.0	248.5	7875.0	26.0
Minimum	8.8	0.5	0.2	115.2	2.1	49.7	0.0	0.4	5.2	159.1	1.6
Median	51.9	13.5	17.4	1153.3	10.1	488.4	0.4	17.3	50.2	500.7	5.9
98th percentile	237.3	66.8	80.6	6101.7	47.9	5380.6	4.1	107.9	208.7	7307.3	13.2

N-number of samples; S.D.- Standard arithmetic deviation.

The average PM_{10} and $PM_{2.5}$ mass concentrations (68.4 μg/m^3 and 61.4 μg/m^3) in Belgrade were higher than in many European cities, although high mean PM_{10} mass concentrations were reported in this region. Thus, Thessalonica, Greece, had mean values of 82, 78, and 84 μg/m^3 [87]. Athens mean values of 75 μg/m^3 for PM_{10} and 40 μg/m^3 for $PM_{2.5}$ [8] , while mean values of 72, 60, 70 μg/m^3 were found in Bulgaria, Hungary and Romania [32]. Our results for PM_{10} and $PM_{2.5}$ concentrations are similar with those reported earlier for the present site [61,62,71].

The results for the total mean concentrations of individual metals presented in Table 12. indicate iron as the most abundant metallic element (1462.9 ng/m^3) in the suspended particles with diameter less than 10 μm (PM_{10}). Zn and Al concentrations in this fraction were very high too. The highest mean concentration in fine fraction of particulate matter ($PM_{2.5}$) was 1998 ng/m^3 for Zn, followed by Al and Fe.

Table 12. Mean daily mass (μg/m^3) and trace metal concentrations (ng/m^3) at Autokomanda (KP site), Belgrade, from June 2003 to July 2005 in PM_{10} and $PM_{2.5}$

Year		PM_{10}	Pb	Cu	Zn	Mn	Fe	Cd	Ni	V	Al	Cr
	N	28	28	28	28	28	28	28	28	28	28	28
2003	Mean	92	26	43.2	655.9	17	1713.2	0.20	26.3	62	830.7	11.7
	N	72	72	72	72	72	72	72	72	72	72	51
2004	Mean	76.4	81.7	131.7	262.9	17.9	1036.2	2.10	19.0	46.4	509.9	5.1
	N	30	30	30	30	30	30	30	30	30	30	30
2005	Mean	92	42.4	48.7	5333.2	46.1	3858.5	0.60	35.7	62	2266.1	12.5

N- Number of samples.

Year		$PM_{2.5}$	Pb	Cu	Zn	Mn	Fe	Cd	Ni	V	Al	Cr
	N	24	24	24	24	24	24	24	24	24	24	24
2003	Mean	77	22.1	19.3	770	7.3	428.5	0.20	36.4	55.5	392.8	6.4
	N	13	13	13	13	13	13	13	13	13	13	13
2004	Mean	68.78	15.5	21.3	261.5	3.8	509.8	1.47	21.6	87.8	416.6	2.1
	N	13	13	13	13	13	13	13	13	13	13	13
2005	Mean	71.72	37.7	18.4	4629.5	38.1	1774.7	0.60	48.8	98.2	2915.0	8.0

N- Number of samples.

The high Zn concentration implies that emissions from anthropogenic sources contributed significantly to its level in the particulate matter. Zinc is a reliable tracer for unleaded fuel and diesel powered motor vehicle emissions [44]. Moreover, Zn can be released, in large amounts, from wear and tear of vulcanized vehicle tires, corrosion of galvanized automobile parts or various industrial activities. From the results of The Principal Component Analysis (PCA) and Cluster Analysis (CA) emission of metals most associated with traffic (Zn,Cu, Fe, Mn, Pb, Al) is probably more related to suspension or resuspension of road dust, which includes soil dust mixed with traffic related particles, than direct exhaust emission [63, 92].

Although Al and Fe are typically crustal elements, if coupled with other elements, they can indicate the presence of anthropogenic sources, such as steel production industry (Smederevo, 50 km in SE direction). Nickel is mainly associated with fossil fuel use, oil burning and emissions from industrial sources. The mean Ni concentration of 28.4 ng/m^3 in the $PM_{2.5}$ fraction was above the critical value of 20 ng/m^3 for PM_{10} [19].

Much higher concentrations for Zn, Al, Mn and Fe, in the year 2005 could be explained, partly, by changed extraction procedure. As dilute acid was used in the first two years for the extraction, the concentrations would be slightly underestimated for some common crustal elements, especially for the coarse particle size range [46]. Another contribution could be markedly increased production in Smederevo steel industry since the beginning of 2005.

Air-back trajectories model was applied to determine the main sources of pollution in the Belgrade area and the results confirmed the assumption that the dominant source of particles was local traffic, individual power stations and the complex of coaled fired power plants 20 km far from Belgrade [43]. Physical and chemical characterization of PM particles done with scanning electron microscope [72,86] have indicated the presence of soot, particles with high content of Si, sulfates, particles enriched with heavy metals and biological particles.

CONCLUSION

Radionuclides Measurements

The study of radionuclides (^{137}Cs, ^{7}Be and ^{210}Pb) in ground level air in Belgrade central city area performed in 2004 and 2005 provided information on the differences in the content of radionuclides in ground level air due to the origin of the radionuclides and the conditions of the microclimate. Generally, the lowest values in radionuclides concentrations were obtained at sampling sites exposed to good ventilation, while higher values were measured at sites with dense vegetation and poor ventilation. The local topology, rate of precipitation, and temperature differences are highly important, as city centers appear to be a sort of islands regarding temperature: in Belgrade, average yearly temperatures between the city center and the surrounding area differ even for 2°C. But as the content of radionuclides in air is governed by the stochastic processes, the two years sampling period could not provide sufficient data for definite conclusions.

^{137}Cs concentrations in ground level air in central city area in 2004 and 2005 were below the lower level of detection (10^{-6} Bq/m^3), while in the surrounding area the concentrations were still within the measurable range (10^{-5} Bq/m^3), mainly due to the local surface dust resuspension effects. As stratospheric contribution of cesium has been less significant since the 90ties, its seasonal variations are less uniform and more dependent on the local weather conditions.

Concentrations of ^{7}Be measured in 2004 and 2005 correspond with the values measured at our midlatitides, with the overall average of 3.04×10^{-3} Bq/m^3 for the period. The pattern of ^{7}Be seasonal variations differ for the sampling sites within the central city area and the one outside the city, but not significantly The pattern exhibit maxima in autumns and a mild increase in summers; high concentrations of ^{7}Be in air in winter periods could be possible due to high mass concentrations of PM measured within the central city area.

The pattern of ^{210}Pb seasonal variations is more unstable due to heavy traffic in the center of the city and the anthropogenic sources of Pb: use of leaded gasoline by passenger cars and city busses, emission from industrial plants and thermoelectric coal burning and crude oil burning plants in the intermediate vicinity of the city, etc. The concentrations of ^{210}Pb in ground level air measured in the central city area are lower that those measured outside the city, but still within the same order of values (10^{-4} Bq/m^3). The reason for this could be because the sampling site outside the city is on the open field, while the site in the city is surrounded by the buildings.

The variations of the $^{7}Be/^{210}Pb$ monthly concentrations ratio obtained in 2004 and 2005 differ both in values and pattern for the sampling sites within the city and outside, while calculated mean RLI for both sites approach zero, as expected. It can be concluded that although the study covers just a period of two years the influence of the local climate on the radionuclides activity ratio pattern could be clearly seen.

Particulate Matter Measurements

PM_{10} and $PM_{2.5}$ measurements were performed in the central urban area of Belgrade in order to provide information about the annual average and daily maximum concentrations and to compare them to the corresponding EU standards. The results may be used in the future to assess the effectiveness of the implemented emission control strategy.

The annual mean PM_{10} and $PM_{2.5}$ mass concentrations exceeded the European Union air quality annual standard. Chemical analysis of the metal content showed that Zn (1998 ng/m^3) was the most abundant element in $PM_{2.5}$. The mean Ni concentration of 28.4 ng/m^3 in the $PM_{2.5}$ fraction was above the critical value of 20 ng/m^3 (for PM_{10}) according to EC (2003) regulations.

The meteorological study revealed some differences in concentrations, stratified by prevailing wind directions. The classification of concentration levels by defined wind subgroups can provide important information on the potential impact of various sources and particle transport between zones of a wider geographical area. Data evaluation showed that higher trace element concentrations were associated with calm conditions suggesting a great influence of local sources. The appearance of stronger W, SE and NE flows is responsible for the transport of particles originating from the thermoelectric plant complex and industrial zones surrounding Belgrade.

The Principal Component Analysis and Cluster Analysis resulted in tentative source categories for each site. Quantification of their contribution showed that the main contributors to heavy metals in PM_{10} and $PM_{2.5}$ generated in Belgrade are combustion sources, e.g. emissions from mobile and stationary units, with the major contribution from traffic emissions, road dust resuspension and industrial activities.

References

[1] Agelaio,G.; Cannizzaro, F.; Greco, G.; Rizzo, S.; Spitale, M.S. 1984. *Health Phys.* 47, 96 –101.

[2] Arimoto, R.; Snow, J.A.; Graustein, W.C.; Moody, J.L.; Ray, B.J.; Duce, R.A.; Turekian, K.K.; Maring, H.B. 1999. *J. Geophys. Res.* 104 (D17), 301-321.

[3] Baeza, A.; Delrio, L.M.; Jimenez, A.; Miro, C.; Paniagua, J.M.; Rufo, M. 1996. *J. Radioanal Nucl. Ch.* 207 (2), 331-344.

[4] Bettoli, M.; Bonasoni, P.; Calzolari, F.; Colombo, T.; Ravanelli, M.; Tositti, L.; Tubertini, O. 1998. *Ann. Chim.-Rome* 88, 769-770.

[5] Bunzl, K.; Hotzl H.; Rosuer G.; Winkler R. 1995. *Naturwissenschaften* 82(9), 417-420.

[6] Cannizzaro, F.; Greco, G.; Raneli, M.; Spitale, M.; Tomarchio, E. 1995. *Nucl. Geophys.* 9, 597-607.

[7] Chaloulakou, A.; Kassomenos, P.; Spyrellis, N.; Demokritou, P.; Koutrakis, P. 2003. *Atmos. Environ.* 37, 649-660.

[8] Chaloulakou, A.; Kassomenos, P.; Grivas, G.; Spyrellis, N.; 2005. *Environ. Int.* 31, 651-659.

[9] Directive on permissable levels, measuring methodology and criteria for immission monitoring sites and collecting data, *Službeni Glasnik* Republic of Serbia 8.08.1992, No. 54, 1998.

[10] Djuric, G.; Petrovic, B.; Smelcerovic, M.; Popovic, D. 1988. *Acta Veterinaria*, 37(4), 555-560.

[11] Djuric, G.; Popovic, D.; Smelcerovic, M.; Sarvajic, A.; Ivankovic S.; Misic, N. 1992. *Vet. Glasnik* (Belgrade), 46/9, 491-497.

[12] Djuric, G.; Popovic, D. 1994. *Ecologica* 1(2), 19-24.

[13] Djuric, G.; Popovic, D. 2000. *Chem. Industry* 54(2), 50-52.

[14] Department of Meteorology, *Annual Reports*, Institute of Nuclear Sciences Vinca, Belgrade, 1996-2006.

[15] Dockery, D. W.; Pope III C.A. 1994. *Annu. Rev. Publ. Health* 15, 107-132.

[16] Duenas, C.; Fernandez, M.C.; Carretero, J.; Liger, E.; Canete, S. 2004. *Atmos. Environ.* 38, 1291 –1301.

[17] Durana, L.; Chudy, M.; Masarik, J. 2000. *J. Radioanal. Nucl. Ch.* 207, 345-356.

[18] EC, 1999. Air Quality Directive 1999/30 EC of the European Parliament and of the Council of 22 April 1999 relating to limit values for SO_2, NO_2 and NOx, particulate matter and lead in ambient air. *Off J. Eur. Communities* L163, Brussels.

[19] EC, 2003. Proposal for a Directive of the European Parliament and of the Council, Relating to arsenic, Cadmium, mercury, *Nickel and Polycyclic Aromatic Hydrocarbons in Ambient Air*. Brussels.

[20] El Daoushy, F. 1988. *Environ. Int.* 14, 305-319.

[21] European Standard 14907.2005. *Ambient Air Quality – Standard Gravimetric Measurement Method for the Determination of the $PM_{2.5}$ Mass Fraction of Suspended Particulate Matter*. European Committee for Standardization, Brussels.

[22] Gaggelar, H.W. 1995. *Radiochim. Acta* 70/71, 345-353.

[23] Gaffney, J.S.; Orlandini, K.A.; Marley, N.A.; Popp, C.J. 1994. *J. Appl. Meteorol.* 33, 869-873.

[24] Gaffney, J.S.; Marley, N.A.; Cunningham, M.M. 2004. *Atmos. Environ.* 38, 3191-3200.

[25] Gehrig, R.; Buchmann, B. 2003. *Atmos. Environ.* 37, 2571-2580.

[26] Gerasopoulos, E.; Zerefos, C.S.; Papastefanou, C.; Zanis, P.; O' Brien, K. 2003. *Atmos. Environ.* 37, 1745-1756.

[27] Groundel, M.; Postendourfen, J. 2004. *Atmos. Environ.* 38, 3723-3728.

[28] Gustafson, P.F.; Kerrigan, M.A.; Brar, S.S. 1981. *Nature* 191, 454-456.

[29] Hartwig, S. 1996. *Z. Naturforsch* A. S1 (10/11), 1139-1143.

[30] Hirose, K.; Honda, T.; Yagishita, S.; Igarashi, Y.; Aoyama, M. 2004. *Atmos. Environ.* 38, 6601-6608.

[31] Hernandez, F.; Hernandez-Armas, J.; Catalau, A.; Fernandez-Aldecoa,J.C.; Karlsson, L. 2005. *Atmos. Environ.* 39, 4057-4066.

[32] Houthuijs, D.; Breugelmans, O.; Hoek, G.; Vaskovi, E.; Mihalikova, E.; Pastuszka J. S.; Jirik, V.; Sachelarescu, S.; Lolova, D.; Meliefste, K.; Uzunova, E.; Marinescu, C.; Volf, J.; de Leeuw, F.; van de Wiel, H.; Fletcher, T.; Lebret, E.; Brunekreef, B. 2001. *Atmos. Environ.* 35, 2757-2771.

[33] Ilyin,L.A.; Pavloskij, O.A.1987. *IAEA Bull.* 29(4), 17-24.

[34] Ioannidou, A.; Papastefanou, C. 1997. *Nucleus* 34, 1-2, 111-115.

[35] Ishikawa, Y.; Murakami, H.; Sekine, T.; Yoshihara, K. 1995. *J. Environ. Radioactiv.* 26(1), 19-36.

[36] Kathren, R.L. 1984. *Radioactivity in the Environment: Sources, Distribution and Surveillance,* Harwood Academic Publishers. New York.

[37] Kyotani, T.; Iwatsuki, M. 2002. *Atmos. Environ.* 36, 639-649.

[38] Larsen , R.J.;.Sanderson, C.G.; Kada, J. *EML Surface Air Sampling Program*, 1990-1993 Data, EML-572, 247 pp., U.S. Department of Energy, Environmental Measurements Laboratory, New York, 1995.

[39] Lighty, J.S.; Veranth, J.M.; Sarofim, A.F. 2000. JAPCA J Air Waste Ma 50, 1565-1681.

[40] Manic,G.; Radojkovic,B.; Manic, V.; Popovic, D. 2005. In: *Proc. »EcoPHysics2005«Conference*, Krusevac, 169-172.

[41] Manic, G.; Petrovic, S.; Manic, V.; Popovic, D.; Todorovic, D. 2006. *Environ. Int.* 32, 533-537.

[42] Marcazzan, G.; Vaccaro, S.; Valli, G.; Vecchi, R. 2001. *Atmos. Environ.* 35, 4639-4650.

[43] Mijic, Z.; Lazic, L.; Rajsic, S.; Tasic, M.; Novakovic, N. In: Book of Extended Abstracts ; First ACCENT Symposium, The Changing Chemical Climate of the Atmosphere, 12-16. Sep.2005. Urbino, Italy, 2006; *in press.*

[44] Monaci, F.; Moni, F.; Lanciotti, E.; Greichi, D.; Bargagli, R. 2000. *Environ. Pollut* 107, 321-327.

[45] Paatero, J.; Hatanka, J.; Mattsson, R. 2001. *Measurements of airborne berillylium-7 in Northern Finland.* http:// www.gr.is/nsfs/paatero1.htm

[46] Pakkanen, A. T.; Kerminen, V.; Loukkola, K.; Hillamo, E. R.; Aarnio, P.; Koskentalo, T.; Maenhaut, W. 2003. *Atmos. Environ.* 37, 1673-1690.

[47] Papastefanou, C. ; Joannidou, A. 1995. *Environ. Int.* 22 , S125-130.

[48] Perkins, D.W.; Roberston, D.E.; Thomas, W.; Young, J.A. In Proc.Int.Symp.on Environmental Contamination Following a Major Nuclear Accident, *IAEA*, Vienna, 1989, 113-139.

[49] Peters, A. J.; Gregor, D. J.; Wilkinson, P.; Spenser, C. 1997. *J. Geophys. Res.* 102(5), 5971-5978.

[50] Pope, C.A. 2000. *Aerosol. Sci. Tech.* 32, 4-14.

[51] Popovic, D.; Djuric, G.; Todorovic, D.1996. *Radiat. Prot. Dosim.* 63/3, 223-225.

[52] Popovic, D.; Djuric, G. Todorovic, D. Chernobyl fallout radionuclides in soil, plants and honey of a mountain region. *IAEA Tech.Reports* No.964, Vol.II, 432-437, Vienna, 1996.

[53] Popovic, D.; Djuric, G. In *"Environmental Impact of the NATO attack in Yugoslavia"* An Independent Expert Group Report. Editor Pavlovic, V.; Eco-center, Belgrade, SCG, 1999; 103-115.

[54] Popovic, D.; Todorovic, D.; Radenkovic, M.; Djuric, G. 2000. *J. Environ. Prot. and Ecol.* 3, 124-130.

[55] Popovic, D.; Djuric, G.; Todorovic, D.; Spasic-Jokic, V. 2000. *Cent. Eur. J. Occup. Environ. Med.* 6 (2/3), 129-134.

[56] Popovic, D.; Todorovic, D.; Radenkovic, M.; Djuric, G. 2000. *J. Environ. Prot.and Ecol.* 3, S134-130.

[57] Popovic, D.; Todorovic, D.; Djuric, G. 2002. *Cent. Eur. J. Occup. Environ. Med.* 8(2/3), 227-231.

[58] Popovic, D.; Todorovic D. 2006. *Facta Universitatis - Series: Physics, Chemistry and Technology,* 4/1, 11-20.

[59] Popovic, D.; Spasic-Jokic, V. *Military Medicine Review* (Vojnosanitetski pregled) (Belgrade) 63/5, 2006, 481-487.

[60] Preiss, N.; Melieres, M.; A., Pourchet, M. 1996. *J. Geophys. Res.* 101(D22), 28.847-28.862.

[61] Rajsic, F. S.; Tasic, D. M.; Novakovic, T. V.; Tomasevic, N. M. 2004. *Environ. Sci.and Pollut. Res.* 11, 158-164.

[62] Rajsic, S.; Tasic, M.; Markovic, D.; Novakovic, V.; Markovic, D.; Tomasevic, M. 2004. *Cent. Eur. J. Occup. Environ. Med.* 10(1) 25-29.

[63] Rajsic, S.; Mijic, Z.; Tasic, M.; Djordjevic, D.; Radenkovic, M.; Joksic, J. 2006. 8th *International Conference on Fundamental and Applied Aspects of Physical Chemistry. Physical Chemistry 2006.* Sept. 26-29., Belgrade, Serbia. Proceedings, 624-626.

[64] Rulik, P.; Bucina, I.; Malatova, I.; Tomasek, M. 1997. *Proc. IRPA Regional Symp on Radiation Protection in Neighbouring Countries of Central Europe*, Praque, CR, 8-12 Sept.

[65] Samara, C.; Kouimtzis, Th.; Tsitouridou, R.; Kanias, G.; Simeonov, V. 2003. *Atmos. Environ.* 37, 41-54.

[66] Schwartz, J.; Dochery, D.W.; Neas, L.M. 1996. *J. Air. Waste.* MA 46, 927-939.

[67] Schwartz, J.; Ballester, F.; Saez, M.; Perez-Hoyos, S.; Bellido, J.; Cambra, K. Arribas, F.; Canada, A.; Perez-Boillos, M.J.; Sunyer, J. 2001. *Environ. Health. Persp.* 109, 1001-1006.

[68] SFOPH- Swiss Federal Office of Public Health, Division of Radiation Protection – Environmental Radiation Section. Geneva 2003. *Environmental Radioactivity and Radiation Exposures in Switzerland in 2002.*

[69] Tasic, D, M.; Rajsic, F, S.; Novakovic, T.V.; Tomasevic, N.M. 2001. *Fresenius Environ Bulletin,* 10 (3), 305-309.

[70] Tasic, M.D.; Rajsic, S.F.; Novakovic, V.T.; Mijic, Z.R.; Tomasevic, M.N. 2004. In *Proc. 7th Int. Conf. on Fundamental and Applied Aspects of Physical Chemistry*; Belgrade, 643-645.

[71] Tasic, D. M.; Rajsic, F. S.; Novakovic, T. V.; Mijic, R. Z.; Tomasevic, T. M. 2005. *Phys. Scr.* T118, 29-30.

[72] Tasic, M.; Djurić-Stanojevic, B.; Rajsic, S.; Mijic, Z.; Novakovic, V.2006. *Acta. Chim. Slov.* 53, 401-405.

[73] Tasic, M.; Rajsic, S.; Novakovic, V.; Mijic Z. 2006. *Facta Universitatis: Series Physics, Chemistry and Technology* 4 (1) 83-91.

[74] Todorovic, D.; Djuric, G.; Popovic, D. In: Strahlenschutz: Physic und Messtechnik« Eds. W.Koelrer and R.Manshart, *IRPA Publ.Ser.Radiation.Protect.* FS 94-71T, 1994, 331-335.

[75] Todorovic, D.; Smiljanic, R.; Manic, S. 1996. *Ecologica* 10(2), 33-40.

[76] Todorovic, D. 1997. The effect of tropopause height on the content of radioactive debris in surface atmopshere. *Environ. Int* .23(6), 815-818.

[77] Todorovic, D.; Popovic, D.; Djuric, G. 1997. Vienna, *IAEA-SM* 349/35P, 849-850.

[78] Todorovic, D.; Popovic, D.; Djuric, G. *Vinca Bull.*, Ed. Antic, D. Belgrade, SCG; 1997, 2/1, 635-638.
[79] Todorovic, D.; Popovic, D.; Radenkovic, M.; Djuric, G. 1999.*Environ. Int.* 25(1), 59-66.
[80] Todorovic, D.; Popovic, D.; Djuric, G.; Radenkovic, M. 2000. *Atmos. Environ.* 34 (19), 3245-3248.
[81] Todorovic, D.; Popovic, D.; Radenkovic, M.; Djuric, G. In:*Applied Physcs in Serbia*, Eds. Koicki, S.; Konjevic, N.; Petrovic, Z.; Bek Uzarov, Dj.; XVCIII (2/1), SANU: Belgrade, SCG, 2002, 63-66.
[82] Todorovic, D.; Radenkovic, M.; Popovic, D.; Tasic M. In: *Recent Advances in Applied Physics,* Ed. Mendey-Vilas, A.; Elsevier, Amsterdam, Nl, 2005, 479.
[83] Todorovic, D.; Popovic, D.; Djuric, G.; Radenkovic, M. 2005. *J. Environ. Radioactiv.* 79, 297-307.
[84] Tokieda, T.; Yamanaka, K.; Harada, K.; Tsunogai, S. 1996. *Tellus* 48B, 690-702.
[85] Tomasevic, M.; Rajsic, S.; Djordjević, D.; Tasic, M.; Krstic, J.; Novakovic, V. 2004. *Environ. Chem..Lett.* 2, 151-154.
[86] Tomasevic, M.; Vukmirovic, Z.; Rajsic, S.; Tasic, M.; Stevanovic, B. 2005. *Chemosphere* 61, 753-760.
[87] Tsitouridou, R.; Voutsa, D.; Koumitzis, Th. 2003. *Chemosphere* 52, 883-891.
[88] UNCEAR Report: *Exposures from Chernobyl Accident* A/AC82/Rep469, Vienna, 1988.
[89] Van Grieken, R.; Delalieux, F. In *Proceedings of the Fifth General Conference of the Balkan Physical Union*, Vrnjacka Banja, Serbia and Montenegro, August 25-29, 2003, Eds: Jokic, S; Milosevic, I.; Balaz, A.; Nikolic, Z. ; Serbian Physical Society, Belgrade, SCG 2003, 234-246.
[90] Vaupotic, J.; Krizman, M.; Sutej, T.; Peternel, M.; Djuric, G.; Popovic, D.; Planinic, J.; Faj, Z.; Klajic, R.; Stegnar, P.; Kobol, I. 1992. *Radiat. Prot. Dosim.* 45(1/4), 487-493.
[91] Vukmirovic, Z. 1990. *Idojaras* 94, 249.
[92] Vukmirovic, Z.; Marendic-Miljkovic, J.; Rajsic S.; Tasic, M.; Novakovic. V. 1997. *Water, Air and Soil Pollution*, 93(1-4), 137-156.
[93] Vukmirovic, Z.; Tosic,I.; Tasic, M.; Rajsic, S. 2004. *Meteorol Atmos Phys* 85(4) 265-273.
[94] ZND89. *Protocol on the coordination on a project of atmospheric radioactivity investigation.* SEV Intercomparision Programme. Sofia, SEV, 1989.
[95] World Health Organization (WHO): *Health aspects of air pollution with particulate matter, ozone and nitrogen dioxide.* Report on a WHO Working Group, Regional Office for Europe; Bonn, Germany 13–15 January 2003. EUR/03/5042688. Available also at http://www.euro.who.int/document/e79097.pdf.

INDEX

B

D

E

F

G

H

I

N

O

P

Q

R

S

T

U

V

W

Y

Z